Human
Genetics

Human Genetics

Elof Axel Carlson

Distinguished Teaching Professor
State University of New York, Stony Brook

D. C. Heath and Company
Lexington, Massachusetts   Toronto

*Dedicated to
Claudia, Christina, Erica, John,
and Anders Carlson*

Preface

Human Genetics is based on a course I have taught at the State University of New York at Stony Brook since 1968. It is intended primarily for use in a one-term course in human genetics, human heredity, or heredity and society. This text, like my course, emphasizes the underlying biology of the human condition. By relating genetic principles to the human condition, this text will prepare students to consider some controversial issues involving the science of genetics in contemporary society.

Coverage of genetic principles includes a balanced treatment of cells and chromosomes, classical genetics, analysis of complex traits, developmental genetics, molecular genetics, and population genetics. In each of these six topic sections, I have made an effort to include applications to the individual and society. When identifying societal and ethical problems, I have emphasized options to which students can apply their new knowledge of genetics. In a society of diverse peoples, there is no way to avoid controversies, and I have deliberately used several as a way of introducing biological principles. Among the controversies covered are questions of embryo transfer, recombinant DNA technology, and environmental mutagens. When discussing current subjects such as IQ testing, eugenics, genetic counseling, genetic engineering, fertility, and cancer, I have also attempted to place these issues in their historical perspective.

No prior science course work is necessary for this text and thus it can be used by a variety of nonscience majors. I have included a number of pedagogical aids to complement the topic discussions and to help the student study the material. Included are an abundance of photographs and illustrations, many clinical in nature; end-of-chapter aids, including questions for review and discussion and a list of key words; a glossary; and a bibliography of related books and readings in human genetics.

I acknowledge with deep appreciation the many helpful suggestions and criticisms offered by those who have reviewed the manuscript for this book. They include: Harvey A. Bender, University of Notre Dame; Claire M. Berg, University of Connecticut; Peter S. Dawson, Oregon State University; Wendell H. McKenzie, North Carolina State University; Muriel N. Nesbitt, University of California, San Diego; Robert M. Petters, Pennsylvania State University; Frank J. Ratty, San Diego State University; and Joan K. Stadler, Iowa State University.

I thank the Dight Institute of Genetics at the University of Minnesota where I began the first draft of the book. I especially appreciate the stimulation provided by Robert Desnick, Gregory Grabowski, Robert Gorlin, Richard King, Carl Witkop, Sheldon Reed, and Elving Anderson.

I am grateful to Bentley Glass and Frank Erk who recruited me to Stony Brook and to Vera King Farris who shared many discussions on the design of the course from which this book was developed. I sincerely appreciate the superb editorial advice of Harvey Pantzis, and the detailed criticisms and encouragement of Cedric Davern. Many of the original sketches for the illustrations for this text were prepared by Christina, Claudia, and Erica Carlson; the balance of the sketches and all the illustrations were attractively rendered by Carmela Ciampa. Cathy Cantin was immensely helpful and talented in producing the book and I thank Janice Wheeler, Eve Mendelsohn, and Kate Bramer for the care and skill they put in the design and photo research for this book.

Most of all I have benefited from the continued interest and criticisms offered by Stony Brook students about the content, emphasis, and issues dealt with in this text. I hope that students who read this text will develop an enriched view of the life sciences and appreciate the central role of heredity in relating all of life, past and present.

Elof Axel Carlson

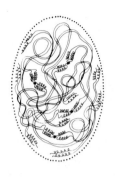

Contents
in Brief

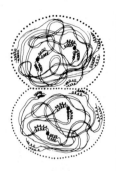

Contents

Chapter 3
Radiation Damage: How Chromosome
Breakage Can Lead to Cell Death   *30*

Chapter 4
The Formation of Sex Cells and the
Consequences of Nondisjunctional Errors   *45*

Chapter 5
Fertility and Infertility   *61*

Part 3
Single Gene Effects   *78*

Chapter 6
Mendel's Laws and Genetic Disorders   *80*

Chapter 7
Indispensable Blood: The ABO and
Rh Systems *106*

Chapter 8
Prenatal Diagnosis *119*

Chapter 9
Genetic Counseling   *130*

Chapter 12
Miscegenation and the Emergence of New Races *177*

Chapter 20
Induced Birth Defects: Teratogenesis   *283*

Chapter 21
Cancer: The Errant Cell   *290*

Chapter 22
Smoking and Lung Cancer: A Self-Inflicted Biohazard *312*

Part 6
Molecular Genetics   *325*

Part 7
Population Genetics   *378*

Chapter 28
Human Evolution *410*

Chapter 29
What Should We Do with Our Genes? *421*

Part 1
Reflections on the Human Condition

What is the human condition? For an English teacher it may be the way we depict life through tragedy, comedy, irony, and romance. For the cultural historian it may reflect the common experiences of a civilization or nation, a ruling class, or even a special population such as a guild or the clergy. We might explore it in a psychology course, using the vaguer concept of "human nature," and assign to it what we believe to be universal traits (often contradictory ones), such as aggression and cooperation, territoriality, male or female dominance with specific gender roles, or a search for values or God. We would learn that no single behavioral attribute associated with our commonality as humans has gone unchallenged.

As a college student you may be requested to take courses in the liberal arts, including the sciences, humanities, social studies, and fine arts. The content and purpose of these courses have changed greatly since the days of Athens and Sparta more than 2500 years ago. Today there are approximately 3000 colleges and universities in the United States, each with its own statement on the significance of the liberal arts. Most of your professors would probably agree that studying the liberal arts frees you from ignorance and narrowness and gives you the chance to see yourself and the world from multiple perspectives. Impaired health, death in the family, infertility, and other stressful or tragic medical problems often have a connection to heredity, although we are usually unaware of that relation. That is precisely why this book deals with human genetics (and biology in its broadest sense) through the human condition.

You will discover that our ideas of the human condition change with our circumstances and values. You can readily demonstrate this by talking with your grandparents and parents about their knowledge of their ancestry, careers, crises, tragedies, and triumphs. If all four of your grand-

1

parents are alive, find out how long their parents lived and what caused their deaths. Ask about uncles and aunts and how many died as infants or young children. You will probably be surprised at the size of families in your grandparents' and great-grandparents' generations. Did any of them have tuberculosis? As children did any of them have diphtheria? What childhood diseases did your parents have? What has your health been like compared to that of your immediate ancestors?

Those are biological differences. You will find other differences too—in education, employment, and military service, for example. Who went to college among your parents, grandparents, and great-grandparents? What were your ancestors doing during the great Depression in the 1930s? Did any of your relatives serve in the armed forces during a war? How many of your ancestors were divorced? Were any of your ancestors raised by relatives other than their parents, or as orphans? Was anyone adopted?

For many of us this exercise in family history reveals that our ancestors had larger families, experienced more severe infectious diseases, and faced more hardships than those who were born after World War II. Yet, in spite of all the progress in medical and social services since World War II, every newborn infant today still encounters some biological factors that can limit his or her potential for normal life.

In the past we have intervened in what was considered the natural world: We would rather prevent plagues than accept them as chastening visitations of divine wrath for our wayward behavior; we would rather promote universal literacy than restrict education to a ruling elite; we would rather regulate the production of goods than face periodic starvation; we would rather have Social Security than ignore the needs of the aged. Each generation has faced some of the problems of the human condition and made changes through advances in technology, medicine, basic science, social legislation, and cultural custom. In retrospect the changes seem rational and orderly, but in reality they were often suspected, resisted, debated, modified, and achieved only after years or decades of conflict. During your lifetime you, too, will have to face and debate such issues. Some of the major problems of the human condition today are introduced in the following chapter.

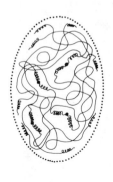

Chapter 1
The Human Condition

Whether we go to a physician for a checkup when we are healthy or go to seek both a diagnosis and a treatment when we are ill, we are likely to ask few questions about the doctor's procedures. Most of us are willing to yield to the medical profession's control of our bodies as well as of issues of human biology about which we know little. Perhaps our willingness to leave our human biology to the medical professions stems from a fear that we cannot cope with the knowledge of our biological selves.

Yet our ignorance robs us of the chance to act, to legislate, to protect our lives. It prevents us from recognizing what can harm us. It leaves us unaware of the damage we inflict on ourselves and our children. If we were enlightened about our own biology, we would find ourselves benefited in the same way that printing and mass public education liberated us by allowing us to be informed of local, national, and world events. Knowledge of human biology would free us to make decisions about the quality of a life we take for granted.

The fear that holds us back may stem from an awareness that life is not perfect. For reasons unknown to us, a stalking sense of tragedy in the form of genetically transmitted defects that may cause infertility, spontaneous abortion, and birth defects may haunt each human family. Fortunately, however, most of us will be among the lucky who will have healthy babies (Figure 1-1).

Infertility or Sterility
Often Frustrates a Couple   About 14 percent of couples are not able to conceive after one year of trying. (See Table 1-1.) Almost half of these infertile couples will succeed sooner or later, often after seeing a medical

Figure 1-1
All other desires seem unimportant as parents await the birth of their child.
Despite all the hazards in the biology of the human condition, most parents
will experience a happy outcome: an alert, responsive, healthy, normal child.
(Anna Kaufman Moon/STOCK, Boston)

specialist, but this still leaves some eight percent of couples who are permanently sterile. Contradictory as it seems, much of this sterility is inherited: The processes that lead to mature sperm and eggs, bring them together, and transport the fertilized egg to the uterus are under genetic control. For many of these couples, adoption is not a useful alternative because healthy children given to adoption agencies are no longer readily available. Once again a biological issue is involved. Widespread use of birth control methods, changes in public attitudes concerning sex education, legislative and court-mandated liberalization of abortion laws, and the public acceptance of the single parent have caused the rapid depletion of the major sources of children available for adoption. In the future, if this trend continues, the outlook can only become more bleak for sterile couples as adoption will become virtually impossible.

Spontaneous Abortion Occurs Frequently, Is Often Unnoticed, and Usually Involves a Chromosomal Defect

Even if the first hurdle of fertility is overcome and pregnancy is achieved, the imperfection of the human condition often reveals itself. About one-fifth of all confirmed pregnant women spontaneously abort or miscarry the developing fetus.

Parents will always feel grief over the loss of a developing baby, but the longer the term of the pregnancy, the greater will be the impact of

Table 1-1   Time of Success in Couples Attempting to Achieve a Pregnancy

Most couples who attempt to have children will achieve a pregnancy within three months. About 85 percent of all pregnancies will begin within the first year of effort. Couples are usually encouraged to seek medical advice if, after one year, no pregnancy has occurred. Note that about two-thirds of those who have not succeeded after the first year will do so eventually without medical attention. Sterility after all medical tests have been done will still exist for about eight percent of these couples. Dr. Guttmacher's survey involved 5574 couples.

Length of Time Required for Fertilization to Occur	Percentage of Couples with Pregnancies	Percentage of All Couples Still Not Pregnant
1 month or less	0.33	0.67
2–3 months	0.50	0.50
4–6 months	0.65	0.35
7–12 months	0.78	0.22
1–2 years	0.86	0.14
More than 2 years	0.92	0.08
Total pregnant	0.92	

Source: Adapted from Guttmacher, 1973.

the abortion on the parents. In the very early stages of the embryo's development, the mother may not even be aware of her pregnancy and subsequent abortion. For example, a fertilized egg that fails to implant in the inner wall of the mother's womb is never even known to exist as an individual. If the early embryo aborts just as its tissues begin to form embryonic organs, long before it is a month old, the mother may not even be aware that the pregnancy existed. Even a three-month-old fetus that aborts may cause less trauma to the parents than a five- or six-month-old fetus. It is often true that the longer the parents invest in the developing child their expectations and hopes for its future, the greater will be their grief if it is lost.

Examination of spontaneously aborted fetuses is very revealing. About half of them contain abnormal numbers of chromosomes. At fertilization, the sperm and the egg each contribute 23 chromosomes (forming a cell with 46 chromosomes), containing all the messages and instructions to enable the cell to generate a human being (Figure 1-2). Aborted fetuses frequently have one chromosome more or less than the normal count. Sometimes there are as many as three sets of chromosomes (69 per cell) in the malformed, aborted embryo (Figure 1-3).

Birth Defects May Be
Present in the Newborn   Sometimes the anxiety about seeing the fetus to its birth is less powerful than the surge of fear that engulfs the parents on the day of the baby's birth. The waiting room pacing, the fright-benumbed hands, and the jumpy, edgy father-to-be seen in cartoons

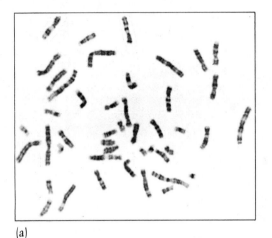

(a)

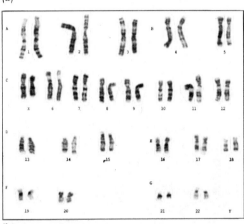

(b)

Figure 1-2
The normal number of chromosomes in humans is 46. (a) These are the chromosomes as they appear under the microscope and (b) as they appear in a karyotype, *an alignment of the chromosomes in order of size. Note that there are two of each chromosome type; such pairs of chromosomes are called* homologs. *(The donor in this case was a female.) (Courtesy of Dr. Carolyn Trunca)*

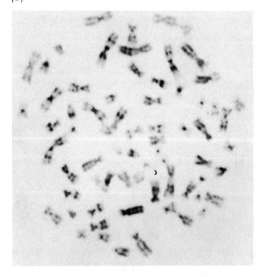

Figure 1-3
A cell from an aborted fetus shows the triploid condition (three of each chromosome type) for the 23 types of chromosomes found normally as homologous pairs. Thus there are 69, not 46, chromosomes. Triploidy is incompatible with human life and most triploid fetuses abort spontaneously. There are many other forms of chromosome abnormality found among spontaneously aborted fetuses. Triploids often arise from the fertilization of one egg (with 23 chromosomes) by two sperm (each with 23 chromosomes). (Courtesy of Dr. Carolyn Trunca)

are part of our tradition because of the possibility of complications for the mother or a child born with a birth defect.

Some five percent of all births involve serious birth defects requiring major medical treatment. Some of the defects are too complex and serious for any hope of the baby's survival. These include those unfortunate babies with an extra chromosome whose excessive instructions cause dozens of organs and tissues to develop improperly. Multiple birth defects may result in a defect of the heart, liver, kidneys, intestines, and/or brain, and sufficient external abnormalities to quickly alert the attending physician and nurses that a tragedy has befallen the parents.

Often the birth defect arises from a defect in a single gene, as in the cases of Tay-Sachs disease, sickle cell anemia, hemophilia, cystic fibrosis, and muscular dystrophy. A generation ago many of these disorders were unknown to the general public, but as concern over birth defects has increased, public campaigns to help the children and parents devastated by the diseases have made their names more familiar.

Many other disorders, whose exact genetic bases are not known, seem to be hereditary, occurring among members of certain families at a very low incidence. Harelip, clubfoot, many blue-baby cardiac defects, and spinal defects such as anencephaly (an absence of most of the brain) and spina bifida (a sac-like extrusion of the spinal cord) belong to this group. A substantial portion of retardation is also genetically linked. Unlike the exact single gene defects, these disorders involve several genes and very little is known about their precise mechanism of inheritance.

We Vary
in Our Genetic Make-up   When the child has escaped these early hurdles of infertility, spontaneous abortion, and birth defect and is on his or her way to a full life potential, yet another hazard must be faced. The particular combination of genes the child receives (which is somewhat dependent upon luck) may impair its survival. Every fertilization (combination of genes) is like a new deal from a well-shuffled deck of cards. Some individuals have combinations that cause them to die younger or to experience more medical crises, surgery, and chronic illnesses than other people. Some geneticists have estimated that about 20 percent of humanity harbors such unfortunate combinations of genes.

The Human Condition Changes
Every Generation or So   Less than two centuries ago, people experienced a far more ruthless and tragic sense of life. Infectious diseases ravaged newborn babies and young children as well as adults (Table 1-2). As recently as 1900 in the United States, the respiratory infections tuberculosis and pneumonia were the two main causes of death, accounting for 11 percent of all mortalities. In third place was a childhood dysentery arising from raw milk and unsanitary food preparation. The

Table 1-2   Incidence of Certain Causes of Mortality in the United States, 1900–1980

Note the rapid decrease in mortality from infectious diseases after 1920. The numbers in parentheses indicate the rank order of the ten major causes of mortality. Any increase or decrease in a given cause of mortality reflects changes in social conditions, medical practices, or lifestyles. Note the decreases in strokes and most accidents, and the increases in cancer, heart disease, and motor vehicle accidents. The number of deaths for each disease is among 100,000 persons in that particular year.

Cause	1900	1920	1940	1980
Influenza and pneumonia*	202 (1)	147 (2)	60 (5)	26 (6)
Tuberculosis*	194 (2)	102 (6)	43 (7)	4
Gastritis*	142 (3)	34 (8)	9	3
Heart disease	137 (4)	210 (1)	289 (1)	334 (1)
Stroke	106 (5)	122 (3)	88 (3)	80 (3)
Kidney disease*	88 (6)	104 (5)	69 (4)	5
Accidents (non-motor vehicle)	72 (7)	62 (7)	49 (6)	27 (5)
Cancer	64 (8)	108 (4)	118 (2)	181 (2)
Senility	50 (9)	—	—	—
Diphtheria*	40 (10)	13	1	0
Typhoid fever*	27	8	1	0
Diabetes	15	21 (9)	25 (8)	15 (7)
Congenital malformations	13	15 (10)	12 (10)	10
Motor vehicle accidents	0	13	22 (9)	27 (4)
Cirrhosis of the liver	16	9	9	13 (8)
Arteriosclerosis	—	—	—	13 (9)
Suicide	12	13	12	12 (10)
Homicide	1	8	6	6
Whooping cough*	8	7	2	0
Measles*	8	6	1	0
All mortalities	1719	1328	1017	883

Source: From National Center for Health Statistics, Division of Vital Statistics (June 1982) and Erhardt and Berlin, 1974.

*Infectious diseases.

high frequency of mortality among the young shortened the average life span. At the time of the signing of the Declaration of Independence (1776), the mean life expectancy in Philadelphia was 28. Today it is about 75.

The very rapid increase in the average length of life was largely due to advances in medicine and public health practices after 1900, which made possible the survival of infants through their late childhood and into adulthood. Milk was pasteurized to kill pathogenic microbes. Visiting nurses instructed mothers on infant care and nutrition. Public health

programs chlorinated water supplies. Mass immunization was introduced for school children to eliminate diseases such as smallpox and
diphtheria. People learned to intervene in the natural world, and to apply
a strategy of defensive living to their lifestyles (Figure 1-4 and Table
1-3). The conquest of disease by immunization was extended to whoop-
ing cough, measles, mumps, and polio. The introduction of antibiotics,
beginning with sulfa drugs in the 1930s and penicillin in the early 1940s,
represented another technique of defensive living. These artificial means
of controlling diseases are widely accepted today as improvements over
the natural lottery of premature death and suffering of previous centuries.

Resistance to Change
Is Usually Encountered We should not forget, however, that public
acceptance of these techniques has been uneven. Vaccination was opposed early in its history. The right of the state to demand immunization
against certain communicable diseases is still resented by some religious
groups and by individuals who feel that they, and not a public health
administration, should make that decision for themselves. Even more
controversial today are the issues of medical abortion, genetic counseling, and artificial insemination. Clearly, it is more than just a desire for
improved health that determines our enthusiasm for some forms of defensive living and our hostility toward others. The issue is complex be-

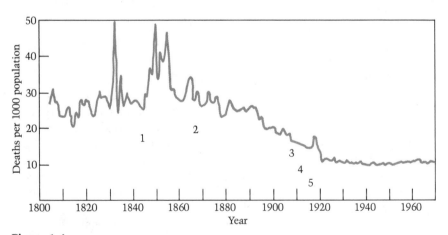

Figure 1-4
*Annual mortality between 1800 and 1890 was about three percent, after
which the rate declined to one percent by 1920, where it has remained fairly
stable. Two-thirds of mortality before 1920 was due to infectious diseases
which were rampant after the opening of the Croton aqueduct in 1842. The
formation of the Board of Health (1866), the control of typhoid carriers (1907),
the chlorination of drinking water (1910), and the pasteurization of milk
(1912) all helped to reduce the mortality rate to its 1920 level. Table 1-3
indicates the epidemics that struck New York City between 1800 and 1920.
(Adapted from Erhardt and Berlin, 1974.)*

Table 1-3 Mortal Epidemics in New York City

Date	Disease	Deaths	Population
1804	smallpox	169	75,000
1805	yellow fever	270	75,000
1822	yellow fever	166	124,000
1824	smallpox	394	165,000
1832	cholera	3,513	202,000
1834	cholera	971	205,000
1834	smallpox	233	205,000
1849	cholera	5,071	519,000
1851	smallpox	562	521,000
1854	cholera	2,509	600,000
1865	smallpox	664	1,000,000
1866	cholera	1,137	1,100,000
1870	yellow fever	9	1,391,000
1872	smallpox	1,660	1,500,000
1873	smallpox	1,859	1,600,000
1881	smallpox	505	1,900,000
1891	smallpox	182	2,300,000
1892	smallpox	302	2,400,000
1892	typhus fever	200	2,400,000
1892	cholera	9	2,400,000
1901	smallpox	410	3,500,000
1902	smallpox	310	3,700,000
1918	influenza	12,562	5,700,000

Source: Adapted from Erhardt and Berlin, 1974.

cause personal freedom to make choices may conflict with family, religious, cultural, or political values which override individual desires.

Our values change and new generations practice habits our fore-fathers would have abhorred. We should expect in our lifetime many new techniques of defensive living to be developed by biologists and physicians. Not all of these will appeal to our present values, and some may even be more undesirable than the conditions they are designed to cure or prevent. Our decision to add a new technique to defensive living or to reject it requires a familiarity with the biological causes of the human condition and a grasp of the biological mechanism involved in each new technique. To some degree we can control nature's claim on our lives through medication, surgery, artificial devices, and other public health measures. There is no question that many of these interventions in the natural state of things involve risks. To decide if the benefits are worth the use of a new technique, we need to know what goes on in our cells and organs. We must become biologically literate in order to intelligently practice defensive living.

Questions to Answer

1. Name four biological crises in the human condition and the stage of the life cycle when they occur.
2. What is your interpretation of the paradoxical opinion that "the more drastic and biologically devastating the defect is to a fetus, the less serious will the blow be to the parents"?
3. What percentage of couples (a) are sterile, (b) experience a spontaneous abortion, or (c) have a child with a birth defect?
4. What are the probable causes of the difference in the mean life expectancy in Philadelphia between 1776 and 1976 (28 years and 75 years, respectively)?
5. What changes took place in urban living between 1800 and 1950 to greatly reduce the occurrence of epidemics?

Terms to Master

Briefly define or characterize each of the following terms.

human condition
human nature
infertility
sterility

spontaneous abortion
chromosome number
triploid
birth defect

Part 2
Cells and Chromosomes

Your concept of self gives you a unique feeling. You are not only a person but you are unlike any other individual alive. You are one among four billion people, and each person is perceived as a unit in a community, a population, and a nation. Whatever language, social customs, ethnic traits, and familial habits you share with others, you know that even if you have an identical twin, you are unique.

Yet, as we descend from the totality of our bodies to our individual organs and tissues, we seem to find less uniqueness, and more similarity, with others. That is, under the microscope, it can be seen that every person's organs and tissues are composed of multitudes of units, called *cells*. This realization did not immediately strike Robert Hook in 1665 when he observed a piece of cork. He called the empty spaces "cells," not realizing that later on it would be found that cells were living units making up all plants and animals. At the time he thought the buoyancy of cork was caused by the cells or empty spaces. It was not until 1838 that cells were thought to be living units composing tissues and that each individual was a community of cells. The cell theory of Matthew Schleiden and Theodore Schwann (1838) has been repeatedly confirmed throughout the entire life cycles of all higher plants and animals. As is true for these organisms, we begin our biological individuality as a fertilized egg and, through numerous cell divisions, become an organized being.

If we understand our cellularity at all, it is at an abstract level, for we cannot see our own individual cells except through the microscope. We cannot experience the death of a single cell. If a cell turns malignant we often do not know it for many years (20 years or more for most cancers). If a sperm or egg is suddenly altered we are unaware of the change—a change that could profoundly alter the life of one of our future children.

Our response is quite the opposite when thousands or millions of cells are killed (as when we scald ourselves with hot water) or toxified (by viruses, bacteria, or chemicals). The nerve cells embedded in our tissues let us know almost immediately when massive numbers of cells are in trouble. Unfortunately, there are two exceptions to this—cancers and mutations—which arise as changes in single cells. It takes many cell divisions to produce an individual with a genetic disorder or a tumor capable of producing symptoms.

Our inability to reach down to our cellularity, or view our own cells, limits our response to events that slice, crush, poison, burn, freeze, and kill our tissues. Even if we are able to scan our individual cells we would probably not see the chemical or physical agents that act on the cells as mutagens or carcinogens. They may be working at the level of the chromosomes or genes within the nucleus of the cell. Such components of the cell may represent only one millionth of the total mass or volume of the cell. They are at the molecular level, and only occasionally are visible under an optical microscope.

In this section on cells and chromosomes you will learn how cells make more cells, how the cell components are multiplied and distributed, and how errors can occur, often with tragic and profound effects on the human condition.

Chapter 2
Cell Structure, Cell Division, and the Cell Cycle

There are three important structures—the cell, the chromosome, and the gene—that we should know about if we are to understand some of the hazards that living things encounter. Unfortunately we live our lives at a level of perception many thousands of times larger than these three structures, and it is only through microscopes that the cells and chromosomes can be perceived. For this reason it is easy to ignore their existence and significance. If we are ill we often think of our body as being sick or a specific organ as being out of sorts. When environmental hazards exist in our foods or our neighborhoods we may not think of their noxious effects on our genes, chromosomes, or cells. As long as we consider our total body or self as the unit being attacked we have a tendency to ignore all but the most toxic and hazardous conditions facing us.

Each Individual Is Composed
of a Multitude of Cells
with a Complex, Microscopic Anatomy   Although each of us is a single being, we are each composed of about 100 trillion (10^{14}) cells. This immense number of cells is the result of approximately 50 rounds of cell divisions, starting with one fertilized egg. Each cell consists of several structures called *organelles* (Figure 2-1). They are given that name because each of these organelles, like an organ in a body, carries out certain functions essential to the life of the cell. A cell is alive if it has a surrounding cell *membrane*; a *nucleus*; a network of inner membranes (the *endoplasmic reticulum*) within which protein synthesis occurs; a number of *mitochondria*, which produce energy by burning the carbon compounds we eat with the oxygen we breathe; and a number of *lysosomes*,

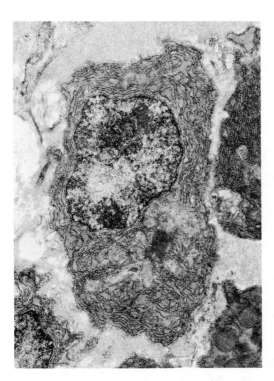

Figure 2-1
An electron micrograph of a eukaryotic cell shows the detail of its organelles. Note the potato-shaped mitochondria, the spongy network of the endoplasmic reticulum, the nuclear membrane, and the cell membrane. (Courtesy of Dr. Ben Walcott)

which digest debris and large molecules in the cell. Figure 2-2 shows a typical cell. The functions of its organelles are more fully described in the legend.

Cells Can Be Physically
Taken Apart and Reassembled   To some degree cells can be manipulated and still function. The nucleus can be plucked from one cell and placed in another that has had its own nucleus removed. Such cells may survive, especially if the nucleus is from the same organism. Similarly, it is even possible in a large cell, such as an amoeba, to isolate a cell membrane from one cell, a nucleus from another, and all of the remaining organelles and fluid contents from a third cell. These three components can then be reassembled into a living single cell, as John Danielli demonstrated in 1970 (Figure 2-3).

Such manipulations of the living cell suggest that the cell is composed of interchangeable parts and enable us to study the functions of these parts in great detail. By no means does this imply that today's technology and understanding of the cell is sufficient to synthesize a cell from ordinary chemicals. For example, each of our cells is millions of times larger than the non-cellular organisms called viruses. The simple composition and the small size of a virus permit scientists to take it apart and even synthesize its components, which then function nor-

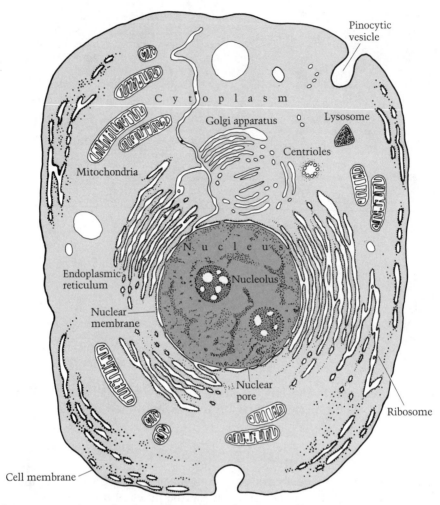

Figure 2-2   The Internal Anatomy of a Eukaryotic Cell
The major organelles of a cell are represented in this diagram. The endoplasmic reticulum *is the location where protein synthesis is carried out, with the individual* ribosomes *serving as the specific sites for this process. The* nuclear pores *permit large molecules (especially RNA) to leave the* nuclear membrane. *The* nucleolus *is formed from a special region of one or more chromosomes, and the vesicle it forms stores RNA molecules. The* cytoplasm *is all of the cell outside of the nucleus.* Lysosomes *digest cell waste and bring molecules into and out of the cell. The* centrioles *form the fibers which move chromosomes during cell division. The* golgi apparatus *stores enzymes which are released into the lysosomes. The* nucleus *is a membrane-bound region housing the chromosomes, which contain the genes or instructions for the components and organization of the cell.* Mitochondria *carry out cell respiration, releasing energy through oxidation of carbon compounds. A* pinocytic vesicle *is formed from the inpocketing of the* cell membrane *(also called the plasma membrane). The liquid suspension within the vesicle is often digested by enzymes present in the lysosomes. The diagram only shows a few lysosomes and mitochondria for clarity, but in a typical cell there may be dozens or hundreds of these organelles.*

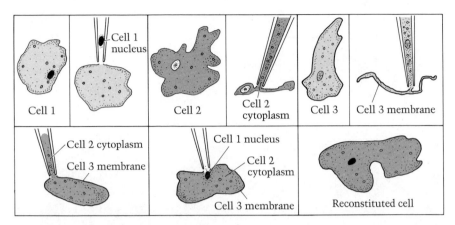

Figure 2-3 Reconstitution of an Amoeba
Danielli used components from three different amoebas. The nucleus is removed from Cell 1 and retained in a micropipette. Most of the cytoplasm from Cell 2 is retained in a second pipette. The third amoeba (Cell 3) is depleted of both nucleus and inner cytoplasm, leaving the collapsed cell membrane. If the cell membrane receives the contents of both pipettes, a reconstituted cell is formed, which is alive and normal. None of the isolated components is capable of independent life, nor can any of the cells depleted of one or more of these three major constituents survive. The surgical removal and replacement of all three components, each from a separate cell, proves that these components are interchangeable among the same species of amoeba.

mally. A bacterial cell (which lacks a nucleus and has a much simpler organization than a eukaryotic cell) is also too complicated at present for us to completely understand how all of its components work together (Figure 2-4).

Is the Study of Life Uniquely Different from the Study of Chemistry or Physics?

Many scientists believe that their inability to take apart a cell down to the molecular level reflects technical difficulties rather than a philosophical impossibility. Some people are troubled by this thought. They would like to believe that the cell, which is the smallest unit of human life and is capable of sustaining itself and forming a human being, is structurally unique and that no matter how much effort scientists put into their research some essential features will remain unknowable. Those who believe this are called *holists* because they claim that the cell is more than the sum of its parts. Those who reject this view are called *reductionists* because they believe that the whole cell can be interpreted from a knowledge of its parts. In general, holists believe that there is a unique, vital, often supernatural, aspect of the living cell, although there are some holists who deny any supernatural features in life. Most reductionists reject the idea that biological life possesses any supernatural components at all.

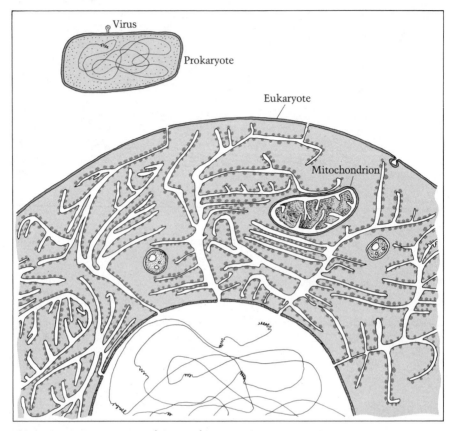

Figure 2-4 Comparison of Sizes of Living Units
*Living organisms consist of viruses (which are not cells), prokaryotes (cells
lacking separate nuclei), and eukaryotes (cells containing separate nuclei). The
virus is so much smaller than a prokaryote (such as a typical bacterium or
blue-green alga) that it resembles a part of an organelle of a prokaryote. The
prokaryote is about 1/100 of the size of a typical eukaryotic cell. Note that a
rod-shaped bacterium is about the size of a mitochondrion in a eukaryotic
cell.*

For many people all this fuss over holism and reductionism may seem
silly because they have no interest in philosophical or religious questions. For others it can be disturbing because the reduction of life to non-
life shakes their religious beliefs about the significance of life. Does the
reductionist view permit a scientist to do research on human embryos
in test tubes? Would a holistic view prevent a scientist from studying
the molecular basis of cancer or other pathologies? There are no simple
answers to these questions. A scientist can be a reductionist and still
feel it is ethically wrong to do research on human embryos. A holist may
do research on cancer at the cell level because the cancer is viewed as
an abnormality imposed on the cell and is thus subject to analysis even
if the cell itself is not comprehensible in its entirety.

**There Are Three Major Ways
in Which the Instructions
for Heredity May Be Organized**   Whatever one's belief about our future understanding of life, there is universal agreement that we are composed of cells, that the cells contain organelles, and that the functions of these organelles are known, at least in part. There is also agreement that living organisms exist on at least three levels of complexity. There are *viruses*, which are not cells, but which are more complex than most of the larger molecules that compose our cells. There are *prokaryotes*, cells like bacteria which lack nuclei, have very few organelles, and have a relatively simple organization. Finally, there are *eukaryotes*—true cells—like those in the tissues of all higher plants and animals (including humans). They contain many types of organelles and a nucleus that is clearly separated from the cytoplasm.

The simplest viruses consist of two components: an inner thread-like or circular molecule of nucleic acid and a coating or protective envelope of protein. The nucleic acid may be ribonucleic acid (RNA) or deoxyribonucleic acid (DNA), but never both in the same virus. Whichever nucleic acid is present, it functions as a chromosome, if we define a chromosome as a sequence of genes or hereditary units.

Prokaryotes are familiar to us as bacteria or as blue-green algae (also called *cyanobacteria*). They consist of an outer wall, a cell membrane, and an inner cytoplasm or collection of living material. The prokaryotic cytoplasm includes one circular (and very twisted) DNA molecule, which forms its chromosome; many thousands of ribosomes (units of RNA and protein which synthesize proteins); and millions of molecules of water, minerals, organic compounds, and enzymes.

**Human Cells Contain
a Constant Chromosome Number**   The *chromosome* is an organelle which carries the information that is stored in genes. It is like a cassette tape, and in its copied and packaged form, it can be distributed to both cells during cell division. Each eukaryotic organism has a constant chromosome number. For humans it is 46; for fruit flies, 8; for sweet peas, 14. The chromosome number may be as small as 2 (in cells of the parasitic thread worm, *Ascaris*) or as high as 480 (in cells of the fern, *Ophioglossum*). In eukaryotes the chromosome number is usually even rather than odd (Table 2-1). When the chromosomes are fully coiled, aligned, and ready to be distributed, they can be photographed. The photo can be enlarged and the chromosomes cut out and lined up in order of size. Such a procedure produces a *karyotype*. If we did this in most human cells we would see that the chromosomes can be made to match up in 23 pairs, totalling 46 individual chromosomes. If we looked at developing eggs or sperm, however, we would usually encounter only 23 chromosomes. When fertilization occurs, the resulting cell, called a *zygote*, is once again restored to 46 chromosomes.

Table 2-1   Chromosome Numbers of Selected Organisms

There is no correlation between the chromosome number and the complexity of an organism. Organisms with larger numbers of chromosomes (several dozen) usually have more than two of each of the chromosomes. This is particularly characteristic of plants. The largest known chromosome number in an animal is found in the crayfish. Among plants the adder's tongue fern holds that record. All the organisms shown in this table have a sexual phase (and thus a haploid state) except *Amoeba proteus*, whose haploid state is inferred because sexual reproduction in this species has never been demonstrated.

Common Name	Taxonomic Name	2N
Human	*Homo sapiens*	46
Chimpanzee	*Pan troglodytes*	48
Spider monkey	*Ateles paniscus*	34
Cow	*Bos taurus*	60
Horse	*Equus caballus*	64
Dog	*Canis familarus*	78
Cat	*Felis catus*	38
Mouse	*Mus musculus*	40
Rat	*Rattus norvegicus*	42
Chicken	*Gallus gallus*	78
Alligator	*Alligator mississippiensis*	32
Frog	*Rana pipiens*	26
Crayfish	*Cambarus clarkii*	200
Bee	*Apis mellifera*	16
Fruitfly	*Drosophila melanogaster*	8
Earthworm	*Lumbricus terrestis*	36
Snail	*Helix pomatia*	54
Hydra	*Hydra vulgaris*	32
Amoeba	*Amoeba proteus*	50
Corn	*Zea mays*	20
Wheat	*Triticum aestivum*	42
Rice	*Oryza sativa*	24
Potato	*Solanum tuberosum*	48
Oak tree	*Quercus alba*	24
Cotton	*Gossypium hirsutum*	52
Adder's tongue fern	*Ophioglossum vulgatum*	480
Mushroom	*Agaricus campestris*	24
Orange bread mold	*Neurospora crassa*	14
Black bread mold	*Rhizipus nigricans*	32

Source: From P. L. Altman and D. S. Dittmer, *Biology Data Book* (Federation of American Societies for Experimental Biology, 1972).

Living things grow. To do so they require a precise process for transferring their essential features from cell to cell. Yet when we see eukaryotic cells divide, there is no exact distribution of the cytoplasmic components. One cell may get more mitochondria than another; one may have a larger portion of cytoplasm than another. However, there is always an exact distribution of the chromosomes within the cell. This precision gives each cell the same instructions for carrying out cell functions that were originally present in the cell prior to its division.

The Cell Cycle Describes the State of a Cell

Most of the cells in our tissues do not have nuclei with sausage-shaped chromosomes as we saw in Figure 1-2. More often the chromosomes cannot be seen at all because they are finely dispersed as almost completely unwound threads (Figure 2-5). During the early part of one's life cycle (from fertilization through formation of a fetus in the mother's womb) all the cells are going through a process of cell division called *mitosis.* The rate begins to slow down for some tissues as the fetus enlarges and is eventually born as a baby. The slowing of the cell division rate varies in different tissues and organs as the child develops. Very few cell divisions occur in most college-age students, especially in their muscle and nerve tissue. Cell growth in adult bony tissue is limited to the replacement of dead cells only.

When cells are studied carefully they can be assigned to different stages of a cell cycle. Most of an adult's cells are in a stage called G_0 (the G stands for a gap between cell divisions and the 0 stands for the inability to divide). Cells that still have a potential to divide are usually in G_1, a

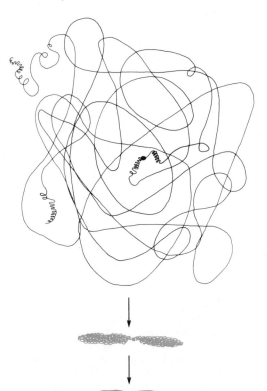

Figure 2-5   Chromosome Coiling

During cell division the long thin genetic thread present in the active metabolic stages, as illustrated at the top, undergoes coiling. This makes the chromosome appear wider and shorter. The length is actually the same, tip to tip, but in the wound-up form it appears shorter. This compact state permits movement of the chromosome without entanglement. During cell division there are two threads for each chromosome. For clarity, only one is shown to emphasize the effect of coiling on apparent size.

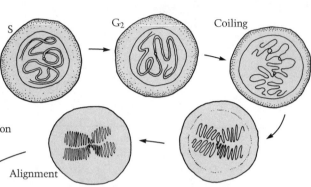

Figure 2-6(a)   Cell Cycle

Active cell division or mitosis (M) is followed by a metabolic phase (G_1), leading to DNA synthesis (S) and a second metabolic phase (G_2), which prepares the cell for mitosis. This is known as the cell cycle.

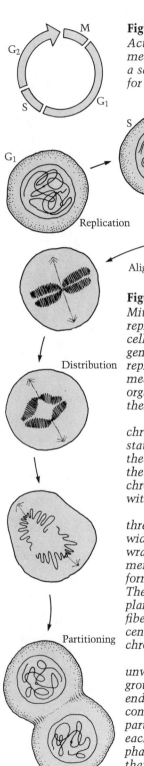

Figure 2-6(b)   Mitosis

Mitosis involves a precise process for chromosome replication, separation, and distribution, with each new cell having exactly the same chromosome number and genetic constitution as the original cell prior to its replication. At the same time, it involves an imprecise means of partitioning off the numerous cytoplasmic organelles, so that each new cell gets about half of them.

In this diagram of mitosis we follow one chromosome of a cell from G_1, the normal metabolic state for that tissue, through replication of the DNA in the S stage, and to the preparation for cell division in the G_2 phase. Note that DNA synthesis converts one chromosome with one chromatid into one chromosome with two chromatids held by the centromere.

The chromosomes, still finely spun out as long threads, begin to coil, causing them to shorten and widen and become more compact, much like string wrapped around a pencil. At the same time the nuclear membrane dissolves and a spindle-shaped apparatus is formed, surrounding the chromosomes like a bird cage. These compact chromosomes are aligned in a single plane with each centromere attached by two spindle fibers facing opposite poles of the cell. When the centromere divides, two chromosomes are formed, each chromatid now having a separate centromere.

As the chromatids are pulled apart they begin to unwind, and a nuclear membrane surrounds the two groups of chromosomes that are gathered at opposite ends of the cell. The cytoplasm begins to partition by a contraction of the region between the two nuclei. This partitioning process, or cytokinesis, yields two cells, each of which is ready to grow during the new G_1 phase. The G_0 stage is not represented here because that is the terminal stage of a nondividing cell.

period of active metabolism characteristic for the tissue. G_1 is also the stage in which the components necessary for DNA synthesis are manufactured in large quantities. The synthesis of DNA, which is the basis for chromosome replication, occurs in the S phase. When the DNA is completely replicated each chromosome contains two threads of DNA instead of one. The genetic thread, tip to tip, is called a *chromatid.* Thus each chromosome contains two chromatids at the end of the S phase. The two chromatids are held together at a special location by a *centromere,* which is formed from the genetic thread. Another long interval, the G_2, occurs in the cell cycle. During this stage the chromosomes actively produce substances needed for cell division. When this is successfully done the cell finally enters cell division.

With your knowledge of the terms chromosome, chromatid, and centromere, it is now possible to describe cell division itself (Figure 2-6).

Cell Division (Mitosis)
Produces Two Cells from One   The major stages in cell division can be summarized as follows. (1) The replicated chromosomes become coiled or packaged into sausage-shaped objects (this prevents the genetic threads from becoming entangled). (2) The nuclear membrane disperses as the chromosomes are aligned or spread out in a single plane. (3) The chromosomes are attached by contractile fibers that separate and pull the two chromatids of each chromosome in opposite directions. (4) When this

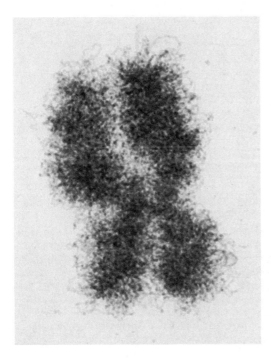

Figure 2-7
During the greater part of the cell cycle, the chromosome remains in an unwound condition. Coiling during mitosis is complex, with several levels of supercoiling, as seen in this electron micrograph of an individual eukaryotic chromosome in late prophase or early metaphase when the chromatids are fully coiled. (Courtesy of Dr. G. F. Bahr)

separation occurs the chromosomes begin to unwind in the confines of new separate nuclear membranes. (5) A pinching in or partitioning of the cytoplasm establishes the physical boundaries of two new cells.

Biologists give technical names to some of these processes. The early coiling occurs during *prophase* (Figure 2-7). The alignment of the tightly coiled chromosomes occurs during *metaphase*. When the two chromatids separate, *anaphase* begins. The distribution and uncoiling of the separated chromosomes takes place in *telophase*, as does the cytoplasmic partitioning, or *cytokinesis*, of the cell. The G_1, S, and G_2 phases of the cell cycle are sometimes collectively lumped together as the *interphase*.

From this knowledge of the cell cycle we can conclude that the cell division process between the time the egg is fertilized and the time the baby is born involves a precise and repeated copying of the 46 chromosomes and a distribution of them to each of the 10^{13} to 10^{14} cells found in the child. Also, we can infer, from the matching pairs of chromosomes revealed by a karyotype, that a special cell division (which actually occurs only in testes and ovaries) reduces the chromosome number from 46 to 23.

The cell division that produces most of our cells is called *mitosis*,

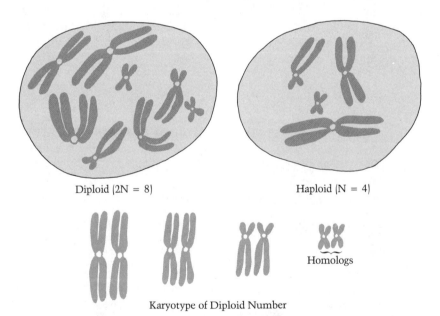

Diploid (2N = 8)   Haploid (N = 4)

Homologs

Karyotype of Diploid Number

Figure 2-8 Diploid and Haploid Chromosome Numbers
When cells of nonreproductive tissues are studied, they show a constant chromosome number in which each type of chromosome is represented twice. This is called the diploid chromosome number. In maturing sex cells, however, especially those about to become gametes (the mature sperm or eggs), only one of each chromosome is found, half the number of the diploid. The gametes contain a haploid chromosome number. The karyotype of the chromosomes from a diploid cell proves the duality of each chromosome in the diploid. Such pairs of matching chromosomes are called homologs.

and the special process that produces eggs and sperm is called *meiosis*. At fertilization the cell containing two sets of instructions (that is, the fertilized egg) is called *diploid*. The individual reproductive cells (sperm and egg) are called *haploid* because they have only a single set of instructions (Figure 2-8), represented by one of each of the pairs of chromosomes present in a karyotype.

The First Meiotic Division Reduces
the Chromosome Number by Half
During meiosis the reproductive cells accomplish a reduction of chromosome number from the diploid to the haploid state (Figure 2-9). This unique form of cell division differs strikingly from a typical mitotic division. The chromosomes do not coil uniformly; they produce a beaded pattern of densely and sparsely coiled regions—a pattern that is unique for each of the matching pairs of chromosomes (called *homologous* chromosomes). These thin, irregularly coiled chromosomes proceed to pair by a process analogous to the closing of a zipper, which brings the homologs together. After pairing and the completion of compact coiling throughout the length of the chromosomes, the pairs of homologs are aligned in two planes. Unlike mitosis, where the diploid number of chromosomes is aligned in a single plane and each centromere is attached to both poles of the spindle, each member of a homologous pair during reductional division has its centromere attached to only one pole of the spindle. The upper haploid set of chromosomes is attached to one pole, and the corresponding haploid set of chromosomes is attached to the opposite pole.

The actual reduction from diploid to haploid takes place when the spindle fiber attached to each centromere contracts. One haploid set of chromosomes migrates to each pole of the spindle and when the cytoplasm pinches in and is partitioned off into separate cells, *reductional division*, or meiosis I, is completed.

A Second Division Is Needed
to Complete Meiosis
Each one of these reduced cells has chromosomes with replicated strands. Thus a second division—*equational division*, or meiosis II—occurs. It is very much like mitosis except that no new replication occurs. The chromosomes coil uniformly and align in a single plane, and each centromere is attached by a fiber from each pole. During distribution the two chromatids are pulled apart to opposite poles.

The two divisions of meiosis yield four cells from one, each of them haploid. The bookkeeping is fairly simple: 46 single-stranded chromosomes replicate, producing 46 chromosomes with a total of 92 strands. These 46 chromosomes pair and complete reductional division, forming two cells with 23 chromosomes each. Since each chromosome is composed of two strands joined by a centromere, equational division merely

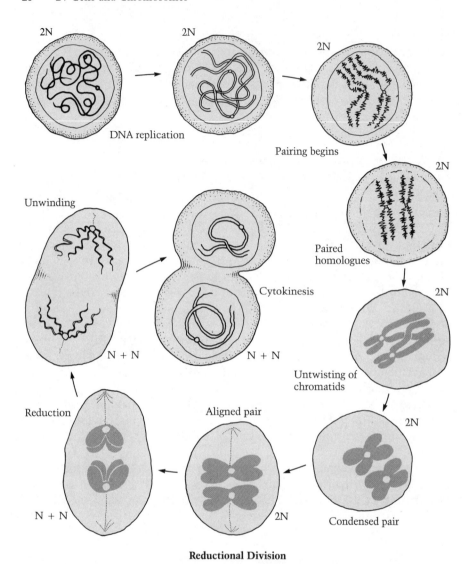

Figure 2-9   Meiosis

In most animals and plants a modified cell division occurs in a specialized organ, the gonad, resulting in gametes, or reproductive cells (the sperm or pollen, and the eggs or ovules). Two functions are served by this process: the chromosome number is shifted from diploid to haploid (from 46 to 23 in humans), and the reproductive cells are modified to carry out the function of fertilization, which reestablishes the diploid number. The G_2 stage of a diploid reproductive cell about to enter meiosis bears the molecules for initiating the two divisions needed to complete the process. It also contains those molecules which modify the cell into a gamete rather than returning it to the original G_1 functions.

The first of the two meiotic divisions, called reductional division, *converts the diploid (2N) to the haploid (N) chromosome number. In this process matching chromosomes (homologs) pair by a process called* synapsis. *They also undergo a thick and thin coiling, which results in knobby chromosomes. When more fully condensed, the chromatids, which are twisted around each*

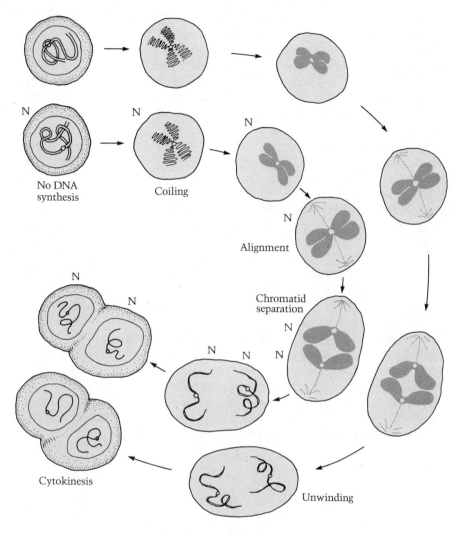

No DNA
synthesis

Coiling

Alignment

Chromatid
separation

Cytokinesis

Unwinding

Equational Division

other, slide free, and the very compact chromosomes of each pair of homologs
are moved to form a double plane at alignment. The centromere of each
chromosome is attached to only one pole, the nearer one. No separation of the
chromatids occurs; rather, the two homologs move apart and the two haploid
cells are partitioned, completing the first division.

The second meiotic division converts haploid chromosomes with two
chromatids to haploid chromosomes with one chromatid. In this respect the
second division, or equational division, resembles mitosis. Unlike mitosis and
the first meiotic division, however, there is no S phase prior to the G_2 as the
cells prepare for a typical coiling and alignment of the chromosomes in a
single plane. The centromeres are now attached to opposite poles, and the two
chromatids of each chromosome move apart. This assures the production, at
cytokinesis, of two haploid cells from each of the haploid cells produced by
reductional division. Altogether, four haploid cells are produced from one cell
during meiosis.

converts it to the single-stranded form. Each of the four haploid cells has 23 chromosomes, and all of those chromosomes arose from the total 92 strands that were present after replication of the diploid, immature sex cell.

It would be puzzling if this elaborate, precise mechanism for replication and distribution of chromosomes had no striking significance. But, of course, it is significant. It is easy to demonstrate, both experimentally and by observation of occasional errors in this process, that this precision is essential for normal life. Any departure from a diploid chromosome number in a newly fertilized cell usually leads to the impairment or death of the individual derived from that cell.

If a few mitochondria or some of the internal membranous network in a developing zygote were removed, the cell would probably go on to form a perfectly healthy organism. But the removal of even a tiny chromosome from a cell would, in most instances, be disastrous to the cell and probably to the organism derived from it.

Chromosomes are important because they contain the genes or individual instructions for virtually all of the activities characteristic of a species (Table 2-2). Genes, as we shall see, carry the information for making proteins, for components of the protein-synthesizing mechanisms in the cell, and for regulation of activity within and between cells. The genes carry out these functions according to the type of proteins synthesized from their stored information. Some proteins are enzymes, which put together complex molecules from simpler ones or shred larger molecules into simpler components. Other proteins are the structural subunits of membranes and other organelles in the cell. Still others act as switches, turning banks of genes on and off so that different cells end

Table 2-2 Gene Functions

Genes are composed of nucleic acids, especially DNA in eukaryotic organisms. When properly decoded by the cell's metabolic machinery in the endoplasmic reticulum, the proteins that are synthesized may serve as enzymes, structural subunits of organelles, or regulatory switches to turn other genes on or off. In addition to decoding proteins, the genes also produce RNA, a nucleic acid involved in protein synthesis. The decoding process for protein synthesis requires an RNA found in ribosomes (*ribosomal RNA*) and an RNA which carries amino acids, the constituents of proteins, to ribosomes. (This form of RNA is called *transfer RNA*.)

Gene Product	Function
Enzymes	Synthesis and digestion of molecules; the production of energy
Subunits	Protein components of membranes and other organelles of the cell
Regulators	Switches that turn various genes in the cell on or off
Ribosomal and transfer RNA	Associated with protein synthesis in the cell cytoplasm

up as different tissues, each highly specialized to perform a limited but essential function in an organ.

It may be difficult to imagine how events at this lilliputian level can affect not only each gene, but each cell, tissue, organ, individual organism, and finally the population of humanity itself, where the effects of these genes may influence public policy! Yet this sometimes occurs, and it is our quest to use these three units—the cell, the chromosome, and the gene—to explore the impact of biology on many controversial and sometimes highly personal issues.

Questions to Answer

1. What is an organelle?
2. What is the relation of the size of a virus to the size of (a) a prokaryote and (b) a eukaryote?
3. Which organelles are common to both prokaryotes and eukaryotes?
4. Can a reductionist believe there is a purpose to life? Defend your viewpoint.
5. What do Danielli's amoeba reconstitution experiments imply about cellular holism?
6. What is a karyotype and what is its relation to the diploid number?
7. Without using the terms prophase, metaphase, anaphase, and telophase, and without the use of diagrams, describe the process of mitosis.
8. What are the major differences between the reductional division of meiosis and the equational division of meiosis?
9. What are the functions of chromosomal genes?
10. What is meant by the term "cell cycle"?

Terms to Master

cell cycle	prokaryote
mitosis	eukaryote
meiosis	virus
reductional division	centromere
equational division	mitochondrion
organelle	lysosome
holism	endoplasmic reticulum
reductionism	diploid
chromatid	haploid
chromosome	homologs

Chapter 3
Radiation Damage: How Chromosome Breakage Can Lead to Cell Death

At 8:15 A.M. on August 6, 1945, a solitary plane, the *Enola Gay,* passed over Hiroshima on a sparsely clouded day. Few of the residents concerned themselves with this apparent reconnaissance flight, or with the small object that floated downward from the plane. But in an instant a brilliant flash blinded those looking skyward, and some 80,000 citizens were either vaporized, blasted to bits, showered with flying glass and other debris,

Figure 3-1
Views of Hiroshima, (a) as it was before the explosion of the atomic bomb on August 6, 1945, and (b) as it appeared shortly after the first use of nuclear weapons on a populated city. (U.S. Air Force photos)

(a)

(b)

or suffocated by the fire storm which sucked up the oxygen around them (Figure 3-1). Many thousands of wounded survivors, moments after the blast, looked with fright at the charred bodies on the ground. It was as if the world as they knew it suddenly ceased to exist.

Most of the survivors who were not much more than a mile from the area of total destruction soon began to experience nausea. Over the next few days they experienced weakness, a measles-like rash over their bodies, and some bleeding in their mouths. They became feverish, racked by the misery of uncontrollable nausea, and began to succumb within a few days or weeks after the blast. Some lingered on and became anemic, often developing peritonitis (an inflammation of the gut cavity) and other infections. Their burns did not heal easily (Figure 3-2). Those who seemed least ill after the first few days gradually began to recover. Sometimes their hair fell out, and grew back after several months. For many, permanent disfiguring scars remained, along with the haunting possibility of cancer later in life.

A second atomic bomb was dropped on Nagasaki a few days later, with similar devastating results.

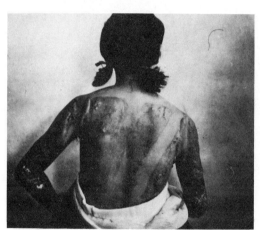

(a)

Figure 3-2
Victims of the Hiroshima nuclear bombing, showing scars produced by slow-healing radiation burns. Note that thicker layers of clothing on the shoulders and back of the victim in (a) and the cap on the victim in (b) gave partial protection against thermal radiation. Special plastic surgery was required to help those victims whose faces were disfigured by these scars. (Courtesy of Defense Nuclear Agency)

(b)

A Few Principles of Biology Can
Explain the Effects of the Atomic Bombs
on Hiroshima and Nagasaki
Many of the striking symptoms of radiation sickness can be explained by a few biological principles. The atomic bomb releases ionizing radiation, mostly highly penetrating gamma rays, the same type of rays that pass through tissues when a doctor or dentist takes a diagnostic x ray. These ionizing rays convert water to hydrogen peroxide, which then reacts with the nucleic acid in the chromosomes of cells. The chromosomes carry the genes or genetic instructions for all of our enzymes and structural proteins and the metabolic activities of our cells.

The radiation may also break a chromosome in a cell. This is likely to occur when the radiation makes a direct hit and its energy becomes absorbed by the chromosome, producing an extremely localized disruption of the giant molecule of nucleic acid which forms the major component of the chromosome. As we shall see, chromosome breakage, and not the alteration or mutation of the individual genes, is the major cause of radiation sickness (Figure 3-3). Although gene mutations are important, to both living individuals and future generations, they had little to do with the illness that struck down tens of thousands of victims in

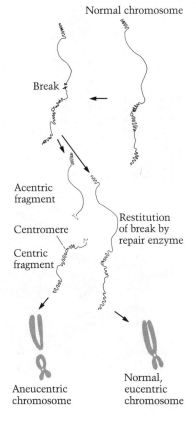

Normal chromosome

Break

Acentric
fragment

Centromere

Centric
fragment

Restitution
of break by
repair enzyme

Aneucentric
chromosome

Normal,
eucentric
chromosome

Figure 3-3   Chromosome Breakage
When a single break occurs in a chromosome it is usually restituted or repaired by a repair enzyme. Such a chromosome functions normally during cell division and shows little, if any, damage to the cell's metabolic functions. If the breakage occurs at a time when the chromosomes are subject to the internal currents of cell fluids, then they may drift apart before the repair enzyme can act. This leads to a nonrestituted break, with the original chromosome now represented by an acentric fragment (lacking the centromere) and a centric fragment (containing the centromere and having its two replicated chromatids united at the site of the original breakage). This cell will eventually die if it enters cell division. If the cell does not divide, the break will not seriously affect the cell. If a chromosome has one centromere and two free ends (telomeres) it is normal or eucentric. If it lacks a centromere or if the number of telomeres is decreased, the chromosomes are called aneucentric.

Hiroshima and Nagasaki who thought they had survived damage from the bomb:

A broken chromosome is distinctly different from a normal chromosome. The normal chromosome is usually a long thread with a single centromere located somewhere along its length. We have already seen that this centromere, when attached to a specialized fiber, drags the chromosome to one end of the cell when cell division occurs.

The broken chromosome automatically introduces two new ends (Figure 3-4). These ends, unlike the normal tips (*telomeres*), are "sticky" and will unite with any other sticky end they meet. They never join with the telomeres. If the broken chromosome fragments do not immediately rejoin, each of the two fragments will have a sticky end to create future problems for it and the cell in which it resides. One fragment will contain the centromere. This *centric* fragment, like a normal chromosome, can participate in cell division because a contractile fiber attached to its centromere can guide it to one end or the other of the dividing cell. The other fragment, which is *acentric,* cannot be guided to either end of the cell when it is dividing and it may then get lost, resulting in the loss of a large number of genes for the cell. Unlike acentric and centric chromosomes, normal unbroken chromosomes are *eucentric,* with one centromere and two telomeres.

Biological Damage
from Chromosome Breakage
Requires Dividing Cells   The cells of our bodies are organized for specific functions. Muscle, nerve, and skin cells, for example, each have different functions. A population of similar cells carrying out a common function is called a tissue. Not all adult tissues have dividing cells. Indeed, after adolescence, very few cells in the tissues in our bodies divide. Most of the cells in our bones, nervous system, and muscles fail to divide throughout adult life. However, our red blood cells are continually replaced through cell division in our bone marrow. So, too, are our white blood cells, the lining of our digestive tract from mouth to anus, the cells that line our blood vessels, and the surface layers of our skin.

Cells that do not divide are not seriously damaged by radiation (Table 3-1). Dividing cells suffer because chromosome breakage creates centric and acentric fragments. Most cells, even when actively dividing, are in the G_1 stage. Thus most broken chromosomes occur in cells that have not yet replicated or copied their genes.

Cell Death Arises from
the Breakage-Fusion-Bridge Cycle   The broken chromosome replicates its acentric and centric fragments. Each fragment now has a copy immediately adjacent to it. Consequently, there are two sticky ends of each replicated acentric fragment, which join to one another. The region

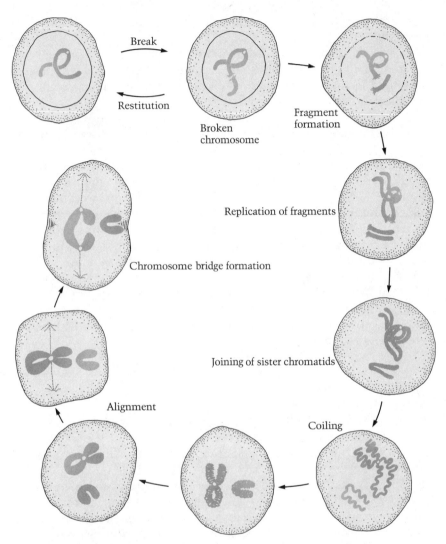

Figure 3-4 Breakage-Fusion-Bridge Cycle

*When a chromosome is broken, as by exposure to ionizing radiation, the
break may be repaired enzymatically, in which case the restituted
chromosome functions normally. If the two pieces formed by the break drift
apart, then the acentric fragment (lacking the centromere) and the centric
fragment (containing the centromere) have unrepaired breaks. When the
chromatids replicate, the broken ends of the centric fragment fuse, as do the
broken ends of the acentric fragment. During mitosis the coiled, compact
fragments are aligned, but only the centric fragment is attached to the spindle
fibers. The separation of the chromatids of the centric fragment leads to a
chromosome bridge.*

*Two things can happen when the chromosome bridge forms. The bridge
may break, permitting each new nucleus to contain a centric fragment with
an unrepaired break. This will generate another breakage-fusion-bridge cycle.
Or, the bridge may not break, in which case the nuclear membrane forms
around it and the cell does not enter cytokinesis. A single giant cell is formed*

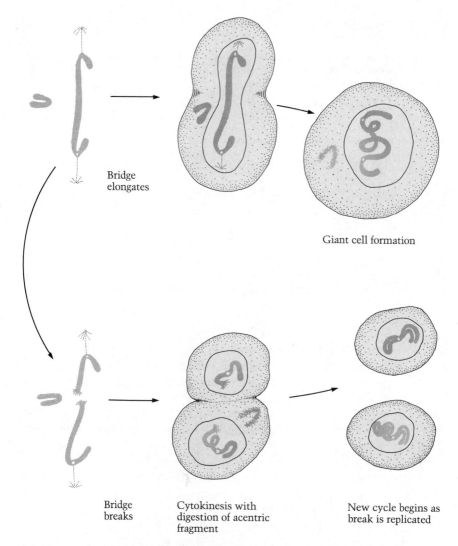

Bridge
elongates

Giant cell formation

Bridge
breaks

Cytokinesis with
digestion of acentric
fragment

New cycle begins as
break is replicated

*with almost double the diploid chromosome number. In either case, the
acentric fragment, going to the poles, may be left out where it will be digested
in the cytoplasm. Note that the giant cell contains an abnormal chromosome
with two centromeres.*

*For dividing cells in animal tissues, the breakage-fusion-bridge cycle is
fatal to the affected cells. Nondividing cells do not form a chromosome
bridge, and thus virtually all the genes needed for functioning are retained.
Radiation-induced breaks in nondividing cells are thus of insignificant
consequence.*

*In embryos all tissues divide and are therefore vulnerable to radiation
damage. The nondividing cells in adult tissue do not show symptoms of
radiation damage, or do so very rarely. However, although they do not
directly express damage from broken chromosomes, they can do so indirectly
if their blood supply is reduced by local damage to the capillaries, or from
other complications initiated by damage to the neighboring dividing cells
(e.g., swelling or edema in the brain).*

Table 3-1   Tissues Whose Cells Are Dividing and Nondividing in the Adult

After fertilization the zygote divides mitotically, and embryonic tissues maintain a rapid rate of cell division (about one division every few days for most tissues after the first month of pregnancy). At birth the rate slows down, and by the end of adolescence, the cells of the tissues can be classified as dividing or nondividing. In general, tissues whose cells wear out from mechanical abrasion, lack nuclei (red blood cells), or are called upon in larger numbers for emergencies (white blood cells) have cells which maintain their cell cycles. Nondividing cells are usually protected from mechanical wear or increase their size by internal synthesis of organelles.

Major tissues with dividing cells	*Major tissues with nondividing cells*
Bone marrow erythroblasts — red blood cells	Neurons
	Muscle
Bone marrow myeloblasts — white blood cells	Bone
	Cartilage
Lymphatic tissue — lymphocytes	Tendon
Intestinal lining	
Skin	
Endothelial cells (lining of blood vessels)	
Reproductive cells	

where the sticky ends meet forms a chromosome bridge, binding these two chromosomes together. Through coiling, the replicated fragments become streamlined and packaged in preparation for the distribution of these precisely copied portions to opposite sides of the cell. Unfortunately, the acentric fragment cannot distribute its copies because there is no centromere to which the guiding fibers can be attached. The centric fragment is properly attached, so that the centromere separates into two pieces, which move to opposite ends of the cell.

The chromosome bridge where the two acentric fragments were joined stretches and either fails to yield, resulting in a failure of cell division, or snaps, generating in each new cell another round of this vicious cycle until the cells eventually die. This is the *breakage-fusion-bridge* cycle which kills dividing cells by producing excess and deficient quantities of genes.

Tissues That Divide the Most Produce the Most Dramatic Symptoms of Radiation Sickness

If these cycles occur in the dividing cells of bone marrow, then red blood cell production will decrease and the patient will become anemic. If the cycles occur in the bone marrow and lymph glands that produce white blood cells, then the body's defense against germs will be impaired and subsequent infections and high fevers may result. If the breakage occurs in the lining of the blood vessels then the wear and tear of blood pounding against the thin walls of the capillaries will cause the damaged and dying cells to dislodge, leading to minute ruptures in the capillary walls, through which the blood will leak into tissue spaces, causing the noticeable measles-like rash of radiation sickness. Similarly, as food passes through the gut and scours the inner linings, new cells must replace those that are sloughed off. If the cells die in the inner lining of the gut they will be digested by our enzymes, just as would any piece of meat we had eaten. The erosion of these dying cells will cause ulceration and the perforation of the gut, leading to peritonitis, or inflammation of the membrane lining the walls of the abdominal cavity and viscera.

It is not surprising, then, that our knowledge of dividing and non-dividing tissues, and the breakage-fusion-bridge cycle, enables us to understand what radiation sickness is. We should also be able to predict some other effects of irradiation. If a pregnant woman were heavily irradiated, we would predict, for the very early embryo, such severe tissue and organ damage that spontaneous abortion would be likely. For more developed embryos (one to three months) we would expect serious congenital malformations at birth, such as microcephaly, the abnormal smallness of the head. The microcephaly occurs because normally the skull and the brain continue to enlarge in the later fetal stages and radiation damage slows down or prevents that growth. For much later embryos, such as those in the sixth to ninth month of gestation, we would find far fewer congenital abnormalities because the major stages of organ formation would have already taken place prior to the exposure to radiation. It does not take too much imagination to visualize why hair falls out following irradiation (the follicles have dividing cells), why skin becomes partially depigmented (the melanocytes killed by the cycle cannot synthesize melanin), or why pinpoint cataracts form in the lenses of the eyes (the dead or abnormal cells form opaque microscopic masses).

The Dose of Radiation and the Percentage of the Body Exposed Are Also Important Factors

We should consider another important factor to guide us in understanding what happened in Hiroshima: the amount of damage is directly related to the dose of radiation (measured in roentgens) received by the entire body (Figures 3-5 and 3-6). If the dose received was very high (200 to 400 roentgens) severe radiation sickness developed. Doses of more than 500 roentgens were usually fatal (Table 3-2). Doses under 100 roentgens produced relatively mild side effects, such as nausea and mild anemia, which were readily treated or cleared up in a few weeks. Very low doses (10 roentgens or less) produced no noticeable clinical symptoms. Any effects produced by these very low doses have only been detected through careful statistical studies of large numbers of exposed individuals, as compared with unexposed individuals. In such cases a slight depression of white and red blood counts can be observed. Doses below one roentgen virtually defy monitoring because so few cells experience chromosome breakage.

There are also long-range effects of radiation. Some of these are due

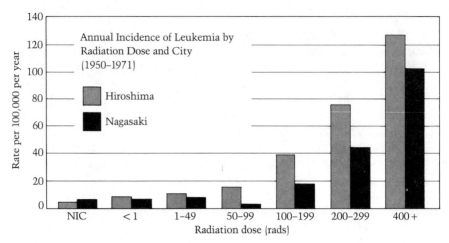

Figure 3-5 Atomic Bombs and Leukemia
The exposed populations of Hiroshima and Nagasaki have been studied since the bombings of August 1945. The estimated whole-body doses, in roentgens, were determined by the distance the survivors were from the hypocenters of the bomb blasts. The incidence of leukemia shows a nearly linear increase with the body dose received. Recent studies suggest that the doses shown here are about twice as high as those actually received. This would make human susceptibility to radiation-induced cancer correspondingly greater than the graph indicates. [From The Radiation Effects Research Foundation: a brief description *(pamphlet) (Hiroshima, Japan), p. 7.*]

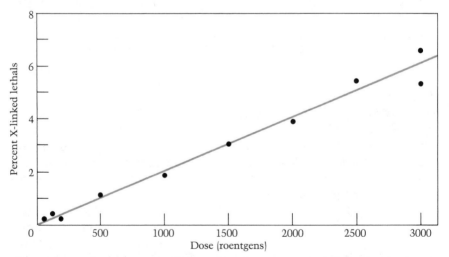

Figure 3-6 Radiation Dose and Gene Mutation in Drosophila (Fruit Flies)
Ionizing radiation, such as is produced by x rays, induces gene mutations. A genetic test used by H. J. Muller in 1927 for the induction of mutations by radiation has been widely used to show the effects of radiation doses ranging from a few roentgens to several thousand roentgens. The curve shows that the frequency of mutations rises as a simple multiple of the dose received. Although mutation experiments in fruit flies involving a total exposure of less than 10 roentgens are difficult to perform, geneticists generally believe that there is no threshold dose—any dose of radiation has some probability of inducing mutations.

Table 3-2

Adult organisms vary in their response to massive doses of radiation. Radiation biologists use a figure called the LD_{50} (mean lethal dose) to measure the lethal effects of radiation. The LD_{50} is the dose at which half of the exposed population will die of radiation sickness. Mammals have an LD_{50} of several hundred roentgens. An adult fruit fly is more resistant to radiation because it has no dividing cells other than in its testes or ovaries, and thus no breakage-fusion-bridge cycles to generate cell death. Insects at immature stages, such as the egg, larva, or pupa (a protective cocoon within which the adult tissues and body organization are formed), are more sensitive to radiation. At a dose of 75,000 R, however, the adult fly's chromosomes are badly fragmented and about 300 gene mutations are induced per cell. This can seriously affect metabolism and organelle function. Paramecia are highly polyploid (they have multiple sets of chromosomes) and can live for many days before dividing; thus their LD_{50} is higher. Organisms that have more catalase (an enzyme that breaks up the peroxides induced by x rays) or fewer dividing cells have a higher LD_{50}.

Organism	LD_{50} for Irradiation with x rays (in roentgens)
Guinea pig	250
Dog	335
Pig	350–400
Human	400–450
Mouse	550–665
Rat	665
Chicken	600–800
Frog	700
Snail	8000–20,000
Yeast	30,000
Drosophila (adult fruit fly)	75,000
Drosophila (pupa formation)	2000
Amoeba	100,000
Bacterium	150,000
Paramecium	300,000

to the breakage of chromosomes and some are due to gene mutations in the exposed cells. One of these effects is called *pseudoaging*, an artificial shortening of life span. It is difficult to obtain precise figures for humans, but studies using mice suggest that one to fifteen days of life are lost for each roentgen received. For Hiroshima survivors who received about 100 roentgens, the estimated mean reduction in life span is about one year. The same causes of death that affect nonirradiated people are experienced by the irradiated population.

Excluded from these estimates of pseudoaging are the leukemias (blood cancers), which appear much earlier than other cancers after exposure to radiation. These leukemias rise proportionately with the dose of radiation received by the body. Most cancers develop slowly and appear 20 or more years after an acute exposure. Cancer is readily induced by ionizing radiation, whether the radiation is administered as a single dose or as repeated doses over a period of time. (The effects of radiation are cumulative.) The mutational mechanism for inducing cancers is not known.

Radiation Not Only Induces Cancer
But It Is Used to Treat Cancer

How is it that cancers can be induced by radiation and yet radiation is used as a treatment for some cancers? We can infer that the reduction in size of a tumor is based on the same principle that enables us to predict the symptoms of radiation sickness (Figure 3-7). The cancer cells and the capillaries nourishing them usually divide more rapidly than the surrounding tissues (indeed, most normal adult tissues do not divide at all). Heavy localized doses of x rays administered to the tumor will induce many chromosome breaks, which frequently enter the breakage-fusion-bridge cycles and kill the cancer cells (Figure 3-8). The mechanism by which x rays induce mutations is different from the way they induce chromosome breaks. Tumor cell production often arises from multiple gene mutations in a cell. The tumor cell then divides for many years before a mass of cells large enough to produce symptoms appears. Through these two different processes, gene mutation and chromosome breakage, cancer can be both induced or destroyed by ionizing radiation. Also note that a localized lethal dose (several thousand roentgens) is only lethal to the dividing cells in that region. It is exposure of the whole body that determines a dose lethal to the individual.

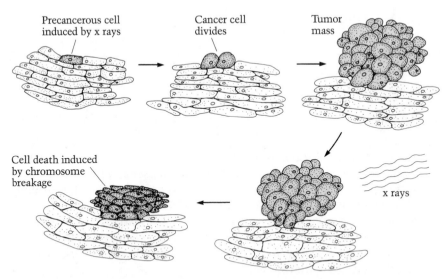

Figure 3-7   X Rays and Cancer
X rays kill cancer cells by preventing the tumor cells from dividing and by killing the capillary endothelial cells so that the blood supply to the tumor is reduced. In the cancer cycle shown here a spontaneous or induced mutation converts a smooth muscle cell into a tumor cell. This tumor cell multiplies, forming a lump or mass of tumor cells. If detected early, x rays in high doses will prevent further cell division, and the starved, dying tumor cells will be resorbed by the body's scavenger cells.

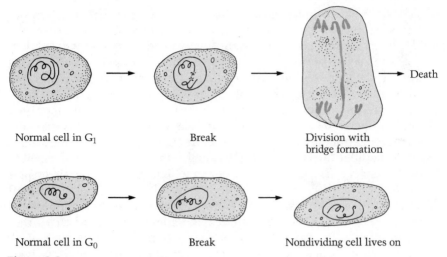

Normal cell in G_1 Break   Division with bridge formation   → Death

Normal cell in G_0 Break   Nondividing cell lives on

Figure 3-8
Note that the breakage in the dividing cancer (or normal cell) kills the cell by the breakage-fusion-bridge cycle, but the nondividing cell with the broken chromosome lives because virtually all of its genes continue to function in the same nucleus. Only one or a few neighboring genes would be impaired at the site of the breaks; the corresponding genes in the unbroken homolog would compensate and permit near-normal function in that cell.

How You Can Use Your Knowledge of Radiation to Protect Yourself

These observations on radiation damage enable us to look at many social controversies in a more informed way. Unfortunately, even this knowledge is not enough to guide all of us in making wise decisions. It can lead different people to very different positions. Nevertheless, there are some safety measures that all of us should be aware of:

1. Whenever you receive a diagnostic x ray from a dentist or a physician, you should request a lead apron to protect your reproductive system (testes or ovaries) from the small amount of radiation scatter that might occur.
2. Whenever an alternative to radiation is medically acceptable, it should be used. Many geneticists believe that any amount of radiation can induce cancers or mutations. Thus mass screening for tuberculosis should be done through the use of a tuberculin skin test. Only if a positive skin test is returned are chest x rays necessary for the person. Similarly, manual examination of breasts and infrared examination are preferable to mammography by x rays, because the number of induced breast cancers among several million screened females may exceed or equal those detected at an early stage by mammography. When millions of people receive low-dosage exposure, some cancers and mutations will necessarily be induced in the population, but usually the benefits to the individual far outweigh the risks.

3. Try to avoid complacency on issues related to radiation. Often our laws are enacted *after* disasters occur and problems arise, rather than *before*. Through careful foresight, many of these can be avoided. It is easy for some professionals to ignore small risks to the individual and to develop a blind faith in the practices of their profession. For this reason we need to have broad representation on regulatory commissions and other agencies that are responsible for consumer safety and public welfare. For example, decisions on the location, size, design, operation, and number of nuclear energy plants should be made with population as well as individual risk figures in mind. Society today is very divided on the extent to which radiation should be part of our technology. Many people oppose all technical and medical applications of radiation in our lives. That may be both impossible and self-defeating. But we can demand standards and attitudes that minimize risks, and risks can be honestly presented and appraised. We must consider these risks when we develop our opinions about radiation. For example, we may oppose all military use (or development) of nuclear weapons, and yet approve of medical use of x rays and radioactive isotopes for diagnosis and therapy.

4. If you have had extensive (abdominal or whole body) radiation therapy to combat a malignancy, and you have not yet started or completed your family, you should seek medical advice on the advisability of using artificial insemination or (less likely because of the expense and surgery involved) inovulation or of not having children of your own because of possible induced mutations in your reproductive cells. If your testes or ovaries were shielded or radiation was not directed toward your reproductive tissue during treatment, then this problem does not apply to you.

The difficulties in reaching decisions about the use of radiation stem from several factors. We do not know how to accurately measure a few hundred induced cancers spread out over 20 or 30 years in a population of several hundred thousand. This small number of cancers would not appreciably increase the total number of cancers that arise normally and involve 20 percent of the adult population. An increase of a few hundred among many tens of thousands of cancers would not be detectable statistically. We do not know the frequency with which gene mutations in humans are induced with very low doses of radiation (Table 3-3). Some of our knowledge comes from disasters, such as Hiroshima, some from industrial and medical ignorance and carelessness. Most of our estimates come from experimental studies on mice and other organisms, the results of which are extrapolated to humans.

Many social decisions are made by practicing the *risk-benefit philosophy*. Decisions are arrived at by weighing the benefits of our actions and choices against the risks. For instance, in deciding whether or not to become parents, we think not only of the benefits of loving and raising a child, but also of the risks involved, such as birth defects, the death of the mother in childbirth, or having a child who is difficult to raise.

Table 3-3   Average Radiation Doses Received by the U.S. Population Annually

Radiation doses are difficult to estimate because many factors are involved. The recommended dose for a dental or chest x ray is a conservative one, but not all machines are maintained to deliver that dose. Furthermore, practitioners may use higher than recommended doses to obtain better contrast and detail, especially for fluoroscopic examination. The doses are given in milliroentgens (mR). A dose of 1000 mR is equal to 1R.

Source of Radiation	Annual Dose (in milliroentgens)
Dental x rays to the jaws, annual checkup	1000–3000
Medical technician	320
Chest x ray, annual checkup	200
Residual radiation in the construction materials of buildings	75
Surrounding soil and rocks	50
Cosmic rays	45
Mean medical exposure per person	35
Radiation from your own tissues	20
Fallout from weapons tests	4
Jet flight, round trip, from New York to Los Angeles	3
Nuclear plants (assuming no accidents or violations of federal standards)	0.5
Color television set, average home viewing (about 5 hours/day)	0.5

A major problem in society today is the inconsistent way we use the risk-benefit philosophy. For example, most of us accept as inevitable traffic mortalities associated with motor vehicles. We justify our motorized lifestyle by saying that it is essential to our culture and we cannot prevent all accidental deaths because there are too many variables involved. However, we do not accept such an attitude in the case of a person walking in a city and being killed by a brick falling from a skyscraper under construction. According to the law, the contractors would be liable for damages in court, and in addition, our city councils require certain construction safety practices to help prevent all such rare disasters. We deplore motor vehicle deaths but accept them as a fact of life, despite many laws to minimize them, but we do not tolerate any deaths from falling bricks because we believe all of them to be preventable. Radiation risks fit neither the motor-vehicle nor the falling-brick model of risk-benefit values because we are not willing to accept some induced cancers and birth defects as a fact of life and we can rarely prove a direct cause and effect relation of the radiation to the individual with an induced cancer or a genetic disorder.

In decision-making we must learn to live with partial knowledge and apparent inconsistencies, often relying on the opinions of those we trust or respect. Yet the more we learn about the scientific facts and principles involved in a social controversy, the better will our choices and decisions be.

Questions to Answer

1. In what two ways does radiation damage the genetic material?
2. How does chromosome breakage compare with gene mutation in producing radiation sickness when an atomic bomb explodes?
3. What are the major dividing tissues in an adult human?
4. Describe the breakage-fusion-bridge cycle.
5. What do the symptoms of radiation sickness reveal about the tissues damaged by massive whole-body radiation?
6. What have been the delayed effects of radiation on the survivors of Hiroshima and Nagasaki?
7. How can x rays be used to treat cancers if they also induce cancers?
8. What are the reasons you would give to your dentist for requesting a lead apron when receiving mouth x rays?
9. Discuss your reactions to the following situations. What sort of stand would you take? (a) A girl is hit on the head by a stone falling from an old apartment building in New York City. Should anyone be sued or is this just a lamentable, but necessary, price for progress? (b) Veterans dying of a rare cancer often linked to atomic bomb fallout asked the Defense Department for permission to sue for their medical expenses. The veterans were exposed to radiation during maneuvers associated with atomic tests in Nevada in the early 1950s. They believe this exposure to radiation induced their cancer.
10. Would you support or oppose the construction of a nuclear reactor ten miles from your residence? What information would help you make a decision?

Terms to Master

acentric fragment

breakage-fusion-bridge cycle

centric fragment

chromosome break

chromosome tip

gene mutation

giant cell formation

LD_{50}

sticky end of chromosome

whole-body dose

Chapter 4
The Formation of Sex Cells and the Consequences of Nondisjunctional Errors

An accurate count of the human chromosomes was not obtained until 1956, when techniques were developed that made it easy to see each of the individual chromosomes in a cell. The growing cells were treated with a mitotic poison such as colchicine. This prevented the aligned, fully-coiled chromosomes from moving to the poles of the spindle apparatus. The cells were also placed in unbuffered water which made them swell as water passed through the cell membrane. This caused the chromosomes to drift apart from one another. Finally the chromosomes were stained and squashed on a glass slide under a thin glass cover slip, where they could be counted. In this study, as in all subsequent ones, the human chromosome number in diploid cells was found to be 46.

When an enlarged photograph was prepared, the 46 chromosomes were cut out and arranged in order of size. One of the 23 pairs of chromosomes always differs between males and females. These are called the *sex chromosomes*. In males the two members of the pair differ in size; the larger one is called an X chromosome and the smaller one a Y chromosome. In females there is no Y chromosome; instead there are two Xs. Every cell contains one pair of sex chromosomes and 22 pairs of nonsex chromosomes, or *autosomes*. The chromosome composition of a female is XX + 22 pairs of autosomes; that of a male is XY + 22 pairs of autosomes.

To represent the karyotype, the geneticist presents the total chromosome number, followed by the sex chromosomes that are present. Thus a normal female is 46,XX, and a normal male is 46,XY. If there are abnormalities present, these are listed after the sex chromosomes. For example, 47,XY,+21 would be a male with Down syndrome. If the sex of the individual is not important in the chromosome abnormality then the XX or XY is omitted from the karyotype. For example, 47,+21 would be a child with Down syndrome.

45

An Extra Chromosome Is Present
in the Cells of Individuals
with Down Syndrome
Shortly after this technique was publicized, scientists looked at the cells of some mentally-retarded children who had a condition called Down syndrome (Figure 4-1). These children had long been of interest to scientists because of their multiple defects and the inexplicable origin of the defects. It was found that victims of Down syndrome had 47, not 46, chromosomes. When the chromosomes were photographed and the photo enlargements used to make a karyotype, matched pairs were obtained for all except one of the smallest chromosomes, which was present in triplicate. Down syndrome is the result of a triplication of chromosome number 21. For this reason, it is also called trisomy-21, and the karyotype of a Down syndrome child is 47, + 21. Where did the extra chromosome come from?

Cell division (mitosis or meiosis) is normally a precise process. Such precision is necessary if each cell is to have an identical number of informational sets to specify all the activities that the organism must carry out to survive and reproduce. Unfortunately, though, living systems are sometimes imperfect, and during mitosis or meiosis a chromosome may fail to separate or fail to move to the proper pole (Figure 4-2). Such an error is called *nondisjunction*. If it occurs during reductional division, then one pair of chromosomes will fail to separate, moving to one pole, while the chromosomes of all other pairs will separate normally and move apart to opposite poles (Figure 4-3). This disturbed distribution will produce gametes that have one extra (N + 1) or one less (N − 1) than the haploid (N) number of chromosomes. Since the nondisjunctional sex cell will most likely experience fertilization with a normal haploid cell, the fertilized diploid egg will be (2N + 1) or (2N − 1) instead of the normal (2N).

For the nondisjunctional human that means 47 or 45 instead of the normal 46 chromosomes in the zygote from which all tissues and organs

Figure 4-1
A child with trisomy-21 or Down syndrome. Note the characteristic features, including the epicanthal eye fold, the shallow nasal bridge, and coarse features. Children with Down syndrome are usually diagnosed at birth. (March of Dimes Birth Defects Foundation)

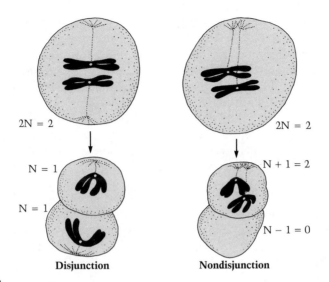

Disjunction   **Nondisjunction**

Figure 4-2
Nondisjunction is the failure of chromosomes to separate during cell division. In the situation shown here, the cell on the left shows a pair of homologous chromosomes during meiosis with the centromere of each chromosome attached to its closest pole. Normal separation or disjunction converts the diploid (2N) to two haploid cells (N). On the right each of the two homologs is attached to the same pole of the cell. This converts the diploid (2N) into nondisjunctional haploid cells—(N + 1) and (N − 1)—containing both homologs and lacking both homologs, respectively.

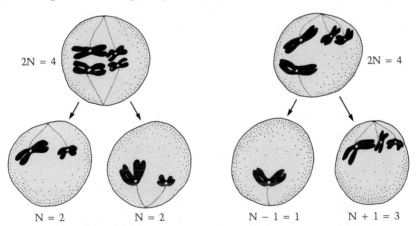

Normal Disjunction in Reductional Division   **Reductional Division Nondisjunction**

Figure 4-3   Meiotic Nondisjunction During Reductional Division
The aligned metaphase chromosomes are normally paired with each homolog attached to the pole it faces. If one pair of homologs is attached to the same pole, then the anaphase movements of the chromosomes lead to nondisjunctional chromosome numbers, one cell bearing (N + 1), and the other (N − 1), chromosomes. A sperm or egg produced from either one of these nondisjunction cells could cause the embryo it forms to abort or develop abnormally. Reductional division is the most common source of nondisjunctional errors in human beings. Reductional division nondisjunction increases in frequency with the age of the individual, especially in the female.

will form. Thus from fertilization on, the developing individual will have to function, as well as it can, with extra or missing genes. Since even a small human chromosome (other than the Y chromosome) contains hundreds of genes, this will create problems for the tissues as they attempt to mobilize the proteins produced in excess or deficit in response to the doses of regulatory signals produced by the abnormal cells.

Down syndrome or trisomy-21 (formerly called Mongolism) is a semilethal condition (Table 4-1). Even when given good medical care for their many illnesses, victims often die young, in their teens or twenties. Many die shortly after birth. A few manage to live normal or near-normal life spans.

The major effect of Down syndrome is mental retardation. About one-third of institutionalized, mentally-retarded children have Down syndrome. Their IQs are usually about 40, but some may be close to 80. Even with a lot of parental and professional effort, the skills of an individual with trisomy-21 are limited.

Table 4-1   Defects Associated with Down Syndrome

The presence of an extra twenty-first chromosome in every cell of the body causes defects in many tissues and organs. Not all the defects in this table are present in each case of Down syndrome. Rather, the frequency varies, depending on the genetic composition of the fetus and its response to the extra gene activity produced by the trisomic chromosomes. For example, cardiac defects are found in 40 percent of babies born with trisomy-21. Intestinal blockage is less commonly encountered (about 10 percent of trisomy-21 cases). The IQ varies with age, being higher in infancy and early childhood than in adolescence or adult life. The reduced life expectancy is due, in part, to a high susceptibility to infectious diseases from defects in the immune system.

Head and Face
  round head
  flat profile
  epicanthal fold
  brushfield spots in iris of eye
  flat nasal bridge
  protruding tongue
  small, round ears

Skin
  rough texture
  elastic, easily pulled

Abdomen
  protruding, often with umbilical
    hernia
  small pelvis
  intestinal blockage

Thorax
  cardiac defects

Limbs
  stubby fingers and toes
  simian crease on palms
  curvature of fifth finger
  wide space between large toe and
    second toe
  double-jointed

Genitalia
  normal, but males impotent

Brain
  retarded, mean IQ = 50 at 5
    years, 38 at 15 years

Life expectancy
  25 percent die in first year of life
  50 percent before fifth year of life
  8 percent survive past age 40
  a one-year-old has a mean life
    expectancy of 23 years

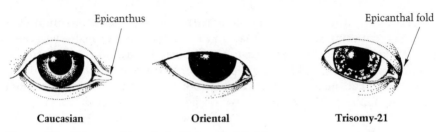

Figure 4-4   Trisomy-21 and the Epicanthal Fold
The junction where the eye folds come together is called the canthus. In trisomy-21 a vertical fold of skin near the nasal bridge (the epicanthus) pulls and tilts the eye slightly toward the nostrils. Note that the upper eyelid in the normal Caucasian eye runs parallel to the skin fold below the eyebrow. In Orientals this skin fold covers a major portion of the upper eyelid. In Caucasians with trisomy-21 the epicanthal fold does not cover a major portion of the upper eyelid, making their eyes distinct from Orientals'. For this reason the terms "Mongoloid" and "Mongolian idiot" are misnomers and are no longer used. The eyes of persons with trisomy-21 frequently show brushfield spots or light patches scattered irregularly on the iris. These spots are not found in the eyes of those with 2N = 46, even if the eye color is a variegated hazel.

Certain physical traits, including short stature, stubby fingers and toes, and distinctive facial features, also characterize the children with trisomy-21. The epicanthus, the part of the eye near the nasal bridge, shows an unusual eye fold which is as distinctive in Oriental trisomy-21 children as it is in Caucasian trisomy-21 children, as seen in Figure 4-4. For this reason older terms such as "Mongolism" and "Mongoloid idiocy" are inaccurate and have become obsolete. Also present are a very depressed nasal bridge, sometimes a protruding, furrowed tongue, and loose skin that stretches easily. The hands usually show a single crease across the palm (a simian crease) instead of the two incomplete creases seen in most individuals. However, about one percent of normal people have simian creases, so don't be frightened if your palms show this feature (Figure 4-5)!

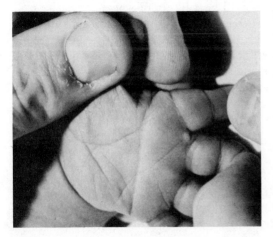

Figure 4-5
A single crease in the palm is usually found in the individual with trisomy-21. The normal individual usually has two palmar creases. Clinodactyly, or curvature of the little finger, often occurs in the trisomy-21 individual. The two creases of the little finger may be so close together that they appear to be a single crease, as if only two, not three, bones (phalanges) formed the finger. (Courtesy of Drs. D. Valle and J. Phillips)

There are other traits that occur with varying frequency among trisomy-21 children. The baby may have a defect in the heart valves or major blood vessels of the heart; the stomach may not open into the small intestine. The child has a poor immune system and suffers from frequent infections. Death from pneumonia is quite common in such children. There is also a higher risk of developing leukemia.

What Do Parents Need to Know If
They Have a Baby with Down Syndrome?
What are some of the difficult issues that parents of trisomy-21 children are faced with? Often the child has a happy disposition and, in a supportive family, can share much love and warmth. Unfortunately, though, many children with trisomy-21 have behavioral problems and can be somewhat difficult to raise. Their emotional outbursts and needs may consume the parents' energies and result in tensions for the other children, who may feel neglected or depressed by the conflicts in the family. Since raising a trisomy-21 child often means large financial expenses for surgery, recurrent infections, and special training or educational programs, the family may not also be able to provide for the economic well-being of the rest of the children. Since children with trisomy-21 are often conceived after the parents already have one or more normal children, the parents are often about 35 to 45 years old when the child with Down syndrome is born. If they decide to raise the child, and, with diligent care, the child lives to be 30 or 40 years old, the parents will be 75 to 85 years old at that time. Thus the middle-aged trisomy-21 child will often either have aged parents (who may not be able to provide proper care) or a brother or sister who takes over that role (perhaps reluctantly) if one or both parents are dead. Most trisomy-21 victims do not have the mental skills to earn a living or manage their health or a household without supervision, so the long-term future for a trisomy-21 who survives the parents and who has no one else to care for him or her may be less than what the parents had hoped for, and institutional care will be a likely prospect.

Trisomy-21 appears later in the birth order because nondisjunction occurs more frequently in older parents than in younger ones (Table 4-2). In young parents (under 30) the incidence is about one in 2000 births; in parents over 40 the incidence may be as high as one in 50. Males contribute an extra twenty-first chromosome as often as do females up to the age of 32. After this age there is a rapid increase in nondisjunctional eggs, but no such increase in the sperm.

Nondisjunction May Also Arise in
Equational Division or During Mitosis
While reductional division is one means by which an extra or missing chromosome can occur in a mature sex cell or *gamete*, nondisjunction is not limited to this stage in

Table 4-2   Down Syndrome: Meiotic and Parental Origin

Most nondisjunctional errors occur during the first meiotic division (reduction). The excess of errors in egg production is age-related (women 32 and over). Since reductional division begins in *all* eggs several months before a female is born, the error must arise in the adult late in the prophase (after the homologs contract) or during the alignment of the homologs and their attachment of spindle fibers to the centromeres.

	Reductional Division	Equational Division	Total
Maternal	23	1	24
Paternal	5	2	7
Total	28	3	31

Source: From R. E. Magenis et al., 1977.

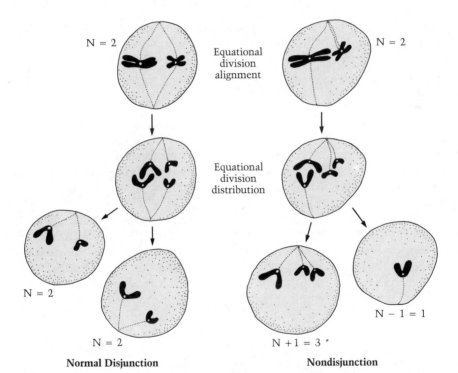

Normal Disjunction   **Nondisjunction**

Figure 4-6   Meiotic Nondisjunction During Equational Division
The haploid cell, bearing one of each pair of chromosomes, aligns in a single plane, but one of the centromeres is attached to only one pole. As a result, there is a movement during anaphase of that chromosome with both of its chromatids separate. This produces (N + 1) chromosomes in that nucleus. The other nucleus, lacking that chromosome, has the (N − 1) chromosome number. In testes, equational division nondisjunction can lead to sperm bearing two Y chromosomes. Special genetic or cytological techniques are needed to demonstrate its occurrence in other chromosomes. There is no relation between equational division nondisjunction and the age of the man or woman.

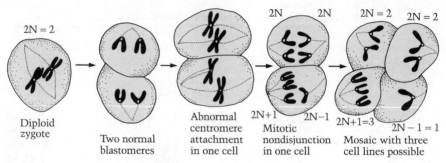

Figure 4-7 Mitotic Nondisjunction and Chromosome Mosaics
The attachment of a centromere to only one pole may occur during mitosis. If this happens shortly after fertilization, there will be a mixture of cell types. In the example used here the first mitotic division of the fertilized egg, or zygote, was normal, producing two normal cells (blastomeres) of the embryo, but during the second division nondisjunction arose, producing a (2N + 1) cell and a (2N − 1) cell. The other, normal cells are both 2N. Thus three cell lines may form (sometimes the (2N + 1) cell line survives, but the (2N − 1) dies out). This can lead to a child who is mosaic for normal and nondisjunctional tissue. Some Down syndrome children (about three percent) result from early mitotic nondisjunction. They are mosaic for Down syndrome; they do not show any (2N − 1) tissue, but do have (2N) and (2N + 1) tissue. The monosomic cells (2N-1) do not usually survive in the early cell divisions after fertilization, so that the embryo forms from the surviving (2N + 1) and (2N) cells.

a cell's development. The two replicated strands of a chromosome in equational division may not separate until that chromosome reaches one of the poles (Figure 4-6). Just as in reductional division, the gamete will be (N + 1) or (N − 1) in content.

Another source of nondisjunction is the mitotically dividing cell. If the two replicated strands fail to separate in the cell shortly after fertilization occurs, the resulting embryo will be a *mosaic*, consisting of two or more cell lines differing in their chromosome numbers (Figure 4-7). It is even possible to have (2N), (2N + 1), and (2N − 1) tissue in the same individual if mitotic nondisjunction happened in the second or later cell divisions of the 2N fertilized egg.

Most Nondisjunctional Fertilizations
Result in Spontaneous Abortion
Trisomy-21 is not the only nondisjunctional disorder in humans. It is very likely that nondisjunction occurs during meiosis because about 40 percent of spontaneously aborted fetuses have abnormal chromosome numbers. Since almost one-fifth of all fertile women experience a spontaneous abortion at some time, this means that as many as five percent of all gametes have an abnormal chromosome number. Fortunately the most serious nondisjunctional disorders (mostly involving the larger chromosomes) kill the embryos at an early stage of development. Only a few nondisjunctional individuals (involving a few of the smaller autosomes) survive long enough to be born.

There Are Trisomic Autosomal Syndromes
Other Than Trisomy-21

Down syndrome is an autosomal nondisjunction of chromosome 21. There are two additional types of autosomal nondisjunction that occur at a much lower incidence among the newborn. Trisomy-13 or Patau syndrome is marked by abnormalities of the forebrain (Figure 4-8). Since the forebrain of the embryo is necessary for the proper development of the face, the failure of the forebrain to develop leads to severe facial defects, such as missing or reduced eyes and bilaterally cleft palate and lip. These babies may also have an extra finger or toe and many internal organ defects, especially heart trouble. Because of multiple birth defects in vital organs, trisomy-13 babies usually die within a few months after birth. The other autosomal trisomic conditions are designated biologically as *lethal*, in contrast to the trisomy-21 condition, which is *semi-lethal*.

Another autosomal lethal condition caused by nondisjunction is trisomy-18 or Edward syndrome (Figure 4-9). These babies have several skeletal defects, including a tiny, receding chin, a short, deformed sternum, fingers which remain clenched in an overlapping pattern, and a bowing out, or reversal, of the arches of the feet. They may also have low-set abnormally shaped ears, a high, arched palate, cardiac problems, abnormal kidneys, hernias which force the gut into the lung cavities, and other serious defects. As in trisomy-13, the children with trisomy-18

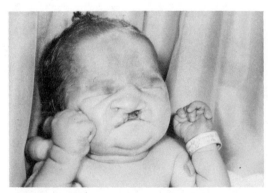

Figure 4-8
Trisomy-13 or Patau syndrome is a lethal condition with multiple gross malformations. Note the polydactyly (extra fingers), the cleft lip, and the microphthalmia (small or missing eyes). Death usually occurs a few weeks or months after birth. (Courtesy of R. J. Gorlin, D.D.S.)

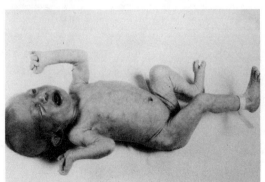

Figure 4-9
The trisomy-18 or Edward syndrome is lethal. This infant shows the rudimentary lower jaw, rotated ears, flexed fingers, and deformed sternum characteristic of most cases. Death usually occurs within a few days or weeks of birth, due to severe internal organ malformations. (Courtesy of R. Desnick, Ph.D., M.D.)

usually die shortly after birth, a few surviving for a full year. The trisomy-13 child is seen about once in 4000 births, trisomy-18 about once in 3000 births.

Klinefelter Syndrome Arises from an XXY Zygote

The normal male is XY and the normal female is XX in chromosome make-up. Normally, the XX female produces eggs with a single X chromosome. The normal male produces some sperm which bear a single X chromosome and some which bear a single Y chromosome (half of one and half of the other). If reductional division nondisjunction occurs the XX females produce either XX eggs or eggs without any X chromosomes (called nullo-X eggs). If it happens in a male, the sperm are either XY or lack any sex chromosomes. All of these gametes are capable of participating in fertilization.

What happens if an XX egg is fertilized by a Y-bearing sperm (or an X egg meets an XY sperm)? Such an individual will be $2N + 1 = 47$, with an XXY chromosome composition, known as the Klinefelter syndrome (Figure 4-10). The karyotype is written as $2N + 1 = 47$,XXY. At birth the baby will be identified as male, because a penis and testes are present. The penis and testes may be small, however. No serious problems occur until puberty, when the boy fails to develop typical masculine hairiness

Figure 4-10
An adult with Klinefelter syndrome (47,XXY) showing breast enlargement. Note that the length from the hips to the soles of the feet exceeds the length from the hips to the top of the head. (Courtesy of R. J. Gorlin, D.D.S.)

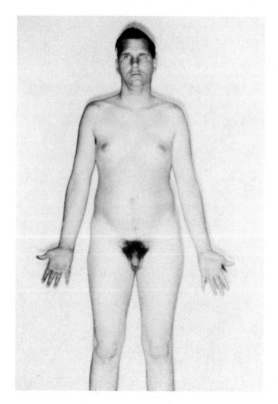

and other secondary sexual characteristics. Often he will develop breasts, which may become large enough to cause embarrassment. These are often surgically removed. The secondary sexual development of these XXY males can be improved by treatment with male hormones, but there are two features that cannot be improved medically. The first is that XXY males tend to be somewhat less intelligent than XY males, and often have mental disturbances. A significant number of men in mental institutions are XXY males. The second uncorrectible defect is sterility. The testes of the XXY males do not produce sperm. About one in every 500 males born has an XXY karyotype. Fortunately most of them are capable of functioning normally in society.

Turner Syndrome Individuals
Are 2N − 1 = 45,X

Whenever two sex chromosomes move to one pole, the other pole fails to receive a sex chromosome. Such (N − 1) sperm or eggs may produce a (2N − 1) fertilized egg, with a single X unaccompanied by another sex chromosome. This results in XO, or Turner syndrome (Figure 4-11). The karyotype is written as 2N − 1 = 45,X. XO cells occur at fertilization, but most do not survive to birth. Those that do have more serious disabilities than do the XXY males. The XO child is female and often shows a webbing of the neck caused by swellings of

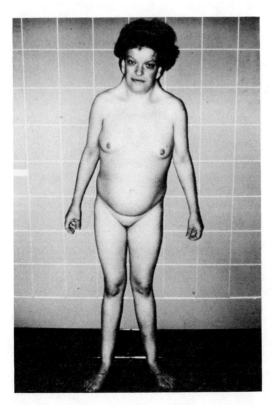

Figure 4-11
Females with Turner syndrome (2N = 45,X) usually have webbed neck, short stature, and elbows that cannot be fully extended. (Courtesy of R. J. Gorlin, D.D.S.)

the skin. Sometimes the XO females have cardiac defects. As adults, they tend to be shorter than average (about four feet tall), and they fail to develop breasts unless they are given hormone therapy. They also lack ovaries (usually present only as streaks of connective tissue) and thus are neither fertile nor capable of menstruation. Most XO females have normal intelligence, but frequently have personality problems. Among newborn females, one in 5000 will have Turner syndrome.

The XYY Condition
May Not Be a Syndrome
Equational division nondisjunction is clearly involved in producing males with two Ys. During sperm maturation the haploid cell bearing a single Y chromosome with two chromatids fails to separate and distribute the chromatids to the two cells formed by the equational division. Thus an egg (bringing in a single X) fertilized by that sperm results in a new (2N + 1) cell that is XYY.

About 1 in 100 males is born with the XYY karyotype. Males with XYY have a highly variable response to the extra Y. Most are tall (over six feet) and heavily acned in adolescence. A few are sterile. Some are mentally retarded, and many have behavioral disorders. In the mid-1960s it was believed that the XYY condition results in violent or aggressive behavior. This has not been substantiated and may be false. Some attempts to detect aggressive behavior in the XYY male have failed to reveal it. A study of 1968 Olympic players showed that there are not an excessive number of XYY males among basketball players. However, several studies have shown that there are a significant number of XYY men among prisoners and in mental institutions. Among tall mental patients the frequency of XYY is 1 in 72, among tall males in prison it is 1 in 56, and among patients in mental-penal institutions it is 1 in 29. In this study, tall is defined as six feet (183 cm) or taller. According to studies in Denmark and the United States, the large number of XYY prisoners may be accounted for, in part, by their diminished intelligence, which may make it easier for them to be apprehended or convicted.

The XXX Condition Is
Also Not a True Syndrome
When equational or reductional division nondisjunction of the X occurs there may be a union of the 2X sex cell with a sex cell bearing a single X, producing an XXX or triple-X female. These females, occurring once in 600 female births, also have a variable expression of symptoms. In a few instances there are visible birth defects, but most XXX females are normal in appearance. Their most common problems are irregularity in menstruation (including precocious menopause at age 20 or 30), occasional retardation, and occasional disturbed personalities. Those with menstrual difficulties are frequently sterile. Among fertile XXX females, a few have produced children with XXY or other nondisjunctional defects, but contrary to expectation, most of them

produce normal XX daughters or XY sons. The reason for this behavior of the sex chromosomes is unknown. It is not present in autosomal non-disjunction. For example, when trisomy-21 females have children, they have about a 50 percent probability of having a child with trisomy-21.

The XYY and XXX conditions are not considered syndromes because only a small percentage of those having the karyotype express one or more of the clinical or behavioral problems described. Recognized syndromes such as XO, XXY, or trisomies 13, 18, and 21 show one or more major clinical effects in each child with such a karyotype.

Sex Chromosome Nondisjunctional Conditions Are More Common Than Autosomal Ones

Why are there so many sex chromosomal, and so few autosomal, defects seen at birth? One likely explanation is related to the role of sex chromosomes. Whatever genes the sex chromosomes carry, the organism has worked out a mechanism to compensate for the double quantity of genes on the two X chromosomes in females and the single quantity of each gene on the one male X chromosome. Extra X or Y chromosomes are more likely to disturb the sexual development of an individual than to disturb the vital organ systems. But one X per cell is essential to life. There is no $2N-1$, YO individual. Since we would expect these to occur nearly as frequently as the XO, XXY, or XXX type, the absence of this class implies that the YO combination is lethal and that the embryo aborts at an early stage of development.

Of the 22 autosomal types, virtually no monosomies $(2N-1)$, lacking one complete autosome, exist. Their near total absence implies that they are aborted naturally. Similarly most of the possible trisomies $(2N+1)$ are probably lethal as well. If any given chromosome has some specific chance of undergoing nondisjunction, then there are 46 potentially abnormal gametes (23 monosomies and 23 trisomies) which can lead to abnormal zygotes. Many of these predicted autosomal trisomies and monosomies are seen when aborted fetuses are brought in for examination.

Some People Have Higher Risks of Nondisjunction Than Others

Nondisjunction during sex cell formation occurs more frequently in older females. In response to this fact there are certain issues that should be addressed. For those who accept abortion for medical reasons, there is a technique of *prenatal diagnosis* called *amniocentesis*. During early pregnancy (about 16 weeks after fertilization) some of the fetal cells floating in the protective fluid around the fetus can be removed and cultured. If such cells show an abnormal chromosome number (Figure 4-12) the parents are informed of this and may choose to abort the fetus rather than see it through to term. For

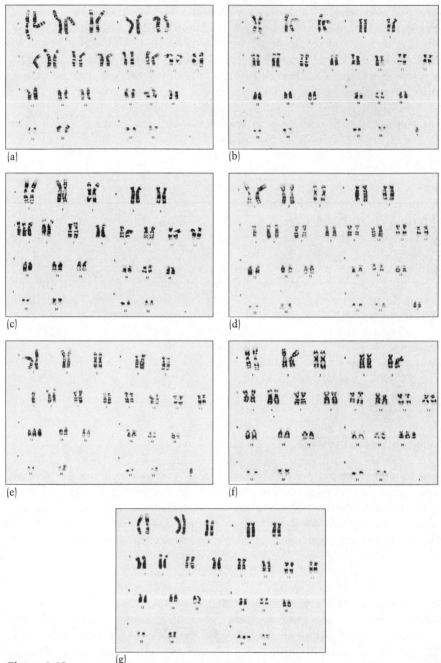

(a)   (b)   (c)   (d)   (e)   (f)   (g)

Figure 4-12

Nondisjunctional disorders involve an extra or missing chromosome. The karyotypes for the sex chromosome aneuploidies shown here are (a) XO (Turner syndrome), (b) XXY (Klinefelter syndrome), (c) XXX female, and (d) XYY male. Also note the autosomal trisomies for (e) 13 (Patau syndrome), (f) 18 (Edward syndrome), and (g) 21 (Down syndrome). The clinical features of these aneuploid cells are described in this chapter. (Courtesy of Carolyn Trunca, Ph.D.)

those women who approve of this procedure prenatal diagnosis of a pregnancy should begin at the age of 32 to 35 when the frequency of nondisjunction begins to increase. For those who would not consider an abortion acceptable under any circumstances, the decision to bear children after age 35 or so should be made only after careful consideration of the possible defects that could occur in the child.

Parents-to-be should also be aware of the following facts:

1. When a child is born with an abnormal chromosome number the probability of any subsequent child of those parents having an abnormal chromosome count is about five percent.
2. A parent who has had an above-average exposure to radiation (medically or occupationally) has a higher risk of having a nondisjunctional child.
3. Certain virus infections (e.g., mumps shortly before fertilization) may increase the risk of nondisjunction.
4. When a child is born with an abnormal chromosome number, the parents would be well-advised to have their own chromosomes checked because certain defects of chromosome organization in one of the parents can increase the risk of producing nondisjunctional gametes.
5. A mosaic child, with cells containing different chromosome numbers, is the result of mitotic nondisjunction of the embryo and usually has nothing to do with either parent's sex-cell-forming process. In a very few cases there may be a hereditary reason why the embryonic cells undergo mitotic nondisjunction.

Familial Down Syndrome Arises from a Different Mechanism

Occasionally there is a form of chromosome abnormality that seems to be hereditary. That is, the risk of a given family having more than one child with Down syndrome or some other chromosome syndrome is high. This familial form of Down syndrome occurs when two nonmatching chromosomes break and rejoin, forming one or two functional chromosomes composed of nonmatching pieces. These are called *translocations*. Often the father or mother who has the translocated chromosomes is normal in all respects, but about one-third of his or her children can theoretically have an excess (or deficit) of one of the chromosomes. This imbalance of genes, if it does not abort the embryo, may cause severe birth defects, including some that are lethal or semi-lethal. About five percent of all Down syndrome children are produced by this translocation mechanism. This is why it is important for parents to have a thorough chromosome examination whenever they already have a child with a known chromosome syndrome or multiple birth defects, and they plan on having another child. Also, a prospective parent who comes from a family with a history of many spontaneous abortions, or an occasional birth with congenital abnormalities, would be wise to request a chromosome study before having children.

Questions to Answer

1. What is nondisjunction? How does it differ from a chromosome break?
2. What are some of the difficulties you would face if you had to raise a child with trisomy-21?
3. Why is it inappropriate to call Down syndrome "Mongolism"?
4. What are the consequences of mitotic nondisjunctional births?
5. Why are sex chromosome nondisjunctional births more numerous than autosomal nondisjunctional births?
6. Why are autosomal monosomies $(2N - 1 = 45)$ virtually nonexistent among newborn babies?
7. Why would an autosomal trisomy be more likely to result in multiple congenital anomalies than in a single-organ defect?
8. Should a physician advise married women who are 35 years old or older that prenatal diagnosis is an available option? Discuss the legal, ethical, and moral reasons for your answer and what you would tell such patients if you were the doctor.
9. What are the major reasons why a parent may have a higher-than-normal risk of having a nondisjunctional child?

Terms to Master

amniocentesis
autosome
monosomies
mosaicism
nondisjunction
trisomy
trisomy-13

trisomy-18
trisomy-21
translocation
XO syndrome
XXX condition
XXY syndrome
XYY condition

Chapter 5
Fertility and Infertility

Most couples who marry have one or more children. They may restrict population growth, consciously or unconsciously, by marrying late, planning their families through artificial or natural methods of birth control, or restricting the number of children to a family size that fits their psychological habits and economic situation. Among married couples desiring to have children, as we saw when discussing the human condition, 25 percent will conceive within their first month's effort, 63 percent will conceive within six months, 75 percent within nine months, and 87 percent within a year. For some 13 percent of married couples who have attempted to achieve a pregnancy, no conception will have taken place after a year of effort.

**Male Infertility Frequently Involves
Inadequate Sperm Production**   If no pregnancy has taken place after a year of trying, most couples will seek medical advice if they have not already done so. Since at least 40 percent of infertility is due to the male (usually inadequate sperm production), and since male infertility problems are less costly to identify than a female's, the male is the first to be examined. In most cases, the male's virility is normal. Sexual intercourse takes place and orgasm is achieved in the vagina. The male's fertility or infertility is usually not related to his sexual performance. Impotence (often confused with sterility) is usually a failure, for psychological reasons, to obtain an erection or to maintain it until orgasm. Impotence is responsible for only a small percentage of male sterile marriages.

The male's sperm count, sperm motility, and sperm morphology are all studied from one or more samples obtained at the doctor's office

(Figure 5-1). Normally about 100 million sperm per milliliter are found in fertile males. About 400 million sperm would be present in an ejaculate although normal ranges show considerable leeway in sperm number and ejaculate volume. A sperm count of less than 20 million per milliliter and a volume of less than 3 ml are indications of low fertility. Additionally, at least 60 percent of the sperm should look normal (Figure 5-2).

Male infertility may arise from infections (mumps), chronic illness (childhood diabetes), occupational exposure to radiation, lead, zinc, copper, or excessive heat, personal habits (infrequent or excessive intercourse, excessive alcohol intake), or impotence. If the male is void of mature sperm, little can be done. Some 29 percent of sterile males produce no sperm cells at all; another 26 percent have cells that begin meiosis but do not complete it; 18 percent have nonfunctional degenerated testicular tissue; and some 27 percent have damaged ducts preventing mature sperm from leaving the testes or various accessory organs. Usually males who cannot produce sperm have a genetic or chromosomal defect that prevents the normal development or functioning of testicular tissue.

To a lesser extent the seminal fluid may be involved in male infertility. The pH of semen is alkaline, 8.0, and it is rich in fructose (2–7 mg/ml). If the pH or fructose level is excessively high or low, the sperm may not function. The sperm must also be *capacitated* or activated by secretions from the vagina in order to be able to penetrate an egg's surface. Failure to respond to such a chemical stimulus or failure of the sperm and egg to be immunologically compatible can lead to sterility.

Figure 5-1
Normal human sperm, as seen under the electron microscope. Note the acrosome which covers the top of the sperm head. The acrosome contains enzymes which digest the egg membrane, permitting entry of the sperm during fertilization. (Courtesy of James K. Koehler, Ph.D.)

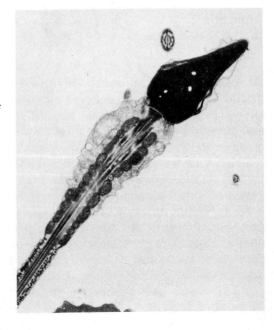

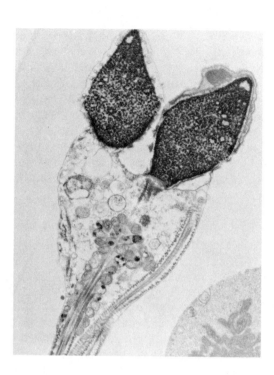

Figure 5-2
The sperm of infertile males may be defective in structure rather than diminished in number. Shown here is an abnormal sperm with two heads. Defective sperm may also show two tails, absence of tail, or other abnormalities. (Courtesy of James K. Koehler, Ph.D.)

Female Infertility Requires More Extensive Examination

If the male is found to be normal in all tests for infertility, then the female is examined. It is easier (and cheaper) to study the male first, although many males are unaware that the incidence of sterility is so high among males. Some males are reluctant to consider infertility as their problem because they falsely equate sterility with impotence or lack of masculine virility, and they err also in considering sterility always to be the female's problem.

The sterile female is more difficult to evaluate but more successfully treated than the sterile male. Unlike males, who produce 10 million sperm per day per gram of testicular tissue from about the age of 13 until they die, females produce only a small number of reproductive cells (one egg per month from the onset of menstruation, or *menarche*, to *menopause*, the cessation of the menstrual cycle, which occurs at about age 45–50). This yields about 400 eggs in the reproductive lifetime of a female. The eggs, unlike the sperm, are not shed externally and thus ovary function must be studied by indirect means. Furthermore, in females there are more reproductive structures and more hormonal controls regulating their functions than in males.

If the female does not menstruate she may suffer from *primary amenorrhoea* (the failure to ever menstruate) or *secondary amenorrhoea* (the premature stopping of the menstrual cycle). Some women have irregular cycles, a fairly common condition which does not interfere with their becoming pregnant but may make family planning difficult.

For most normal women, a cyclical ripening of one egg takes place each lunar month (about 28 days). This does not guarantee that the follicle will rupture, releasing the egg from the ovary, nor that the egg will end up in the oviduct, or later in the uterus. Among sterile females, about 50 percent have blocked or poorly functioning oviducts; about 30 percent have cervical or uterine defects, making them hostile to the sperm or implanting embryo; and about 20 percent have hormonal problems. Blocked oviducts can often be corrected by simple surgical procedures. Some uterine problems can be corrected by hormonal therapy, and some failures of ovulation can be overcome by substances that stimulate fertility.

Clomiphene Is Used to Induce Ovulation in Sterile Females

A synthetic drug, clomiphene, is sometimes used to correct ovulation problems. This compound stimulates the anterior pituitary gland to release two hormones which stimulate egg maturation and the rupture of the follicle that bears the egg. These two hormones are FSH, or follicle-stimulating hormone, and LH, or luteinizing hormone. About 90 percent of nonovulating females will respond to clomiphene therapy and, of these, half will achieve successful pregnancies. One side effect of clomiphene is that it sometimes causes multiple ovulation, resulting in twins or, less commonly, three to six conceptions within one uterus. Normally (that is, without the use of clomiphene) two-egg twins occur about once in 80 pregnancies, but clomiphene-induced twinning occurs about once in 16 pregnancies. The drug does not affect the frequency of identical or one-egg twins. Women whose pituitary glands do not produce FSH or LH may not respond to clomiphene. In such cases, injections of FSH or a hormone from the urine of menopausal women may be used to stimulate follicle development.

Not too long ago, before hormonal therapy for nonovulation existed, infertile women were treated with heavy doses of irradiation to the pituitary gland and to the ovaries (about 80 roentgens to each ovary). The hypothesis for this treatment was based on the belief that radiation stimulated the pituitary gland and cortical ovarian tissue to release hormones. This treatment was used for thousands of women from the 1920s through the late 1960s, until clomiphene and other hormonal treatments were found to be effective. Today the radiation treatment is considered hazardous because both chromosome breaks and gene mutations could be induced in the eggs. After Hiroshima, when concern about radiation hazards became more widespread within both the medical profession and the public, this procedure was criticized. Its proponents justified its use by either denying or minimizing radiation hazards in humans.

In addition to the hazards associated with irradiation, there is considerable doubt that the x rays themselves induced ovulation. Tests were performed which demonstrated that ovulation would sometimes occur even though a lead disc prevented the x rays from reaching the ovaries.

Such women were responding to the placebo effect; they were psychologically assured that something was being done to help them and this released the psychological blocks that prevented the anterior pituitary from releasing adequate gonadal-stimulating hormones. Nearly half of all infertile couples eventually conceive even if they fail to see a physician.

Such psychological factors exist because the anterior pituitary is regulated by hormones released by the hypothalamus, a portion of the brain adjacent to the pituitary gland, which can respond to our emotional states. In normal females a hormone called estrogen is released by the ovaries, which makes the hypothalamus turn off its hormonal stimulation of the anterior pituitary. In response to this, the follicle-stimulating and follicle-rupturing hormones will diminish in production. An emotionally stressed woman may alter the quantity of estrogen she produces or the threshold to estrogen in her hypothalamus. The exact physiological relationship between psychological stress and fertility is not known. When clomiphene is given to a sterile woman, her estrogen is blocked by it and cannot function properly when it reaches the hypothalamus. This permits the anterior pituitary to release FSH and stimulate egg maturation.

As in males, sterility in women sometimes may be due to genetic and chromosomal abnormalities, which cannot be corrected. Also, if meiosis cannot occur at all, or if egg maturation does not respond to hormonal therapy, then sterility will be permanent.

Sperm Production Is Continuous
from Puberty to Death
The testes contain special tissues, the seminiferous tubules, which produce sperm. There are additional tissues in the testes which secrete a masculinizing hormone, testosterone, which assists the maturing sex cells as they are transformed into sperm (Figure 5-3), a process called *spermatogenesis*.

At puberty the adolescent male begins meiotic production of sperm in response to hormones from the anterior pituitary gland. These same hormones are found in females—follicle-stimulating hormone (FSH) and luteinizing hormone (LH). In males FSH is called ICSH (interstitial cell-stimulating hormone), and it causes specialized somatic cells of the testes to release male hormones. These hormones from the testes stimulate sex cell formation and development. The immature diploid reproductive cells are called *spermatogonia*; they divide by ordinary mitosis and retain that capacity indefinitely. Some of them, in response to the ICSH and LH, enter reductional division as *primary spermatocytes*. Each diploid primary spermatocyte completes reductional division, producing two haploid *secondary spermatocytes* (Figure 5-4).

A secondary spermatocyte, although haploid, has two chromatids per chromosome, and thus a second meiotic division must take place. The equational division converts each haploid secondary spermatocyte into

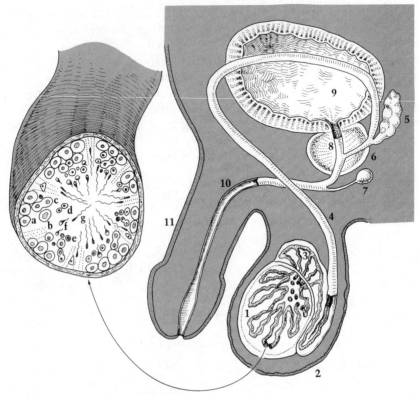

Figure 5-3   Sperm Production in the Adult Male
*The seminiferous tubules of the testes produce haploid mature sperm
(spermatozoa). The testes (1) are housed in the scrotum (2) and transfer the
sperm to the epididymis (3). After undergoing modification there, the sperm
travel along the sperm duct (4) and receive contributions to the seminal fluid
from the seminal vesicle (5), the prostate gland (6), and Cowper's gland (7).
The sperm duct joins the urethra (8) of the bladder (9) and continues through
the penile urethra (10) of the penis (11). An enlarged segment of one of the
seminiferous tubules shows diploid spermatogonia (a) and primary
spermatocytes (b). These undergo reduction (c) forming haploid secondary
spermatocytes (d). Equational division yields spermatids (e) which are
modified (f) into spermatozoa (g).*

two haploid *spermatids*. Thus, the two divisions of meiosis convert one
diploid spermatogonium into four haploid spermatids.

Each spermatid retains a large amount of cytoplasm until a post-
meiotic process, *spermiogenesis*, takes place, eliminating much of the
bulk of the cell and converting the spermatid into a *spermatozoan*. This
process occurs in the testes, and the spermatozoa are then carried by a
fine network of ducts to a storage organ which rests on the surface of
the testis, called the *epididymis*. Within the epididymis some further
changes in sperm morphology take place. The sperm then move to the
seminal vesicle, where they encounter fluid produced by the prostate and
other accessory glands. Altogether it takes about eight weeks for a diploid
spermatogonium to become mature sperm ready for ejaculation.

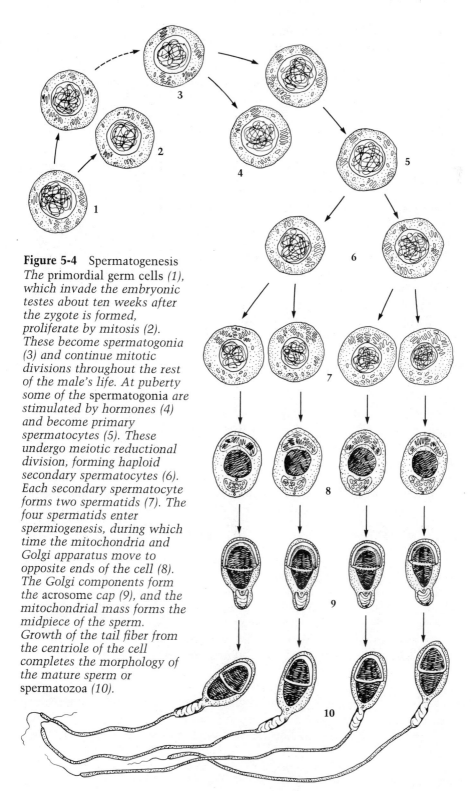

Figure 5-4 Spermatogenesis *The primordial germ cells (1), which invade the embryonic testes about ten weeks after the zygote is formed, proliferate by mitosis (2). These become spermatogonia (3) and continue mitotic divisions throughout the rest of the male's life. At puberty some of the spermatogonia are stimulated by hormones (4) and become primary spermatocytes (5). These undergo meiotic reductional division, forming haploid secondary spermatocytes (6). Each secondary spermatocyte forms two spermatids (7). The four spermatids enter spermiogenesis, during which time the mitochondria and Golgi apparatus move to opposite ends of the cell (8). The Golgi components form the* acrosome *cap (9), and the mitochondrial mass forms the midpiece of the sperm. Growth of the tail fiber from the centriole of the cell completes the morphology of the mature sperm or spermatozoa (10).*

What Should Males Exposed to High Doses of Radiation Do?

What Should Males Exposed to High Doses of Radiation Do? If a male is exposed to ionizing radiation there may be many induced breaks in the chromosomes of cells undergoing spermatogenesis. However, eight weeks after exposure to radiation sperm that are produced will not contain unrepaired breaks. Also, gene mutations are likely to be induced at a higher frequency if the radiation hits tightly coiled chromosomes rather than finely spun out, metabolizing chromosomes. Thus mature sperm at the time of exposure will have more mutations induced in them than those that have not completed meiosis or entered meiosis.

If a male is exposed to a high level of radiation from a nuclear accident, from therapy or fluoroscopic examination near the testes, or during an atomic war, he should not engage in sexual intercourse for two or three months following that exposure without taking careful measures to prevent fertilization. This two-to-three-month delay would not eliminate all the chromosome breaks induced by the radiation, but it would substantially reduce the amount of chromosome damage passed on to the zygote or, by subsequent cell divisions, to a surviving infant.

There is no evidence for cyclicity in sperm production. Unlike the female's hypothalamus, the male's hypothalamus maintains a nearly constant stimulation of pituitary gonad-stimulating hormones.

Egg Production Has a Discontinuous and Complex History

Egg Production Has a Discontinuous and Complex History In the female *all* of the diploid immature, reproductive cells or *oogonia* in the ovaries enter reductional division when the fetus is about six months old. They proceed to the late pairing (diplotene) stage of reductional division. In this suspended state, as *primary oocytes*, they remain for a dozen or more years, and then, at puberty, a single primary oocyte per month will begin its enlargement within a *Graafian follicle* (Figure 5-5). The follicle fills with hormones, particularly estrogen and progesterone, in response to FSH and LH, which are released by the pituitary. These pituitary hormones also rekindle the meiotic process which is held in abeyance throughout pregnancy.

The Graffian follicle forms a blister on the surface of the ovary, and the egg within it completes reductional division. Instead of two *secondary oocytes*, however, only one haploid secondary oocyte is formed. The other nucleus of reductional division gets wrapped in a small amount of cytoplasm and is discarded as a *polar body* on the surface of the egg (Figure 5-6).

The mature egg enters equational division and reaches the point of chromosome alignment (metaphase), at which time meiosis stops. Shortly before reaching this stage, the egg is released from the ovary and falls into the funnel-shaped end of the oviduct, which in response to the hormones of the follicle, has come close to the blistered surface of the ovary.

If sperm are not present in the oviduct when the mature egg is at

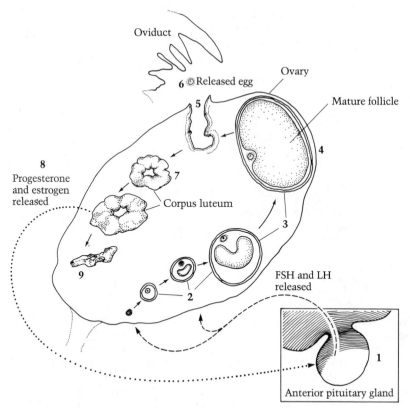

Figure 5-5 Hormonal Control of Egg Production
In the female, hormones from the anterior pituitary gland (1) of the brain—
FSH (follicle-stimulating hormone) and LH (luteinizing hormone)—are
released. FSH stimulates the diploid egg cell (a primary oocyte) *to increase in*
size (2). As it does so a hormonal fluid accumulates in the developing follicle
in response to LH (3, 4). The mature Graafian follicle (5) responds to a surge of
LH and releases the egg (6) which is a haploid secondary oocyte *in the midst*
of equational division. The remaining follicle becomes a corpus luteum (7)
which releases the hormones progesterone and estrogen (8). The estrogen
lowers the production of FSH by the pituitary, and the progesterone lowers the
LH production. If no pregnancy takes place the corpus luteum decreases in
size and leaves a scar (9). The decreased progesterone and estrogen leads to
renewed FSH and LH production by the pituitary gland and the ripening of
another egg.

this point, the egg never completes equational division. It becomes over-
ripe and disintegrates in the uterus. However if a sperm does stimulate
the haploid egg surface at this time (Figure 5-7), the nucleus of the egg
completes equational division, and one nucleus is discarded as a second
polar body. If the first polar body also completes equational division (as
it does in many species but not in humans) the initial diploid primary
oocyte will have produced one haploid egg and three haploid polar bodies.
The nuclei of the sperm and egg will then unite to form the *zygote,* or
fertilized egg, which constitutes a potential new human being.

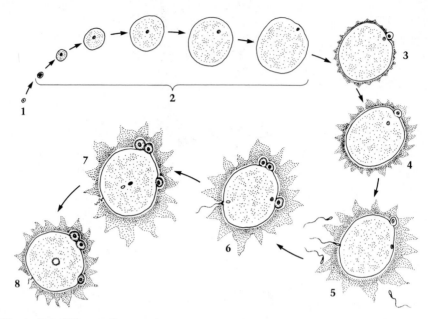

Figure 5-6   Human Oogenesis

*Unlike sperm, which are produced by the hundreds of millions from puberty
to the death of the male, eggs are formed discontinuously. All of the
mitotically dividing* oogonia *shift to form* primary *oocytes while the female
fetus is still in its mother's uterus. These oocytes develop until they reach the
stage when the chromatids are fully paired and partially coiled. They remain
in that condition (1) from birth until puberty, at which time reduction
division occurs, usually in a single cell, about once every 28 days. This is
accompanied by enormous enlargement of the oocyte (2) within a Graafian
follicle. The two nuclei do not share equal parts of the cytoplasm. Instead, the
diploid nucleus moves toward the periphery of the cell (3), and there
completes the reduction division, with one haploid nucleus remaining in the
cell and the other being extruded in a small bleb of cytoplasm. This tiny cell,
which serves no further physiological function, is called a* polar body *(4).*

The equational division of the secondary *oocyte or ovum reaches the
alignment (metaphase) stage (5) and will not proceed further unless a sperm
stimulates the cell cytoplasm. The presence of a sperm signals the haploid
nucleus to complete its division, also near the cell surface, resulting in the
extrusion of another polar body (6). If the first polar body also completes an
equational division, a total of three polar bodies and one ovum constitute the
four products of meiosis (7). The sperm nucleus, free of its midpiece and tail,
moves toward the center of the egg, as does the egg nucleus, where they fuse
to form a new diploid nucleus (8) and the first cell of a new potential
individual.*

Human Reproduction May Differ Substantially
from That of Other Animals   It is both the glory

and the frustrating aspect of biology that we cannot readily generalize
from one organism to all others. While our primary interest here is in
humans, we must realize that what is true for us is not true for all other

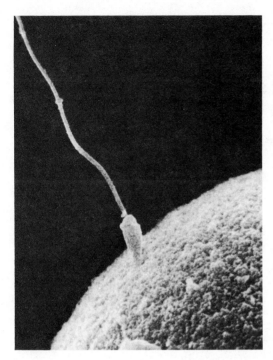

Figure 5-7
The moment of fertilization, showing the contact of a human sperm with an egg. Changes in the chemistry of the egg surface prevent additional sperm from penetrating the egg. (Courtesy Drs. Dan W. Fawcett and Everett Anderson)

living creatures. Very few biological functions have universal application. The general principles governing the functioning of living things have many variations.

Although all eukaryotic, multicellular animals produce sperm through meiosis, there is considerable diversity in ejaculate volume, sperm density, frequency of intercourse, and length of the spermatogenic cycle (Table 5-1). A typical human ejaculate contains 3 to 5 ml of semen, and most human males have intercourse from once a week to daily, but other mammals can be strikingly different. An active tomcat can inseminate ten females in one hour; a sexually active bull can ejaculate 36 times in an hour; and a buck rabbit can inseminate 40 times in eight hours.

The fact that the human female and the apes are receptive to intercourse throughout most of the menstrual cycle is unique in the animal world. Most mammals have only one or a few estrus cycles each year during which intercourse and fertilization can take place. Rhesus monkeys and humans have 28-day cycles; chimpanzees have 35-day cycles. Female receptivity in mammals may be restricted to once a year (as in dogs) or several times during one estrus cycle. Estrus does not involve menstrual bleeding. For many mammals elaborate hormonal and behavioral mechanisms exist for initiating courtship and reproductive activity. How much genetically programmed behavior (if any) exists among humans and determines sexual differences which we refer to as "masculine" and "feminine" has not been established.

Table 5-1   Sperm Number in Some Familiar Animals

The diversity of life is demonstrated by the more than one million different species of plants and animals. Sperm samples from males of different species illustrate a wide variation in sperm density, volume, and number. One human male, in his lifetime, produces about 50 to 75 billion sperm (about 15 times greater than the world's present population).

Organism	Density of Sperm	Volume of Ejaculate	Number of Sperm in Ejaculate
Human	70×10^6/ml	3–5 ml	300×10^6
Rabbit	150×10^6/ml	3 ml	450×10^6
Ram	3500×10^6/ml	1 ml	3500×10^6
Bull	100×10^6/ml	5 ml	500×10^6
Boar	80×10^6/ml	500 ml	4000×10^6
Dog	60×10^6/ml	5 ml	300×10^6
Stallion	113×10^6/ml	56 ml	6000×10^6
Cock	3500×10^6/ml	0.5 ml	700×10^6
Turkey	7000×10^6/ml	0.2 ml	1400×10^6

Egg sizes vary greatly among different species, from microscopic, internal cells in mammals to large, shell-protected eggs in birds and reptiles (Table 5-2). The numbers of eggs also vary, from the solitary egg of larger mammals to the hundreds or thousands of eggs released by amphibians and fishes.

These variations in gamete size, number, and shape, and in the circumstances leading to the generation of new progeny, reflect an immense

Table 5-2   Diameter of Eggs

A mammal's egg is not much larger than the dot of this letter *i*. Frog and fish roe (caviar) are readily visible to the eye because they contain an immense amount of yolk, which supplies the embryo with food until it is free to fend for itself. Reptiles and birds have huge eggs whose masses are millions of times greater than their nuclei, allowing for long gestation periods and relatively large, complex offspring at the time of birth. The unfertilized ostrich egg is considerably bigger than a person's fist and yet, like all eggs, it is a single cell until it has been fertilized.

Species	Size of Egg
Human	0.089–0.091 mm
Cow	0.135–0.157 mm
Frog	0.7–10.0 mm
Fish	0.4–150 mm
Ostrich*	155 × 130 mm
Python*	120 × 60 mm

Source: Adapted from Austin and Short, 1972.

*Mammals, fish, and amphibians have spherical eggs whose diameter range is given. The ostrich and python have oval eggs. For an oval egg there are two diameters, one for length and the other for width.

genetic diversity among species. The differences in their reproductive systems, however, should not obscure the fact that in each species two meiotic divisions occur in spermatogenesis and oogenesis, and fertilization of the egg by a sperm reconstitutes the diploid condition essential for the start of a new life.

Similar variation exists for the size of the male's genitalia (Table 5-3) and for the duration of pregnancy (Table 5-4). There does not seem to be a relationship between penis length and testis weight, but there is an increase in gestation length with the size of the mammal.

Table 5-3 Comparative Size of Male Genitalia

The external genitalia of males vary in size within each species. The penis length and testis weight relative to the size of the mammal also may vary from one species to the next (compare the full body size and penis length of a human and of a rabbit).

Species	Penis Length	Testis Weight
Human	13 cm	20–35 g
Cow	91 cm	300 g
Horse	51 cm	225–300 g
Rabbit	4 cm	2.6 g

Source: From W. S. Spector, *Handbook of Biological Data* (Philadelphia: Saunders, 1956).

Table 5-4 Gestation Length

The duration of pregnancy for a particular species is related to size and longevity. In general, larger animals have longer gestations.

Organism	Days
Opossum	13
Rabbit	31
Cat	63
Dog	63
Guinea pig	68
Pig	114
Sheep	150
Rhesus monkey	163*
Baboon	187*
Chimpanzee	227*
Human	280*
Cow	280
Whale	365
Elephant	623

Source: Adapted from Austin and Short, 1972.

*In primates, the onset of the last menstrual cycle is used as the starting date for calculating gestation.

Sterile Couples Often Must
Face Difficult Options
Some five to ten percent of all married couples will remain childless despite all their efforts and despite medical help. For these couples, adoption is the most frequently sought means of satisfying their desire to be parents. Unfortunately for these people, however, adoptions are becoming more and more difficult to arrange.

Society has changed dramatically since the 1950s, in the industrialized countries, and social services for economically stressed families make it more likely that children, who might formerly have been given up for adoption, will be kept with the parents. Birth control, liberalized abortion laws, and family planning have reduced the number of large families and pregnancies out of wedlock. These factors have resulted in fewer children available for adoption. Many sterile couples must wait several years to adopt a child, and many communities do not have any children available for adoption.

Often, infertile couples want to adopt a newborn child of the same race or ethnic background, and who is also free of serious birth defects. When children were more readily available for adoption, couples had considerable choice, including sex or the resemblance of the child to the adopting parents (through the adoption agency's knowledge of the biological parents). With the dramatic reduction in the number of children available for adoption choices are much more limited. The sterile couple today often selects an older child, a child of a different race, or a child with a birth defect.

Artificial Insemination Is Often
Used in Cases of Male Sterility
Male sterile couples have an alternative to adoption—*artificial insemination*. A considerable number of conceptions are achieved through this technique; in the United States about 10,000 births are attributed to artificial insemination each year. The process may vary with the physician, but certain procedures are universally followed. The sterile couple is informed by the physician that artificial insemination from a fertile donor (AID, artificial insemination by donor) has a high probability of achieving a successful pregnancy. If the couple is receptive to the idea, they are asked to consider the moral issues involved and then return for an appointment if they decide to go ahead.

Most couples who do accept AID for achieving pregnancy usually prefer that no one in their families or among their acquaintances know that the pregnancy was achieved by AID. They prefer complete confidentiality so that their children can be raised without bias on the part of other relatives and without psychological insecurities, which are occasionally present among adopted children. Thus the identity of the donor is unknown to the sterile couple and is known only to the physician (who may destroy the record of the donor after a period of time to assure that anonymity).

The physician selects a donor who matches the sterile male's major blood groups, ethnicity, and general features. The donor must be of proven fertility, and the physician usually (but unfortunately, not always) rejects individuals with poor health, a family history of hereditary diseases, and other serious impairments. Intelligence, longevity, and personality (with the exception of insanity) are not usually considered in selecting a donor, although they are occasionally.

Sometimes the physician will mix some of the sterile male's semen with the donor's. This is done for two reasons. It gives the sterile male some psychological possibility, however remote, that one of his sperm was activated and achieved the fertilization. Second, it gives the child legal protection with regard to his or her right to estate inheritance and parental support, in the event of a later divorce. The physician may also have the sterile couple sign a statement binding them to the support and legitimacy of their child.

State laws vary on the subject of legitimacy, and many legal precedents are based on older values which identified AID use as adultery. AID raises difficult questions: Is the donor liable for child support if the sterile father files for divorce? Does a physician have the legal right to destroy the file on the donor's identity? Should the AID baby be adopted by its sterile parent? AID also raises questions about the rights of sterile couples. Can they specify a donor known to them, selected for qualities they desire? If they wish to know who the donor is for purposes of future medical evaluation (e.g., in genetic counseling) should they have access to this information? A healthy donor may be the donor of a harmful mutation (e.g., Tay-Sachs) or a newly arising mutation (e.g., achondroplastic dwarfism). Should records be kept on donors for that reason? Should the users of AID have the right to sue if the baby has a birth defect?

The semen for AID is most often donated shortly before the woman arrives for her appointment. Some physicians use semen that was donated and then frozen as much as ten years earlier. Whatever the means of AID, the health of such children has been excellent because the semen used represents a segment of the population screened by the physician to eliminate gross abnormalities. The age of the donor selected (usually in his 20s) is also optimal for low nondisjunction and low mutation frequencies. This is true even for the frozen semen. Frozen semen has an advantage in that there is a chance for retrospective evaluation of the donor's health and accomplishments. For this reason frozen semen has been recommended for *eugenic* purposes (which allegedly improve the chances of having healthy or gifted children). The ethical problems and controversies associated with eugenics will be discussed in a later chapter.

Infertile males who have reduced, but not zero-level, sperm counts may use another technique called AIH (artificial insemination of the husband's semen). Several ejaculates of the husband are combined, concentrated by centrifugation, and then administered. Clearly AIH avoids the legal snarls concerning illegitimacy and adultery, but it still raises some moral problems. If a person's religion teaches that masturbation is a sin or that semen should not be ejaculated except in the wife's vagina, then this technique may not be an approved method for achieving a conception.

Present and future research will no doubt increase the medical options (and ethical problems) for sterile couples. Another subject of controversy today is artificial inovulation, the implantation of fertilized or unfertilized eggs in a sterile woman's reproductive tract. However, it will be some time (if ever) before embryos can be developed in artificial wombs with gestation carried out completely outside the mother's body.

For Blocked Oviducts
a Surgical By-Pass Procedure Exists   A form of artificial inovulation was achieved in 1978 when a British medical team, P. C. Steptoe and R. G. Edwards, performed an oviduct by-pass operation in a woman whose oviducts were irreversibly damaged. The woman's egg and her husband's sperm were used to generate a cleavage stage embryo of a few dozen cells. This embryo was implanted in the mother's uterus at the appropriate time in her menstrual cycle. After pregnancy was achieved the fetus was monitored by amniocentesis to make sure its chromosome number was normal. The procedure is not routine, but it is becoming more common. As we learned earlier, half of all sterile females have blocked oviducts, a considerable portion of which cannot be corrected surgically or by being inflated. Potentially there are tens of thousands of women yearly who could elect the oviduct by-pass procedure. At present, few do so because it is expensive to remove an egg, fertilize it in a laboratory, culture the embryo in its first few days of existence, and implant it in the uterus. The fetus also has to be monitored to make sure no gross abnormality is present.

The Steptoe-Edwards technique has been publicized worldwide, and is often referred to as a means of producing "test-tube babies." Actually, the procedure should be considered an oviduct by-pass. The term test-tube baby implies a science fiction quality to the life of that child. For many couples the birth of a healthy baby through this procedure is a "miracle" which does not violate their religious beliefs or personal values. For many other people the procedure of in vitro fertilization raises several ethical problems. The male donates his own semen by masturbation, not intercourse. The female requires surgery to remove the egg, and amniocentesis is needed to make sure the embryo has developed normally.

As in many medical innovations (e.g., vaccination, use of the stethoscope, open heart surgery, and organ transplants) the first successes in oviduct by-pass surgery have been regarded with suspicion by some people, as if some defiance of nature has taken place. As new procedures continue to be used, however (several oviduct by-pass babies have been born since 1978, including some in the United States), they generally become more and more widely accepted. The miraculous or unique qualities associated with them rapidly disappear. Whatever decision a couple makes, they should always insist on knowing the details of the procedures involved as well as the medical risks they face.

Questions to Answer

1. What is the difference between infertility and impotence?
2. How rare or common is sterility in marriages?
3. What are the major reasons for male sterility? For female sterility?
4. What medical options are available for infertile or sterile females?
5. What evidence can you cite to support psychological causes of infertility?
6. Describe spermatogenesis in human males. When does the process begin and at what stage of the life cycle does it cease? Do the same for oogenesis in females.
7. What are the risks for a male having intercourse within a week after exposure to a whole-body dose of 100 R compared to three months later?
8. What options are available to the sterile male?
9. Why is it misleading to refer to a child conceived by means of the Steptoe-Edwards operation as a "test-tube baby"?

Terms to Master

AID
AIH
estrus
Graafian follicle
impotence
infertility
menstruation

oogenesis
oviduct
oviduct by-pass
ovum
polar body
spermatogenesis
sterility

Part 3
Single Gene Effects

Each of us can trace our biological origin to a single cell, which, nine months later, was a newborn baby. How much of you was originally in that one cell? How much of you is biologically determined? Your blood type, skin color, and hair texture are certainly determined. So too are your eye color, facial features, height, and general body build. Less certainty of biological determination exists for your weight (which you can regulate by diet), talents, intelligence, personality, overall health, and life expectancy.

What are the biological determinants that reside in that initial cell? Gregor Mendel was the first to infer that units, later called genes, were the components that determined the inherited characteristics of an organism. But what do we really inherit? We inherit the instructions for how cells divide, how they differentiate into tissues, how they become organized into organs and embryo, fetus, baby, child, adolescent, and adult. The genes dictate how cell organelles are put together and how the cell works. They produce enzymes to digest molecules and to assemble more complex molecules from simpler ones.

All of these activities, collectively, define the living organism either as a cell or a community of cells. The gene, Hermann Muller claimed, is the basis of life. All other components of the cell are synthesized by the genes. The coordination of all gene activities, as one initial cell divides, multiplies, and becomes a fully functional organism, is very complex.

If we concentrate our attention on the individual gene, we can learn a lot about the range of gene activity and the Mendelian law of segregation which governs the transmission of gene instructions from one generation to another. By comparison of how this same law affects humans, garden peas, and fruit flies, we can learn a lot. From one fertilized pea plant several hundred progeny arise. So, too, will a pair of fruit flies produce abundant offspring (in a period of only two weeks). Humans, however, normally produce only two or three offspring, making their traits more difficult to analyze in terms of numerical ratios. Unlike flies and peas, we are not mated to satisfy the curiosity of geneticists. The approach to human heredity is therefore indirect, more laborious, and less certain.

Despite the limitations imposed by our values and culture, we know quite a lot about single gene mutations that affect our health. Some gene mutations are relatively innocuous, others can kill, cripple, deform, and debilitate. Chapters 6–9 will explore the range of single gene defects and how geneticists, biochemists, and physicians have devised strategies to detect and treat these defects or to prevent babies with these disorders from being born. As with all acts of intervention—from the mythological Prometheus enraging the gods by giving fire to humanity to the present-day use of prenatal diagnosis—there is some controversy about altering what fate has assigned to us or what we believe to be the prevailing human condition.

Chapter 6
Mendel's Laws and Genetic Disorders

Andy is a sick baby. He is nearly eight months old but he looks much smaller than most babies his age. He does not cry or stir about much. When he does move he breaks out in perspiration. His dark eyes still shine, hauntingly, but his unsmiling face is washed-out-looking. In a few more months Andy will die, probably of congestive heart failure. His heart is enlarged—immense—but the chambers of his heart pump too little blood to his lungs and body. Surprisingly, when you squeeze his forearm it is not soft and emaciated; it is hard and firm, yet it looks limp. His liver is also enlarged and causes the right side of his abdomen to swell.

Andy is missing an enzyme, acid maltase, which means he cannot convert glycogen, a body starch, into sugar. The glycogen is synthesized in his muscles and liver but instead of using this up between meals, his cells ignore the glycogen as if it were inert. The cells become distended with the massive quantities of glycogen that are produced and built up. That makes Andy's heart muscles get fatter, and the walls of his heart squeeze together, producing an ever narrower passageway for his blood. When Andy dies, his death certificate will mention *Pompe syndrome* (pronounced pom-pay) as the cause of death.

The enzyme Andy needs is produced by a gene on one of the autosomes. Both of Andy's parents are normal, and there is no family history of the disease. For several generations, however, a mutant gene had been passed along, its effects remaining hidden. When a sperm and an egg, each carrying the mutant gene, were united, the long-silent mutation came into expression.

Since children with Pompe syndrome (Figure 6-1) develop heart trouble, and since physicians and scientists cannot get the cells to digest glycogen, the condition is lethal. The normal health and life expectancy that Andy's parents enjoy tell us that the lethal condition only expresses

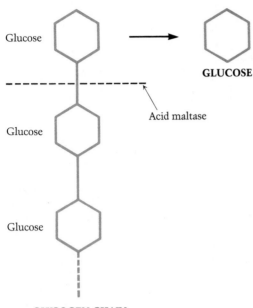

Glucose

GLUCOSE

Acid maltase

Glucose

Glucose

GLYCOGEN CHAIN

Figure 6-1   Pompe Syndrome
*Normally the enzyme acid
maltase digests glucose
molecules from the chain of
glucose residues in glycogen,
releasing the glucose for the
cell's use. If the enzyme is
missing, the glycogen is stored
in the lysosomes. This causes
the disorder known as Pompe
syndrome, a condition that
results in death in infancy.*

itself when the cells lack both of the normal genes. Apparently, one normal gene is all that is necessary to make that enzyme in the cell which mobilizes the glycogen and converts it to glucose when needed.

To understand the way Andy's parents transmitted the factors causing Andy's disorder when they themselves did not show any symptoms, we must introduce a genetic vocabulary and symbolism to describe the mutant gene in Andy and his parents. Once you have mastered that terminology you will be able to apply it to the chromosomes that carry the normal and mutant genes, and you will be able to understand how heredity works.

An Understanding of Genetics
Requires the Use of Genetic Symbols
and Terminology

The inheritance of traits can be interpreted by the movements of chromosomes during meiosis and fertilization. While the process is understood in our minds by reference to the chromosomes, it is symbolized in different ways (Table 6-1). In human genetics the symbols used are capital and lower-case letters.

The normal gene and its mutant variant are called *alleles*. Since the normal gene usually permits a cell to carry out a specific function, the gene for that function is represented by the upper-case (i.e., capital) letter. When a gene cannot carry out that function (usually by not making an enzyme) it is represented by one or more lower-case letters, depending upon the syndrome or defect involved. If the gene for Pompe syndrome is represented by the letter *p*, its normal allele is *P*.

Table 6-1 Symbols Used in Genetic Crosses

Species	Notation	Examples	Advantages	Disadvantages
Humans	Capital letter is used for dominant allele and lower-case letter for recessive allele.	Huntington disease victim is Hh (h is normal allele, and H is Huntington disease allele). Cystic fibrosis (recessive disorder) victim is cc. Normal individual is usually CC, but parents of children with cystic fibrosis are Cc.	Capital letter immediately indicates dominant allele.	Normal allele may be capital or lower-case letter, depending on whether it is dominant or recessive with respect to mutant allele.
Most other eukaryotic organisms	By the convention adopted by the International Congress of Genetics in 1957, recessive mutant is lower-case letter and normal allele has superscript plus sign. Dominant mutant is capital letter or superscript D. Normal recessive allele has superscript plus sign.	Green peas are recessive (g) and yellow peas are dominant (g^+). Yellow peas are either g^+g^+ or g^+g. Green peas are always gg. Fruit fly wings are normally oval and flat. Dominant curly mutant is Cy and its normal allele is Cy^+. Series of mutant alleles may arise or be induced in given normal gene. Recessive symbol is then used with superscript D if dominant mutant allele also arises: dp = dumpy wings, dp^+ = normal wings, dp^D = dominant allele.	Normal function is identified, whether it is dominant or recessive with respect to mutant.	Easier to make errors when plus signs are used.

Because individuals are diploid, their cells contain two of each gene. Thus a person must be either *pp, PP,* or *Pp* for the pair of homologs harboring these genes. When a person has one of each gene, *Pp,* that person is a *carrier* for the mutant gene. Geneticists use the technical term *heterozygous* to represent the carrier state. If the same allele is present in both homologous chromosomes, the individual is described genetically as *homozygous.* Note that Andy is homozygous *pp* but most people are homozygous *PP.* The heterozygous person, *Pp,* looks like the normal, homozygous person who is *PP.*

The mutant allele *p,* which is carried, but not expressed, in the heterozygote *Pp,* is called *recessive.* Similarly, the normal allele, *P,* in the heterozygote, *Pp,* is called *dominant* because only one P is necessary for its expression in the individual. The appearance of the trait is called its *phenotype.* The genetic make-up, which we symbolized as *pp, Pp,* or *PP,* is called the *genotype.* To symbolize the phenotype we place the symbol for the normal or the mutant allele in parentheses. Thus Andy is (*p*) and his parents are (*P*). You too are (*P*), because the Pompe syndrome is lethal, and the fact that you are alive and an adult defines you as phenotypically normal for this trait. What you do not know, however, is whether you are genotypically *PP* or *Pp.* For most genetic traits we do not know if we are homozygous or heterozygous normal because only in a very few cases can biochemical testing reveal whether we are carriers or not.

We can now understand why Andy has Pompe syndrome, and yet has two normal parents. Since Pompe syndrome arises from a homozygous recessive state of the mutant allele, *pp,* each parent had to contribute one *p* allele. This means that both parents must be heterozygous, *Pp.* During meiosis Andy's father will produce 50 percent of sperm with the normal allele, *P,* and 50 percent with the mutant allele, *p.* The same ratio will result from the meiosis of the eggs of Andy's mother. If we assume that any one of these sperm is likely to fertilize an egg, the egg bearing the allele *P* can produce a *Pp* or *PP* zygote. Similarly the egg bearing the allele *p* can produce a *Pp* or *pp* zygote. Of the four possible zygotes, three will be normal (*P*) and one will lead to Pompe syndrome (*p*). This means that any offspring of Andy's parents have a 1 in 4 chance (25 percent) of having Pompe syndrome. The fact that Andy was born with the disorder has no effect on subsequent fertilizations. Neither parent can influence which sperm encounters which egg.

Note also that among the three potentially normal individuals, two are heterozygous, *Pp,* and one is homozygous, *PP.* Thus two out of the three normal children born to the heterozygous parents will pass the mutant gene on to later generations.

Mendel's Law of Segregation
Predicts How Genes Are Passed On   Suppose Andy's parents asked their physician what their chances would be of having another child with Pompe syndrome. To answer that question the physician would rely on

Gregor Mendel's *law of segregation*. This principle of heredity was first described over a century ago when Mendel used the garden pea to study the transmission of contrasting traits, such as yellow and green peas. From the results of his crosses Mendel obtained the law of segregation.

Mendel began with homozygous green peas and homozygous yellow peas. We can call that the *parental generation* (technically, the P_1 or first generation of parents used). The offspring of such a cross between homozygous parents are all heterozygous and constitute the F_1 (for the Latin name for offspring, in this case the *first filial generation*). Since the F_1

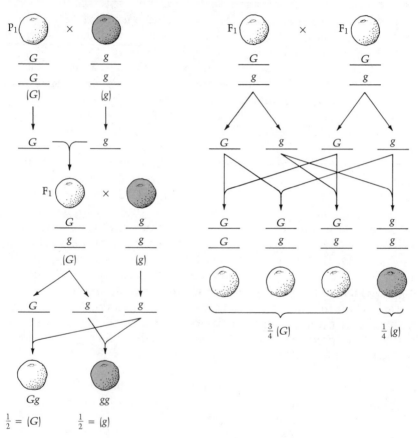

Figure 6-2   Mendel's First Law: Segregation
Mendel, using sharply contrasting traits such as yellow and green pea color, found that the F_1 showed a dominance of the yellow pea color, rather than a blending of the two colors. These F_1 plants, when tested by crossing them to the recessive form, yield a ratio of 1 yellow to 1 green among the F_2 progeny. This procedure is called a test cross.

If the F_1 plants are self-pollinated, they produce both yellow and green peas in the pods of the F_2 plants. The yellow peas are three times as numerous as the green peas. The green, which was recessive or hidden in the F_1 hybrid, had segregated from that association and reappeared in the F_2. Such F_2 homozygous green peas produce green peas indefinitely when self-pollinated and followed for several generations.

turn out to be all yellow, we can conclude that yellow is dominant and green is recessive. Thus we can use the capital and lower-case convention and symbolize the gene for green as g and the gene for yellow as G.

In Figure 6-2 the P_1 and F_1 generations are shown phenotypically and genotypically, with each allele placed on a separate chromosome. The F_1 individuals can be self-pollinated (because flowers have both male and female gametes). The F_2 or second generation of offspring then arise. The two alleles in a heterozygous stamen produce pollen in a ratio of 1 G to 1 g. Similarly the two alleles in heterozygous cells of the plant ovaries produce ovules in the ratio 1 G to 1 g. Note the way the pollen and ovules come together randomly to form four zygotic combinations: 1 GG, 2 Gg, and 1 gg. Thus 75 percent of the F_2 peas are yellow and 25 percent are green.

The chromosomal basis of segregation is shown in Figure 6-3. The pair of homologs for the heterozygote *must* contain one mutant and one normal allele. Thus G and g are aligned on their paired chromosomes in *two* planes, and reductional division segregates G from g, guaranteeing a ratio of 1 G to 1 g for the haploid cells.

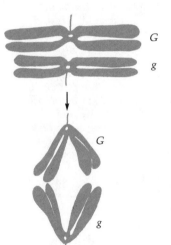

Figure 6-3
The basis for segregation is meiotic reductional division. If the genes G and g are represented as being on a pair of homologous chromosomes, they will segregate from each other as each chromosome moves to an opposite pole. This leads to the segregation ratio of 1 G to 1 g for the gametes produced.

The Founder of Genetics Was a High School Physics Teacher
Gregor Mendel was an Augustinian monk (Figure 6-4). He was born a peasant, and as a youth showed an aptitude for schoolwork. Despite his circumstances and limited education, he became one of the major scientists of the nineteenth century. Mendel came to his interpretation without a knowledge of mitosis, meiosis, or even fertilization. At that time no one even knew that only a single sperm penetrates an egg to start a new individual. Mendel was a substitute physics teacher in a high school for all of his teaching career because he had failed twice to pass the examinations (especially in zoology) for certification as a fully qualified teacher. He had chosen to join the mon-

Figure 6-4   Gregor Mendel (1822–1884)

Although born a peasant's son, Mendel overcame the difficulties of obtaining an education and made one of the great contributions to science through his experiments on garden peas. The laws of heredity governing individual traits were the foundation of genetics. Mendel did not live to see his work recognized and celebrated. His contemporaries could not relate the inheritance of individual traits to more complex character traits (life span, intelligence, body height, overall health) and believed Mendel's results were of minor importance. Unfortunately for Mendel, some of those who tested individual traits in other plants did not confirm his laws. More unusual mechanisms were at work in those plants, but it took another 50 years before these exceptions to Mendel's laws were explained. (Courtesy of Moravské Muzeum V Brne)

astery in Brno, Czechoslovakia because that was the only way a peasant's son in mid-nineteenth century could get a college education.

While he was at the University of Vienna, learning physics and taking courses in biology, Mendel learned about some of the new ideas in science, such as the cell theory and the atomic theory. All living things were thought to be made of small units. Somehow he extended this idea to the study of heredity, and shortly after he began his work as a teacher in Brno, he started the experiments on garden peas which lead to his laws of heredity.

Mendel studied individual plants. He chose very clear and easy-to-follow traits for breeding analysis: pea color (yellow or green), plant size (giant or dwarf), and leaf texture (smooth or hairy). By very carefully cutting off the reproductive organs of the plant after pollination, Mendel guaranteed that unwanted pollen would not stray onto his flowers. He kept precise records and followed each line of his initial cross for three or four generations.

Mendel's results were published in 1866. About 100 copies of the journal that printed his scientific paper were sent to libraries around the world, including the United States and England. He also had 40 copies of his article reprinted and sent to several scientists studying heredity. One of the best of these scientists, Carl Nägeli, corresponded with Mendel for several years. He sent Mendel some seeds of another plant (*Hieracium*, the hawkweed) which did not show Mendelian inheritance properties (it was not known at that time that reproduction in hawkweeds did not involve fertilization). Mendel was disappointed that he could not confirm his earlier work with peas and he dropped his plant breeding experiments for other activities, especially when he was honored by being made Abbot, or head, of the monastery. His work on heredity was rarely cited for 35 years.

Mendel's paper was recognized fully in 1900 when his work was confirmed and extended by three different European botanists. Within six years after the rediscovery, many traits of plants and animals were carefully put to test and found to follow Mendelian laws of inheritance. The field of heredity was given the new name *genetics* in 1906 to replace the phrase "heredity, variation, and evolution." In 1910, the term *gene* was introduced as the name for the unit of Mendelian inheritance.

Some Recessive Disorders
Involve Lysosomal Enzymes
Genes are units of hereditary information that is stored in unique message-bearing sequences along the length of a large nucleic acid molecule, forming the chromosome thread. When their message is successfully decoded, most genes produce proteins that have a corresponding unique sequence of amino acids. A major portion of all proteins are enzymes. Thus numerous genetic disorders occurring in humans can be traced to defective enzymes. These disorders may be lethal at an early age, like Pompe syndrome, or may only slightly impair the individuals, as in albinism. Some are exceedingly rare and others are relatively common.

One major class of enzymatic defects whose biochemical basis has been successfully determined is the group of diseases called *lysosomal storage diseases*. Within the eukaryotic cell the lysosomes serve a digestive function (see Figure 6-5). They break larger, partially broken-down molecules into smaller pieces that can be recycled for use elsewhere in, or outside of, the cell. Lysosomes also digest molecules that are trapped on the surface of the cell and carried into the cell within small bubbles or vacuoles. (The vacuoles merge with the digestive lysosomes.) There are other lysosomes that carry enzymes or waste products to the cell surface and excrete them from the cell. For these reasons lysosomes bear numerous enzymes, and the absence or malfunction of any one of the enzymes may prevent the successful degradation of a cell product into simpler pieces. If the product cannot leave the lysosome, it remains and accumulates, as similar products are brought in by the cell. The lysosomes, engorged with the product, swell to a massive size. In excess the product may be toxic, or the cells jammed with swollen lysosomes may no longer function properly, or the tissues and organs may become severely deformed as the cells become enlarged with the stored material.

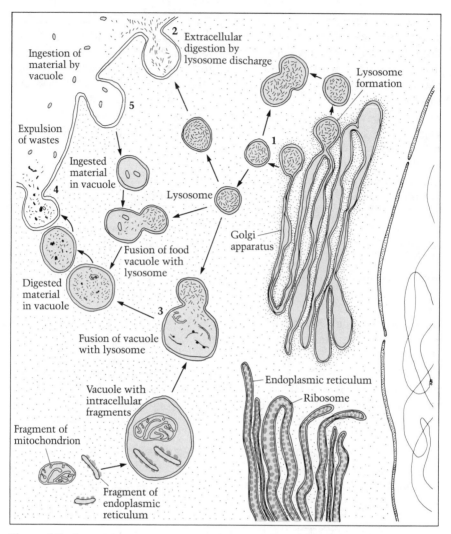

Figure 6-5 Lysosomes

Lysosomes are cell organelles that carry out digestive functions. They are budded off from the Golgi apparatus (1) and contain numerous digestive enzymes (more than 40). Some of these lysosomes leave the cell (2) and carry out digestion outside the cell. Others digest worn-out membranes and organelles within the cell (3), thereby recycling the molecules to make new organelles. Some of the indigestible wastes of the cell are expelled (4) by lysosomes. Still other lysosomes fuse with vacuoles that bring molecules or small objects into the cell (5), where they may be digested.

Lysosomes lacking one of their enzymes may accumulate waste molecules. The waste products may remain engorged within the cell in such cases, producing lysosomal storage disease. Depending on the molecule being digested (which may only be found in restricted tissues) symptoms will appear as the specific tissues involved fail to function properly. These malfunctions may be present in an infant (Tay-Sachs, Pompe, Hurler syndromes) or they may not manifest themselves until puberty when the storage has become excessive (Fabry syndrome).

Tay-Sachs Syndrome Is
a Lysosomal Storage Disease
Affecting the Nervous System   The best-known lysosomal storage disease is the Tay-Sachs syndrome. In this syndrome the product that should be broken down but is not is called ganglioside GM_2, a complex lipid found in nerve cells. In normal individuals this product is cleaved by the enzyme hexosaminidase A. If this enzyme is lacking in an individual, ganglioside GM_2 cannot be broken down, and this condition is called the Tay-Sachs syndrome.

The lysosomes in a Tay-Sachs child at birth are already abnormally swollen, but no symptoms appear until six months of age, when the baby fails to sit up (Figure 6-6). In those early months the baby is listless but responds with a startled reaction to a sharp sound, like the clapping of the parents' hands. The eyes fail to fix on objects and may not follow a moving hand waved in front of the face. If an eye examination is carried out by a physician, a cherry-red spot will be seen on the retina. Progressively, the child will lose nervous function, suffer from seizures, and eventually die, usually before it is four years old.

Tay-Sachs is found more frequently among Ashkenazic Jews (descendants of Yiddish-speaking ancestors who lived in Europe) than in other populations. About 1 out of 27 such Ashkenazim are carriers for the recessive gene that causes the syndrome. The reason for the high incidence of Tay-Sachs among this group of people is not known, but may be related to historical factors, such as the initial founding of a population by a small band of settlers, one or more of whom carried the mutant gene, or some unknown adaptation, such as resistance to an

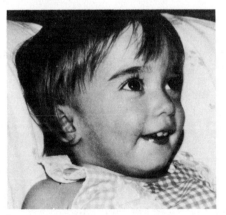 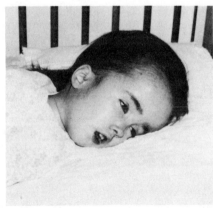

(a)   (b)

Figure 6-6   Tay-Sachs Syndrome
The child with Tay-Sachs syndrome in (a) is 17 months old, has a slightly enlarged head, and is partially blind but can still respond to sound. In (b) the same child at 4 years and 9 months is unresponsive, has a much enlarged head, and is totally blind. The child died a year later. (Courtesy of Robert and Marlene Lippman)

Table 6-2 Lysosomal Storage Diseases

The lysosomal storage diseases vary in severity depending on the tissues that are involved. Thus Pompe syndrome disturbs muscle cells; Fabry syndrome, the endothelial (lining) cells of the blood vessels; Tay-Sachs syndrome, the neurons; and the Morquio syndrome, the connective tissues. A few lysosomal storage diseases are shown in this table. All are autosomal recessive conditions except for the Fabry and Hunter syndromes. The I-cell and Chediak-Higashi syndromes may be defects of cell regulation or membrane defects rather than enzyme deficiencies. Most are rare (incidence is about 1 in 25,000 births) but ethnic prevalence is present in the Tay-Sachs, Gaucher, and Niemann-Pick syndromes for Ashkenazic Jewish populations. A list of some 2800 single gene mutations in humans, with useful information and references on each, is available in V. A. Mckusick, *Mendelian Inheritance in Man, Sixth Edition* (Baltimore: Johns Hopkins University Press, 1983).

Name of Disorder	*Enzyme Defect*	*Major Symptoms*
Lipid (fat) metabolism		
Fabry	α-Galactosidase A	Juvenile onset; purple pimples; burning sensation in hands and feet; kidney and heart disease; death about age 45, X-linked
Gaucher	Glucosylceramidase	Juvenile onset; enlarged spleen and liver; defective clotting; bone erosion, survival variable in adults
Tay-Sachs (GM$_2$ gangliosidosis)	β-N-Acetyl-hexosaminidase A	Infantile onset; paralysis; blindness; seizures; startle response to sound; death by five years of age
Mucopolysaccharide metabolism		
Hurler	α-1-Iduronidase	Infantile onset; coarse features; mental retardation; bone deformities; enlarged liver and spleen; opaque corneas; death by age 20
Hunter	Sulfoiduronate sulfatase	Same as Hurler, but corneas clear; X-linked
Morquio	Chondroitin sulfatase	Growth retardation by two years of age; severe skeletal deformities; normal intelligence; opaque corneas
Carbohydrate metabolism		
Mannosidosis	α-Mannosidase A and B	Juvenile onset; mental retardation; deafness; skeletal deformities; coarse features
Pompe	α-Glucosidase	Infantile onset; enlarged heart; muscular weakness; failure to thrive; death by 1–2 years of age

Other metabolic defects

Nieman-Pick	Sphingomyelin phosphodiesterase	Infantile onset; mental retardation; enlarged liver and spleen; blindness; death by 4 years of age
Metachromatic leukodystrophy	Arylsulfatase A	Onset within 1–2 years; progressive paralysis; weakness; spastic movements; death by 15 years of age
I-cell or Leroy	Unknown; lysosomal enzymes leak into plasma or escape from cell	Infantile onset; skeletal deformities; dwarfism; mild to severe mental retardation; death in childhood
Chediak-Higashi	Unknown; white blood cell granulocytes and melanocytes devour other cells but lysosomal digestion diminished	Infections; pain in hands and feet; seizures; albinism; death before adulthood

infectious disease, that carriers had which was lacking in normal homozygotes. Many ethnic groups or populations show such prevalences of rare genetic diseases. For this reason the diseases are called *ethnic diseases.*

There is no cure for Tay-Sachs syndrome, but the presence or absence of hexosaminidase A can be detected in the fetal cells of the amniotic fluid through amniocentesis. Parents can also know, through tests of their hexosaminidase A levels in blood or tears, whether or not they are carriers, because carriers make only half as much of the enzyme as do homozygous normal persons.

Pompe and Tay-Sachs syndromes kill at an early age, but there are other lysosomal storage diseases that are more insidious and cause severe impairments that persist through life, sometimes reducing life expectancy. Mannosidosis, for example, involves the storage of compounds that have the sugar mannose. The lysosomes of people affected with mannosidosis are particularly engorged in nerve cells, producing mental retardation. The enzyme is also involved with the formation of connective tissue and bones, so there are other abnormalities that can be seen. Persons with mannosidosis have coarse-looking faces, pug noses, thickened bones in the skull (causing deafness), and deformed vertebrae along their spines. While some children with mannosidosis die before their teens, others live on past their twentieth birthday. Because of their severe retardation and physical defects they require lifelong institutional care. Although there are 30 or more known lysosomal storage diseases in humans, most of which affect the degradation of fats and carbohydrates, these make up only a small portion of all recessive disorders (Table 6-2).

Albinism Arises from a Failure
to Synthesize or Store Melanin
in Skin Cells

In albinism one of the genes essential for pigment production may be mutant, resulting in cells that lack the enzyme tyrosinase. Pigment in our skin and hair comes primarily from the compound melanin. This very large molecule is composed almost entirely of the amino acid tyrosine. A tyrosinase-negative albino has white hair, pale skin (even in an African native), eyes lacking pigment in the choroid layer, and usually a quivering motion of the eyes (Figure 6-7). The albino is sensitive to light and squints or wears sunglasses outdoors. If exposed to the sun (or to ultraviolet lamps) on a regular basis, severe sunburn can occur. Also skin cancer is much more frequent among albinos than among those with normal skin pigment because the melanin in a normal person filters out the ultraviolet rays before they reach the nucleic acids of cell nuclei in the skin.

It takes more than tyrosinase, however, to synthesize melanin and distribute it within pigment cells. The melanin has to be stored in special

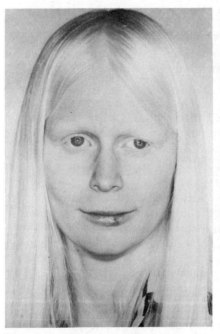
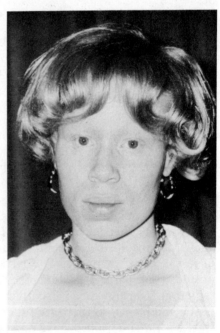

(a)   (b)

Figure 6-7

(a) An oculocutaneous albino Caucasian showing the characteristic white hair, pink skin, and squinting eyes (photophobia). The eyes of such albinos quiver back and forth (nystagmus) due to a neurological defect in the anatomy of their visual tracts to the brain. (b) An oculocutaneous albino Negroid from Nigeria. The skin is a pale tan (straw-colored), and the hair color is a pale yellow. Squinting and nystagmus are also present. Albinos occur in all human races. (Courtesy of R. J. Gorlin, D.D.S.)

organelles and these have to be moved from the cells that produce them to other cells in the skin that accumulate them. Some albinos are even tyrosinase-positive. It is easy to assay the enzyme if it is active. A few hairs plucked from the head will contain hair bulbs which can be examined microscopically after exposure to a tyrosine-rich solution. The staining reaction of the actively growing hair bulb is used to diagnose the state of the enzyme.

In contrast to the digestive enzymes in lysosomal storage disease, tyrosinase is an enzyme that synthesizes complex molecules from simple ones. Since melanin is found in only a few tissues, most of the organs in an albino are normal in function. Curiously, there is one relation of albinism to brain morphology that is not related directly to melanin synthesis. The paths of the nerves leading from the eyes to the brain of an albino person do not form normally. This results in quivering eyes and the inability to perceive depth or three-dimensionality as efficiently as non-albinos.

Some Genetic Disorders Are Expressed in the Heterozygote As Dominant Traits

Recessive disorders usually involve enzyme defects of metabolism. *Dominant mutations* usually result in defects of the processes of embryonic or adult tissue development. It is not known, at the molecular level, how these dominant mutants alter the regulation or normal timing, differentiation, and movements of cells. Unlike metabolic enzymes, *regulatory proteins* serve as switches, turning on or off different "batteries" of genes. Clearly if a group of genes is "turned on" when they should not be functioning, the cell's activities may be grossly changed.

Unlike recessive disorders, dominant disorders are almost invariably found in heterozygotes. When the affected person reproduces, the mutant allele that causes the defect will segregate from the normal allele, and thus half the gametes will contain the dominant (mutant) allele and the other half will carry the normal allele. Most likely the person with a dominant disorder will marry a normal person. The normal individual is homozygous for the normal allele. Thus all zygotes formed in the marriage will be either heterozygous, resulting in the expression of the dominant trait, or homozygous for the normal gene. The gametes of the heterozygous carrier, according to Mendel's laws, are segregated in a ratio of one dominant to one normal. Thus the zygotes (receiving only normal genes from the other parent) must also show a phenotypic ratio of one dominant to one normal. We see then that the risk of a child being born with a dominant disorder when one of the parents carries the mutant gene is 50 percent.

Huntington Disease Is a Classical Dominant Disorder

A good example of a dominant disorder is Huntington disease. It is a degeneration of the nervous system that occurs somewhat late in the life of the individual, at about the age of 30 to 40. For this reason a child who has the gene for Huntington disease may not know if it is present or not until he or she has already had one or more children, since most people start their families in their mid-twenties.

We can represent the mutant gene by H and the normal allele by h. Most persons who have Huntington disease are genotypically Hh. Thus a typical marriage would involve a P_1 of $Hh \times hh$. While two persons with Huntington disease might marry, this is rare, mainly because it is not a common defect and, to a lesser extent, two individuals at risk for the disorder might not wish to face the 75 percent chance of having a child with Huntington disease.

A person with Huntington disease develops neurological defects which progressively worsen over a span of about ten years, at which time paralysis is so severe that death occurs. The disease has an insidious start. The person becomes forgetful and may be fired because of numerous mistakes made while working. A tremor develops, gradually leading to jerky motions of the arms and legs. There may be personality changes, with explosive temper tantrums or paranoid accusations directed at friends, acquaintances, or family members. Walking becomes more difficult and feeding may become a chore because of the palsied hands and arms.

Huntington disease is rarely due to new mutations. Virtually every case of Huntington disease in the United States can be traced to one of two families. One of these families settled in Quebec in 1644 and moved across Canada, heading south into the Midwest during the nineteenth century. The other family came from England, reaching North America in 1637. For 16 generations members of these families have continued to marry and produce offspring. There are about 2000 descendants who have been traced through pedigree studies. Because of the absence of symptoms at the time of marriage, many individuals at risk for Huntington disease decide to take the chance and have children.

The brains of Huntington disease patients show considerable degeneration of the motor control region and the cerebral cortex, the region that controls conceptual skill. The molecular basis for the disease is unknown and no treatment is available to prevent the progressive paralysis and degeneration of the nervous system.

Most Dominant Disorders
Arise As New Mutations
Dominant diseases are numerous. About 1400 are known but fortunately all of them are rare. In general a dominant disease that causes severe medical problems, grotesque disfigurement, mental retardation, psychosis, or unusual appearance may prevent that person from finding a marriage partner. For these reasons persons with disorders like achondroplasia (congenital dwarfism, represented in the movie versions of L. Frank Baum's *Wizard of Oz* and Roald Dahl's *Charlie and the Chocolate Factory*) often remain single and do not have

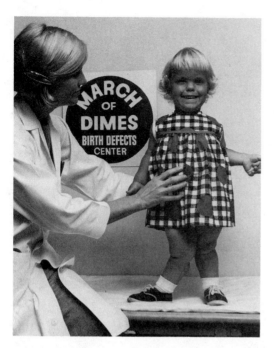

Figure 6-8 Achondroplastic Dwarfism
This child has a 50 percent chance of transmitting her dwarf stature to her children. The gene is an autosomal dominant, producing shortened limbs, spinal curvature, and an unusually shaped skull with a prominent forehead. (March of Dimes Birth Defect Foundation)

children (Figure 6-8). The same is true for disorders like neurofibromatosis, a severe disorder of connective tissue lining the bones and nerves (Figure 6-9), or retinoblastoma, a cancer of the eye in children. Because of the early onset of these disorders, they are less often transmitted to descendants than are late-onset disorders such as Huntington disease.

Normal genes mutate at a constant rate and thus many harmful dominant diseases arise as new mutations in the sperm or egg of a normal parent. Such newly arising dominant disorders are called *sporadic* disorders. If the child with the sporadic disorder reaches maturity and marries, his or her child has a 50 percent risk of inheriting the mutant gene for the defect.

X-Linked Traits, Like Autosomal Traits, Can Affect Any Tissue or Organ

X-linked disorders usually affect males rather than females. In both autosomal recessive and autosomal dominant syndromes there is a one-to-one sex ratio of males to females among the affected persons. Since flowers usually are *bisexual hermaphrodites*, plants rarely have separate sexes. Animals are often organized into separate sexes, and the basis for being male or female is usually chromosomal. In humans, we saw, males are XY and females XX for their sex chromosomes. When a mutant recessive gene is located on the X chromosome in the male, the male invariably expresses this condition because the Y chromosome contains very few genes (and these are chiefly associated with testes formation and sperm production). In a fe-

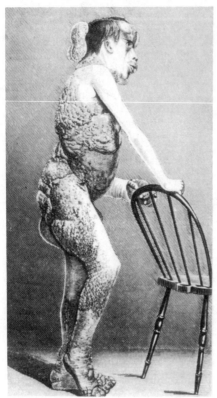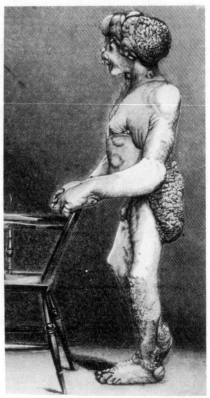

Figure 6-9
Neurofibromatosis or von Recklinghausen disease is an autosomal dominant condition. In its most extreme form the connective tissue of the bones and the lining of the nerves produce disfiguring tumorous growths. A devastating expression of the gene occurred in John Merrick, a late nineteenth-century Englishman, whose tragic life became the subject of a contemporary play and movie, The Elephant Man. *(From* The British Medical Journal, *Vol. 2, December 1886, p. 1189)*

male both Xs must carry the recessive gene in order for her to express the disorder to the same degree as a male.

The X-linked mechanism of inheritance favors the appearance of a mutant condition in males. This is as true for lethal conditions as for less serious ones. Some X-linked traits are common, like colorblindness (about six percent of all males). Others are rare, like Fabry syndrome, a lysosomal storage disease, or Lesch-Nyhan syndrome, a nucleic acid metabolism defect that results in a self-mutilation. Perhaps the X-linked disorder that is most familiar to us is hemophilia, also known as factor VIII deficiency. This disorder was passed on by Queen Victoria to several of her grandsons in the Spanish, German, and Russian royalty. It is a clotting defect of the blood. Without regular transfusions of blood containing the missing factor, or concentrated factor VIII extracted from blood, the person's chances for survival are slim. Untreated, it is a lethal disorder.

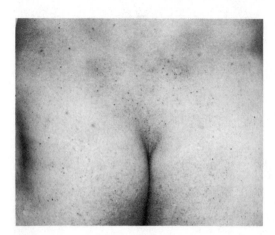

Figure 6-10   Fabry Syndrome *An unusual purplish-red rash is seen on the skin of victims. Fabry syndrome is an X-linked trait and has serious effects in middle age when kidney failure and heart disease are likely to occur. The rash is produced by the rupture of swollen capillaries in the skin. The lining of the capillaries (endothelial cells) store excess lipids in their lysosomes. (Courtesy of Dr. Charles C. Sweeley)*

Fabry syndrome victims are missing an enzyme necessary for breaking up a fat found in the membranes of most cells of the body, but chiefly in the kidneys and the lining of the blood vessels. The disease causes excruciating, burning pain in the fingers and toes. A large number of tiny, dark red skin lesions (pimple-like swellings of the capillaries) form along the lips, thighs, and genital region (Figure 6-10). The most serious symptom, however, is the lipid accumulation in the kidney, sometimes causing kidney failure. Also dangerous is the lipid storage in the heart and blood vessels, leading to murmurs, heart attacks, and strokes. Death usually occurs between the ages of 40 and 50. Although there is no treatment for this enzyme defect, some Fabry patients can prolong their lives through a kidney transplant if they are suffering from renal failure.

In contrast to lethal or semi-lethal X-linked disorders, colorblindness is without medical consequences. The defect is restricted to the cone cells of the retina, which are responsible for color vision. Rod cells, the other visual cells of the retina, are not involved in color perception, and so the person with malfunctioning cone cells will be colorblind.

Colorblind individuals usually confuse red and green, but yellow and blue perception is normal for them. There is no cure or treatment for colorblindness. Because colorblind males have normal life spans and the mutant gene is present in six to eight percent of the population, homozygous recessive colorblind females also occur (they usually have colorblind fathers and carrier mothers) with an overall frequency of 0.4 percent.

The Inheritance of X-Linked Traits
Was Demonstrated Initially in Fruit Flies   X-linked inheritance was worked out by T. H. Morgan in 1910. His study of fruit fly eye color provided the mechanism for interpreting human X-linked disorders. The details of X-linked transmission are shown in Figure 6-11. Although the ratios of X-linked disorders are not typically those of Mendelian segregation the same basic principle is involved. There is a segregation of the

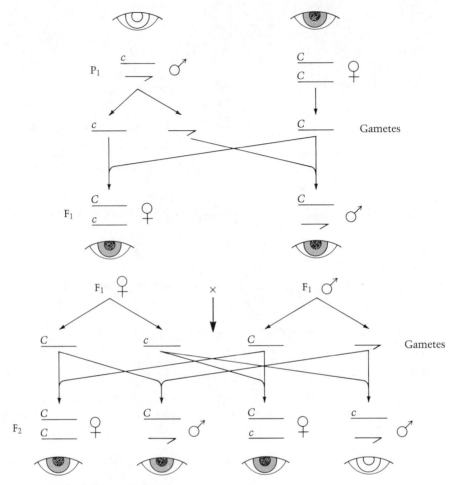

Figure 6-11   X-linked Inheritance

In mammals and many other forms of animal life, including flies, the sexes differ in the make-up of their sex chromosomes. The female has two X chromosomes, represented by XX, and the male has only one X, usually accompanied by a Y chromosome and represented as XY. In the diagram the X is a rod and the Y is J-shaped.

The genes of the X chromosome, for the most part, are not present on the Y, and the Y has very few genes of its own. Thus a male usually expresses whatever is on the X chromosome, regardless of its dominant or recessive condition. For this reason the male is said to be **hemizygous** *rather than homozygous or heterozygous. The fact that the male has only a single X causes a marked departure from the ratios of Mendelian segregation, but the mechanism—meiotic reduction—remains the same.*

This difference is seen dramatically in the case of a **reciprocal cross,** *where the sex of the colorblind partner in a cross of a colorblind person and a person with normal color vision makes a difference in the outcome. (In autosomal traits, which obey Mendel's law of segregation, no such difference occurs.) The colorblind male mated to a noncolorblind female yields progeny who all have normal color vision. When a son from this couple has a wife*

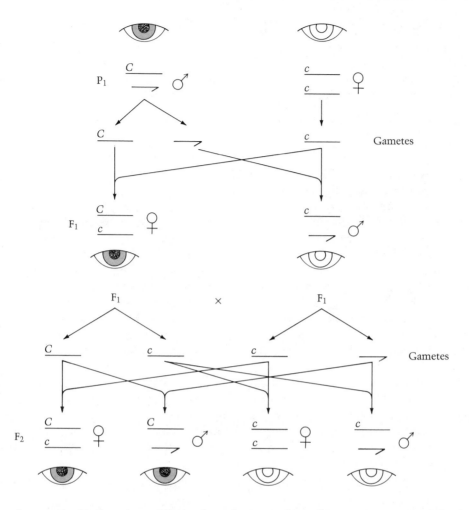

from a similar marriage, they produce the equivalent of an F_2 generation, with a modified 3:1 ratio, with all of the colorblind individuals being males. The rest of the offspring from this union have normal color vision in a ratio of two females to one male.

When the cross is carried out using a P_1 male with normal color vision and a colorblind female, the F_1 are crisscrossed with regard to their sex; the sons resemble their mothers (i.e., they are colorblind) and the daughters resemble their fathers (having normal color vision). Crosses involving F_1-type individuals, who would be heterozygous females and hemizygous colorblind males, produce a 1:1 ratio of normal to colorblind offspring, with equal numbers of males and females of each type.

For human X-linked traits homozygous females are rare (although about 0.4 percent of females are colorblind). In most instances X-linked disorders are transmitted from heterozygous mothers to their sons, and if the disorder is life-threatening, the mutation has probably been of recent origin (not more than 3 or 4 generations old), because hemizygous males with the disorder do not reproduce.

two X chromosomes in the female and of the X and Y chromosomes in the male. The X-linked traits in the male are expressed regardless of whether they are recessive or not. The recessive trait is usually not expressed in the female heterozygote. The XY condition is neither homozygous nor heterozygous in genotype. It is called *hemizygous*. The XX females, of course, can be heterozygous or homozygous for a pair of alleles.

A Female's Two Xs Do Not Produce
Twice the Activity Obtained
from a Male's Single X
The female, as we have learned, is XX and the male is XY, with the Y having very few of the genes that are present on the X. Very clearly the genes on the Y are not essential for an organism to live, for all normal females contain no Y chromosomes in their cells. Theoretically a female should have double the gene products that a male has, since she has two Xs and he has only one. In fact, this does not occur. The male and female have the same quantity of enzymes or gene products from their X-linked genes. This equivalence results from *dosage compensation*, which, in humans, involves the inactivation of one of the two X chromosomes in the female's cells.

In 1955 Murray Barr noted a peculiar blemish or chromatin spot in mammalian cells. This spot, called the *sex chromatin*, was found to be present in the female's cells but not in the male's. Males were thus defined as chromatin-negative and females as chromatin-positive. How-

1   2   3   4

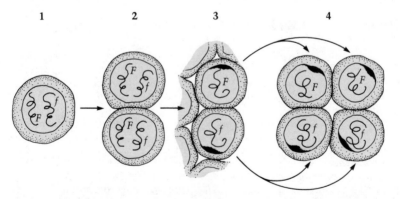

Figure 6-12   Dosage Compensation by X-Inactivation
When the zygote of a potentially normal female is formed, both of its X chromosomes are unwound and functional in the G_1 stage of the cell cycle (1). Cell divisions in the early stages of development produce cells (2) with both X chromosomes functional. At the 1000-cell stage (3) random inactivation occurs and one of the two X chromosomes forms the sex chromatin. From then on, the mitotic descendants of such a cell (4) retain the same X in the sex chromatin. Thus, a female carrier for Fabry syndrome, Ff, whose f-bearing chromosome is inactivated, produces two daughter cells, each of which has the f-bearing X as sex chromatin and a fully-functioning F-bearing X.

ever, it was later noticed that Klinefelter males were chromatin-positive and Turner females were chromatin-negative. Therefore, the chromatin spot was not strictly correlated with sex.

Mary Lyon proposed a theory of *X-inactivation* to interpret these observations (Figure 6-12). At about the time when a female embryo contains 1000 cells one of the two X chromosomes in each cell is inactivated. The inactivated chromosome is tightly coiled and remains attached to the nuclear membrane, forming the chromatin spot. When this first occurs the process is random, occurring on either (but never both) of the Xs. From then on cells always form, among their descendants, a population of cells with the same X inactivated. If a female is heterozygous for the presence of the X-linked enzyme glucose-6-phosphate dehydrogenase (G6PD), the G6PD gene may be inactivated in one of her cells. Since the other X lacks the ability to make the enzyme, it will be unwound and functional in the G_1 stage of the cell cycle, and all mitotic descendants of this cell will thus lack G6PD.

Inactivation Allows Only One
X Chromosome to Function
The Klinefelter male is XXY. Thus we find only one chromatin spot, because the Y plays no role in X-inactivation, and only one of the two X chromosomes is in excess. A Turner female, being 45,X in karyotype, has only one X and thus there is no excess to be inactivated. The triple-X female has two chromatin spots because there are two X chromosomes in excess.

Dosage compensation only occurs among X chromosomes. It does not involve autosomes. Thus autosomal trisomies (e.g., trisomies 13, 18, and 21) are not dosage-compensated and the dose from the excess genes in such individuals creates abnormal metabolism and development.

Not everything is known about dosage compensation. We do not know why 45,X females show Turner syndrome features such as webbed neck and short stature. We do not know why XXY individuals differ in their skeletal and behavioral development from XY males. In the Turner syndrome each cell has one functional X and no Y, as do the cells of normal females. In Klinefelter cells there is one functional X and one functional Y, as in normal males. We do know that X-inactivation does not fully turn off the functions of the second X chromosome; part of the short arm of the X remains unwound during the G_1 stage. The chromatin spot does unwind and gets replicated, but this occurs in G_2 and not in the S phase. What those unwound genes do during the G_2 stage is also unknown.

X-Inactivation May Cause
Genetic Expression
of Some X-Linked Traits
If a female is heterozygous for an X-linked recessive trait, half her adult cells should express the mutant trait and half should express the normal trait. Why doesn't this result in illness

for carriers of Lesch-Nyhan syndrome, Fabry syndrome, hemophilia, or muscular dystrophy? In many disorders, such as Lesch-Nyhan syndrome, the product not digested by the mutant cells will be digested by nearby normal cells. It is the product—the purine in this case—which readily diffuses from cell to cell and gets properly metabolized, permitting the female to enjoy normal health. But if a female is a carrier for Fabry syndrome her cells having the normal gene inactivated within a chromatin spot cannot digest the ceramides accumulating within the lysosomes. The normal enzyme functioning in a neighboring cell cannot help out because the enzyme cannot leave the normal cell and enter the mutant one. Nor can the ceramide float out of the lysosome, leave the cell, and enter a normal cell. The ceramide and the enzyme are both too large and remain within the cell. Fabry carriers thus vary in severity of expression, ranging from some who are symptom-free to some who are afflicted with severe pain and illness. Hemophilia, like Fabry, carriers show variable clotting defects. Carriers for muscular dystrophy, like Fabry carriers, are also variable.

How to Use Pedigrees
for Human Genetic Studies
When a physician or geneticist takes a family history the conventional symbolism is reduced to phenotypes, using a square to designate the male and a circle for the female. This phenotypic representation is called a pedigree. Parents are connected by a crossbar, and a perpendicular line descends to the next generation. The children are attached to that perpendicular line by a horizontal crossbar. Any person in the pedigree showing a trait being studied is represented by a filled-in box or circle (Figure 6-13).

In the case of Andy we would represent him and his parents as follows:

In a family, if the Pompe syndrome appears in a second-born daughter, and the first-born is a normal son, the family is represented thus:

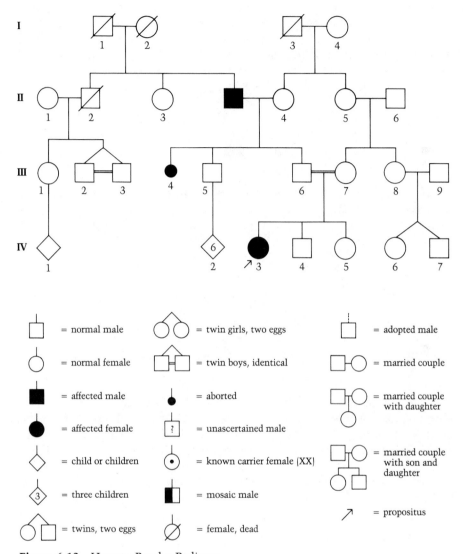

Figure 6-13 How to Read a Pedigree

*In a human pedigree each generation is assigned a Roman numeral beginning
with the oldest traceable generation. The male is represented as a square; the
female as a circle. The filled-in circle or square indicates an individual
affected with the particular trait whose inheritance is being analyzed. A
horizontal line between a circle and a square indicates a mating. A vertical
line indicates children from that union. Twins are connected by a bifurcated
line (IV-6 and IV-7); when two horizontal bars are present in the twin pair
(III-2 and III-3) they are one-egg (identical) twins. When parents have two
horizontal bars between them, they are related by a common relative as are
cousins (III-6 and III-7). A diamond indicates other children (sex(es) not
given), and the number of such children is placed in the diamond (e.g., IV-2
are the six children of III-5). An arrow pointing to an individual is the
propositus, the person whose case is being followed by pedigree analysis. A
small black circle is a spontaneous abortion. Deceased persons are indicated
by a diagonal slash through the circle or square.*

In disorders showing clear-cut dominant inheritance, such as Huntington disease, the trait will be seen in at least one of the parents and in one or more of the children:

In X-linked inherited disorders, like Fabry syndrome, the trait will be passed by female carriers to male children. When the female parent is known to be a carrier, she is represented by a circle with a dot inside it.

The pedigree is a valuable summary, showing the physician or genetic counselor how the trait has been passed on and which relatives are at risk for passing the defect on to their own children. The counselor can then advise members of the family about resources that are available for treatment or prenatal diagnosis. Information about the trait can be presented to family members to help them in making decisions about their families.

Questions to Answer

1. What is a lysosomal storage disease?
2. What are the characteristics of a recessive lethal condition?
3. What features of experimental design were employed by Mendel in revealing the law of segregation?
4. What historical circumstances would you consider likely if an inherited disorder is an ethnic disease?
5. Why are most persons with Huntington disease traceable to ancestors who had that disease some 350 years ago, but retinoblastoma victims rarely can find an affected ancestor more than 1 or 2 generations back?

6. If your spouse has Huntington disease and you do not, what is the probability you will have a normal child? A normal son?

7. If you and your spouse are tested for Tay-Sachs and both of you are carriers, what is your probability of having a normal child? A Tay-Sachs fetus?

8. What is the most likely genotype of the mother of a colorblind son?

9. What are the differences between a genotype and a phenotype?

Terms to Master

allele

carrier

sex chromatin

dominant

dosage compensation

genotype

heterozygous

homozygous

hemizygous

law of segregation

lysosomal storage disease

pedigree

phenotype

recessive

sporadic dominant

X-inactivation

X-linked inheritance

Chapter 7
Indispensable Blood: The ABO and Rh Systems

Blood has long been recognized as a vital fluid. Until the nineteenth century it was considered the hereditary substance, and such phrases as "blue blood," "blood will tell," "blood brother," and "it's in his blood" still reflect that belief. The popular idiom, "blood is thicker than water," identifies a kinship relation which demands a loyalty beyond merit. This is especially seen in the widespread practice of nepotism (job favoritism for relatives).

We know now that it is not our blood, but our genes, that define relatedness and serve as the basis of heredity. But blood, like all other components of our bodies, is composed of substances under genetic control. Fortunately there are only a few different types of blood among humans, making it possible to transfuse blood from one person to another.

Transfusions of blood were rarely done before the twentieth century. Blood was successfully transfused from one dog to another in 1665 in England. Several attempts shortly after that success to transfuse animal blood (usually from a lamb) to humans failed, and laws were passed to prevent such experiments. The first human-to-human transfusion took place in 1818, but during the nineteenth century too many failures occurred to justify its use.

Either Whole Blood or Components of Blood
May Be Used from Donated Blood An adult human contains about 12 or 13 pints of blood (about 6 liters), 45 percent of which is water. The cellular portions of the blood include red blood cells (RBCs or erythrocytes, Figure 7-1), white blood cells (several types, derived from the thymus gland, the lymphatic glands, and the bone marrow), and platelets (which participate in clotting). When blood clots, the remaining fluid

Figure 7-1
The electron micrograph of red blood cells (erythrocytes) shows the typical biconcave appearance of such mature cells. The biconcave appearance of the cell results from the absence of the nucleus. Also, few organelles are present; the erythrocyte is essentially a membrane stuffed with lots of hemoglobin. The cell membrane contains proteins which may have antigenic properties when present in the blood plasma of another individual. These blood antigens are used to type blood and to permit successful blood transfusions. (Courtesy of Ruth Blumershine, Indiana University School of Dentistry)

is called *serum*. However, in blood banks, when the blood is centrifuged so that the cellular portion is removed from the liquid portion, the liquid is called *plasma*. The plasma still contains certain clotting factors that are no longer present in serum.

Donated whole blood must be used for transfusion within 21 days. Whether the blood is refrigerated or not, the red blood cells continue their maturation and wear out. Red blood cells are continually replaced in the body at a rate of one percent per day. They are produced in the bone marrow by erythroblasts, and the nuclei are squeezed out as the cells pass through narrow membranous pores. The life of an erythrocyte in the blood is about four months. Blood banks separate the plasma from whole blood after 21 days, dehydrate it, and store it as a frozen, dried powder. Such plasma concentrate can be kept for years and reconstituted with sterile water. Plasma can be used to compensate for a severe blood loss until whole blood becomes available. Thus battlefield casualties may receive plasma from a medic at the site of battle, but will be given whole blood as soon as they reach a field hospital. Sometimes the red blood cells are separated and concentrated for transfusion into anemic patients who have sufficient plasma but require additional cells to carry their oxygen.

When you donate a pint of blood the cellular loss will be fully re-stored within six weeks. Blood donation policy varies from country to country (Table 7-1). In the United States about 100 pints of old blood are pooled for plasma extraction. In Great Britain this is restricted to ten pints because of the risk of spreading hepatitis virus. In the United States blood may be obtained from volunteers or from paid donors. In Great Britain, blood is obtained only from volunteers. In general, blood from volunteers is less likely to carry the hepatitis virus or parasitic infections. When blood is given freely, it is usually by persons motivated by the desire to save lives. In the case of paid donors, personal needs may some-times outweigh the moral responsibilities of revealing their medical his-tories. Thus a skid row alcoholic with little or no money might put the purchase of alcoholic beverages above all other considerations in donat-ing blood. If such a person were carrying hepatitis, it would be passed on to the person receiving the blood.

When blood is donated, voluntarily or for pay, the donor is tested for syphilis. Those with a positive test are not allowed to donate blood. There are no mandatory tests for malaria or certain tropical parasites which may be present in whole blood for many years. Donors are required to fill out a detailed medical history, however, to protect not only recip-ients, but also themselves. Persons with heart trouble, anemia, or hy-pertension, for example, are not permitted to donate blood without a physician's consent.

Table 7-1 Blood As a Commodity

The legal guidelines and public traditions regarding blood donations are highly variable. In Great Britain (which has the lowest hepatitis infection rate in the world) all the blood is voluntarily donated and carefully regulated. There are no commercial blood banks in Great Britain. In the United States, paid donations (mostly through profit-making commercial blood banks) are as frequent as voluntary donations (as in Red Cross drives). The purchased bloods have higher hepatitis infection rates than the voluntary bloods. In Sweden all donors are paid by hospitals and commercial blood banks. In general, paid donors have higher hepatitis rates than unpaid volunteers. Once a country is on a paid-donor basis, it is very difficult to convert to an all-voluntary basis.

Country	Percentage Paid for Donation
Britain	None
Ireland	None
France	Few
USA	50
USSR	50
Greece	66
West Germany	40–85
East Germany	85
India	83
Japan	98
Sweden	100

Source: Adapted from Titmuss, 1978.

The ABO Blood Types
Are the Most Important
in Determining Transfusion
The early failures with blood transfusion occurred for two major reasons. Sterile techniques were not used until the last quarter of the nineteenth century when the germ theory of infection was firmly established. But even germ-free transfusions would sometimes fail because the blood would clump (agglutinate), causing severe shock reactions. The nature of this incompatibility was solved by Karl Landsteiner in 1901. He showed that red blood cells sometimes contained surface proteins or antigens, which he designated A or B. The plasma or cell-free portion of blood contained antibodies complementary to those found on the cells (Figure 7-2). Thus type A red blood cells floated in anti-B serum. Type B red blood cells were bathed in anti-A serum. Besides the A and B blood types, Landsteiner detected two additional types (now designated AB and O). The AB individuals have both A and B antigens on the surface of the cells, but neither antibody in their plasmas. Individuals whose red blood cells lack both the A and B surface antigens are called type O; they have both anti-A and anti-B present in their plasmas.

The transfusion of type A blood into the veins of persons with type B or O would lead to clumping because both type B and type O recipients

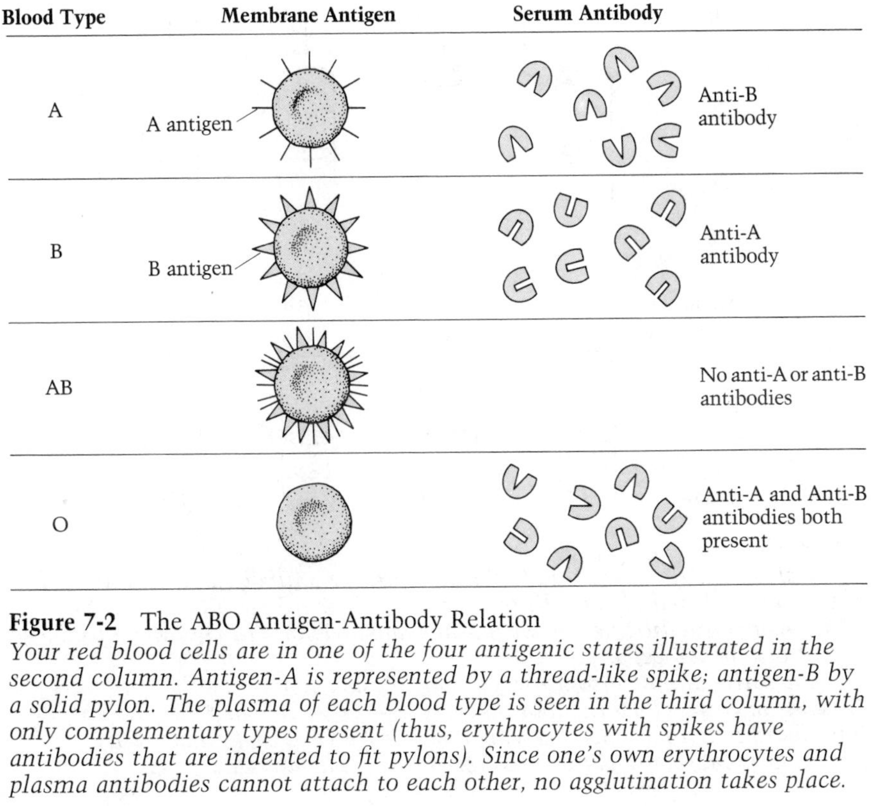

Blood Type	Membrane Antigen	Serum Antibody
A	A antigen	Anti-B antibody
B	B antigen	Anti-A antibody
AB		No anti-A or anti-B antibodies
O		Anti-A and Anti-B antibodies both present

Figure 7-2 The ABO Antigen-Antibody Relation
Your red blood cells are in one of the four antigenic states illustrated in the second column. Antigen-A is represented by a thread-like spike; antigen-B by a solid pylon. The plasma of each blood type is seen in the third column, with only complementary types present (thus, erythrocytes with spikes have antibodies that are indented to fit pylons). Since one's own erythrocytes and plasma antibodies cannot attach to each other, no agglutination takes place.

have massive amounts of anti-A antibody in their plasmas (Figure 7-3). Similarly, type B blood transfused into type A or type O recipients will encounter the antibody anti-B present in the plasmas (Figure 7-4).

Only type AB individuals can receive blood from any donor without a clumping reaction, because such recipients have no anti-A or anti-B in their plasma. The type O individuals, with both anti-A and anti-B in their plasma, can only receive donations from other type O donors.

The type of blood an individual has is genetically determined. A specific gene—the I gene—exists in three different forms. The variant forms of a gene, we have learned, are called alleles. When several alleles are associated with one gene, they constitute a *multiple allelic system*. The ABO system of multiple alleles determines the A, B, AB, and O blood types (Table 7-2). The I gene may carry the A, B, or O allele. It never carries both A and B. Zygotes may be $I^A I^A$ or $I^A I^O$ for type A, $I^B I^B$ or $I^B I^O$ for type B, $I^A I^B$ for type AB, and $I^O I^O$ for type O. Note that type AB individuals are codominant; they contain two alleles, neither of which is exclusively dominant because both the A and the B types are expressed in the cells. Type O individuals are always homozygous for the I^O allele. Type A and B individuals may be either homozygous ($I^A I^A$ or $I^B I^B$) or heterozygous ($I^A I^O$ or $I^B I^O$).

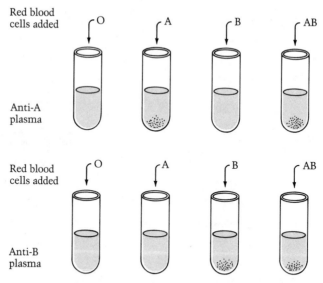

Figure 7-3

The vials in the top row contain plasma with the antibody anti-A. Into each of the four vials red blood cells of type O, A, B, or AB are introduced. Note that the anti-A antibodies in the plasma cause agglutination of the introduced type A or type AB cells.

Similarly, in the bottom row of vials, plasma containing the antibody anti-B is present. When the four blood types—O, A, B, and AB—are introduced note that vials receiving the B or AB cells agglutinate. This agglutination response is the basis for effective blood typing.

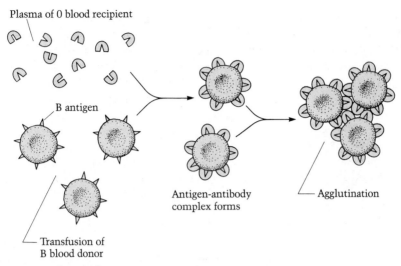

Plasma of 0 blood recipient

B antigen

Antigen-antibody
complex forms

Agglutination

Transfusion of
B blood donor

Figure 7-4   The Agglutination Response
*When blood from a donor is incompatible with the recipient's blood, the
surface antigens of the recipient's blood (shown as pylons on the periphery)
will be covered with antibodies. In this case the antigen B (pylon) is encased
by the antibody anti-B found in type O plasma. Type O individuals, lacking
the A or B antigen on their own cells, have massive quantities of anti-A and
anti-B antibodies. The coated type B cell encounters other coated type B cells
and the altered antibodies adhere to one another, causing the cells to clump
or agglutinate.*

Table 7-2   Donor and Recipient ABO Relations

The ABO blood types involve a dual relationship present throughout the life of the
individual. Proteins on the surface of the red blood cells (antigen A or antigen B) may be
present alone, together, or not at all, depending on the blood type. Complementary
proteins (antibody A or antibody B) are present in the plasma. The table illustrates the
normal cell and plasma contents for the A, B, AB, and O types. In column three are
shown those potential donors whose red blood cells will agglutinate with the recipient's
plasma. In column four are those donors whose red blood cells would survive without
provoking agglutination in the recipient's plasma. Note that the surface antigens in
column one define the blood type of the individual.

Antigen on Your Red Blood Cell Membrane	Antibody in Your Plasma	Incompatible Blood Donors	Acceptable Blood Donors*
A	Anti-B	B, AB	A, O
B	Anti-A	A, AB	B, O
AB	None	None	A, B, AB, O
O	Anti-A and anti-B	A, B, AB	O

*In practice, the whole blood transfused into the recipient is of the same type; the permissible donations
of O for A or B, and of A, B, or O for AB can only be used in extreme emergencies because the donor
serum's antibodies can react with some of the recipient's red blood cell antigens. Thus, a type O donor's
anti-A and anti-B will cause some damage to the recipient's red blood cells if that recipient is type A,
B, or AB.

The Rh Blood Groups
Normally Lack Serum Antibodies
It was not until 1940 that another major cause of blood transfusion failure was confirmed. Blood from rhesus monkeys was found to cause a clumping reaction in rabbit plasma. This gave rise to the name Rh (for rhesus) for the antigen on the erythrocytes. The same Rh factor was found in humans, associated with a disease of newborn children, *erythroblastosis fetalis*. The disease usually involves Rh-negative mothers and Rh-positive fathers. Persons are Rh-positive if their red blood cells have a surface antigen. They are Rh-negative if the red blood cells lack the surface antigen. Couples with other combinations are not at risk. Males who are Rh-negative (whether their spouses are Rh-positive or Rh-negative) do not produce children with erythroblastosis fetalis. Nor do Rh-positive males who marry Rh-positive females.

Unlike the ABO system, the serum of any Rh blood type lacks circulating antibody. Only when Rh-positive blood enters the blood stream of a person who is Rh-negative will serum antibodies against the Rh factor develop. Rh-negative blood, if introduced into a person who is Rh-positive, will not provoke antibodies against the Rh negative condition. This is because an Rh-negative cell *lacks* the Rh factor. The Rh-negative blood cell does not have an antigen of its own.

Erythroblastosis Fetalis Arises
in Rh-Negative Females
Who Marry Rh-Positive Males
A mother who is Rh-negative (that is, her erythrocytes lack the Rh-positive antigen) and whose fetus is Rh-positive is at risk of developing an anti-Rh antibody in her blood (Figure 7-5). Such antibodies in a mother will destroy the Rh-positive cells of the fetus, leading to jaundice, mental retardation, or the death of the newborn child. In Europe a nomenclature different from the one used in the United States is used to designate the Rh series of alleles. Rh-negative individuals in Europe are usually represented as cde (with d being the chief Rh-negative factor). Rh-positive individuals have a D in their specific allelic designation. Thus Rh-negative (cde) individuals can potentially form anti-D (anti-Rh antibodies) when exposed to the D antigen.

Unlike the ABO multiple allelic system, which has complementary antibodies present in the plasma throughout the life of the individual, the Rh allelic system has only the antigens present in the blood of an individual. The antibodies must be artificially provoked by introduction of genotypically different cells. Thus an Rh-negative individual will develop antibodies against the Rh antigen.

There are three methods of introducing the Rh antigen into an Rh-negative person. The person may have had a prior blood transfusion with Rh-positive cells. The person may have received (in the days before viral measles vaccines existed) a blood component, gamma globulin, from a cured measles donor. Such injections, in addition to conferring immunity to measles, may have transmitted the Rh antigen. The most

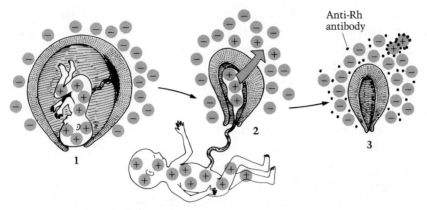

Figure 7-5 Birth and Rh Immunization
The near-term fetus (1) of Rh-positive blood type in an Rh-negative mother faces no risk if the mother's plasma contains no anti-Rh-positive (also called anti-D) antibodies. When birth occurs (2), the placental damage may cause some fetal cells to enter the mother's blood (prior to delivery, this is prevented from happening by the separation of fetal and maternal blood by the placenta). The mother's antibody-forming tissues will produce the anti-Rh antibody (3) represented by the small black dots.

common method by which the Rh antigen is introduced is the production of antibodies by an Rh-negative mother when she gives birth to her Rh-positive child. For an Rh-negative mother there is about one chance in 40 that at delivery some of the fetal Rh-positive cells will enter her blood. If this happens the mother's blood may develop anti-Rh antibody.

Ever since the discovery of the Rh factors, pregnant women have been blood-tested for their Rh alleles to prevent immunization through pregnancy (Figure 7-6). If the woman is Rh-negative and her husband is

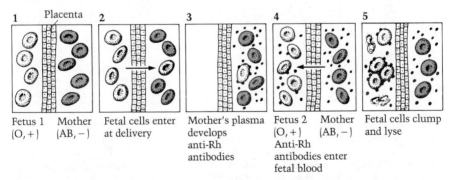

1 Placenta	2	3	4	5
Fetus 1 Mother (O, +) (AB, −)	Fetal cells enter at delivery	Mother's plasma develops anti-Rh antibodies	Fetus 2 Mother (O, +) (AB, −) Anti-Rh antibodies enter fetal blood	Fetal cells clump and lyse

Figure 7-6 How Later Pregnancies Are Affected by Rh Incompatibility
The Rh-positive first pregnancy (1) occurs in an Rh-negative mother whose plasma does not contain anti-Rh antibody. At birth (2) some of the fetal cells may enter the mother's blood and provoke the formation of anti-Rh antibody (3). When the mother carries another Rh-positive fetus (4), the maternal antibodies can cross the placenta (most proteins cannot) and produce clumping and lysis of the fetal blood (5). This may have serious consequences to the fetus, including abortion or birth with erythroblastosis fetalis.

Rh-positive, then more thorough tests are carried out. The mother is checked for anti-Rh antibody and for her prior exposure, through pregnancy or transfusion, to Rh antigen. If the fetus is at risk because the mother has had some prior exposure to Rh antigen, the pregnancy can be monitored, and if there is a rise of anti-Rh antibodies in the mother's blood, transfusion can be carried out in utero, by amniocentesis. This transfusion is rarely used today. The newborn child is more likely to receive an Rh-negative blood transfusion immediately after birth.

The Damaging Effects of Rh Blood Incompatibility Are Now Preventable by a Simple Treatment

Fortunately, in the mid-1960s, a means of preventing Rh blood incompatibility was discovered. The Rh-negative mother is given a special blood test after delivery of her Rh-positive child. If fetal Rh-positive cells are present in the mother's blood she is given injections of *rhogam,* a preparation of anti-Rh antibody obtained from immunized donors (Figure 7-7). The rhogam coats the fetal red blood cells in the mother's blood so that there are no Rh-positive antigens available to stimulate the mother's circulation to form any anti-Rh antibody of her own. Until the mid-1960s, a woman who bore a child with erythroblastosis fetalis often ran the risk of earlier and more severe damage to the fetus in her next pregnancy, often resulting in spontaneous abortion.

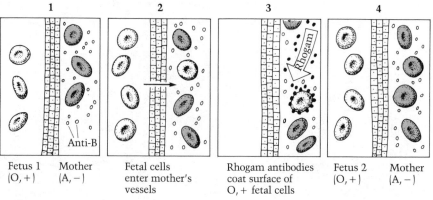

1	2	3	4
Fetus 1 (O, +)	Mother (A, –)	Fetal cells enter mother's vessels	Rhogam antibodies coat surface of O, + fetal cells	Fetus 2 (O, +)	Mother (A, –)

Figure 7-7 The Rhogam Method of Preventing Erythroblastosis Fetalis
The Rh antigens in a type O fetus (1) do not create problems if this is the first such pregnancy. But at birth (2) the fetal cells enter the mother's circulation where they are not disturbed by the anti-B antibodies present in her blood (because she is type A, her antibodies are anti-B, and because the fetus is type O there are no type B antigens on the cell surface). To prevent the formation of anti-D (anti-Rh) antibodies, the mother is given enough rhogam (anti-D antibody) to destroy the fetal cells (3). Thus, in her next Rh-positive pregnancy (4), there are no anti-D antibodies of the mother's own making; the amount supplied artificially was enough to destroy the Rh-positive cells of the first fetus but not enough to remain in her blood to affect the second fetus.

The children with erythroblastosis fetalis often suffer mental retardation from the breakdown products of the disintegrated (lysed) blood cells. The neurons of such infants are stained with the porphyrin (iron-bearing) component of the hemoglobin released by the destroyed cells. This condition, *kernicterus,* is toxic to the neurons and leads to mental retardation.

In the United States (and Western Europe as a whole) about 83 percent of the population is Rh-positive and 17 percent Rh-negative. The Rh-negative blood type is less common among blacks (7 percent have it), relatively rare among Orientals (1.5 percent have it), and almost nonexistent among American Indians.

Most couples who are Rh incompatible do not experience the predicted maternal stimulation of anti-Rh antibody following delivery of an Rh-positive baby. Their protection most often involves *natural immunity* (Figure 7-8). If the Rh-negative mother, for example, is type O, and her Rh-positive husband is type A, an Rh-positive type A fetus could introduce type A cells (also bearing the husband's Rh antigens) into the mother's blood. However, the mother's overwhelming amount of anti-A antibody would coat the surface of such fetal cells so heavily that they would not be able to stimulate anti-Rh antibody. It is this natural pro-

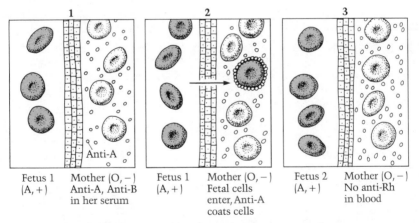

Fetus 1	Mother (O, −)	Fetus 1	Mother (O, −)	Fetus 2	Mother (O, −)
(A, +)	Anti-A, Anti-B	(A, +)	Fetal cells	(A, +)	No anti-Rh
	in her serum		enter, Anti-A		in blood
			coats cells		

Figure 7-8   The Natural Protection Against Rh Incompatibility
Every person's red blood cells have both an Rh blood type and an ABO blood type. If the father is A, Rh-positive and the mother is O, Rh-negative the fetus (1) may be A, Rh-positive (the fetus can also be O, Rh-negative if the father is heterozygous for A and the Rh-positive genes). At birth, such fetal cells may enter the mother's blood and encounter her naturally-occurring anti-A and anti-B antibodies (type O persons always have plasmas with anti-A and anti-B antibodies in massive quantity). These antibodies overwhelm the fetal type A cells, coating them with anti-A antibody. Such inactivated cells are then engulfed by white cells and digested. They do not have a chance to provoke an anti-D (anti-Rh) antibody in the mother. Thus, in her next Rh-positive pregnancy (3), the mother's plasma remains free of anti-D antibodies.

tection that is imitated when at-risk Rh-negative females are given rho-gam injections after delivery of their babies.

Some Minor Blood Groups in Humans
Are Less Likely to Be of Clinical Significance
There are many additional antigens on the surface of red blood cells which are rarely involved with transfusion reactions. They are useful, however, to anthropologists in following the passage of genes among populations. They are also valuable in certain medical situations, such as disputed paternity and identical twin determination. Those antigens that occasionally do cause transfusion reactions or threaten the life of the fetus are the MNS, P, Lutheran (Lu), Lewis (Le), Kell (K), Kidd (JK), Duffy (Fy), and Xgª antigens. When a person donates or requires blood these antigens are typed along with the ABO and Rh systems. The minor blood groups either produce antibodies with difficulty, or the antigens capable of doing this are rarely found in the population.

Clotting Disorders Usually
Involve the Absence
of a Needed Protein in the Serum
If you scratch a mosquito bite or nick your face or leg with a razor blade, blood will trickle from the wound for about a minute and then stop as the wound is covered by a clot. The clot is a mass of cells entrapped and embedded in a matrix of blood platelets and fibrous factors. The formation of a clot involves several stages. Shortly after the wound occurs the capillaries constrict. The platelets, tiny cells lacking nuclei, form a plug in the cut capillaries. Numerous proteins in the blood (designated factor VIII, factor IX, factor XI, and so on) form a complex material called thromboplastin. This is converted to thrombin which, in turn, produces fibrin. The fibrin makes a stable spongy clot of the cellular components of the released blood. As the liquid serum is reabsorbed or evaporated, the clot becomes a hard scab.

Some individuals lack one or more of the components for coagulation. In such individuals bleeding from minor cuts may continue for many minutes or hours; more serious spontaneous bleeding may be indicated by bruiselike marks near body joints, measles-like rashes (petechiae) under the skin, frequent nosebleeds, or other symptoms. Males with this X-linked disorder, which is called hemophilia A, lack factor VIII. Because marriages between surviving male hemophiliacs and carrier females are rare, no cases of homozygous females with hemophilia are known. The process of X-inactivation, however, does result in some heterozygous females having a mild hemophilia. Hemophiliacs can survive to maturity if they receive transfusions of concentrated factor VIII or fresh plasma. The factor VIII treatments are costly (about $20,000 per year) and not all hemophiliacs respond effectively to it. About 20 percent of hemophiliacs develop antibodies against the injected factor VIII.

Clotting disorders can arise from defects of any of the many enzymes or structural proteins involved, including the precursors of fibrin, the vitamin K that is associated with some of these enzymes, and the platelets themselves.

White Blood Cells Are
Part of Our Immune System
Red blood cells carry oxygen to the body tissues and remove carbon dioxide, which is carried by the veins and released in the spongy capillary network of the lungs. Another function of the blood is to protect the body from infectious bacteria and viruses. The cells participating in this role, the white blood cells, are part of the *immune system*. The infectious organisms are destroyed directly either by cells that engulf and digest such foreign material or by cells that form antibodies to immobilize and inactivate foreign proteins. The white blood cells that eat foreign matter (bacteria, cell debris) in the blood plasma or body fluids are produced in the bone marrow. They are called *myelocytes* and there are several types (neutrophils, eosinophils, and basophils) named for their responses to cellular stains. Myelocytes surround and ingest bacteria by a process called *phagocytosis*. The ingested bacteria are degraded by lysosomes.

A second category of white blood cells are synthesized chiefly by the lymph nodes, spleen, and tonsils. These are the *lymphocytes* and *plasma cells*. They produce antibodies which immobilize viruses, foreign protein debris, and bacteria.

Normally the white cell components of blood have relatively short lives (platelets live only 9–11 days, myelocytes about 10–20 days, and lymphocytes up to 100–200 days). Thus synthesis of these cells is a lifelong activity.

The white blood cells may be deficient in number or function because of genetic defects, cancer, or other diseases. Such losses of cell number or function lead to increased and longer-lasting infections.

Fallacies About Blood
Are Widespread
As we have seen, blood is not the basis of heredity. This is not the only fallacy associated with blood. Many advertisements imply that diet affects our blood. Actually, diet has only minor effects on the health of the blood. Except during pregnancy, iron supplements are not usually needed. Nor does an ordinary diet deprive the blood of vitamin K, which is mostly synthesized by the liver.

Another fallacy involves blood transfusions between races. Such transfusions do not alter our personalities or our heredity. In countries where racism is a strong social force, blood is racially segregated, a practice discontinued in most of the United States after World War II. The only states that still legally require segregation of blood according to the donor's race are Arkansas and Louisiana. In South Africa blood is legally segregated into White, Bantu, Indian, Oriental, and Colored (various shades of skin color in persons of mixed racial background with some black ancestry, however remote).

Questions to Answer

1. What are the major components of blood?
2. What antigens are present in the blood of individuals of (a) type A, (b) type B, (c) type AB, (d) type O?
3. What antibodies are normally present in the plasma of individuals of (a) type A, (b) type B, (c) type AB, (d) type O?
4. Distinguish between the terms antigen and antibody.
5. What antibodies are normally present in the plasma of an individual who has an Rh-positive (D or Rh +) blood type?
6. How may anti-D antibodies be produced in an Rh-negative individual?
7. What is the usual Rh blood type of (a) the mother and (b) the fetus in most cases of erythroblastosis fetalis?
8. How does ABO incompatibility naturally protect Rh-positive babies of an Rh-negative mother from being born with erythroblastosis fetalis?
9. What are the advantages and disadvantages of an all-voluntary program of blood donation in the United States?
10. Discuss the differences between clotting and agglutination; between blood plasma and serum.

Terms to Master

ABO series	hemophilia
antibody	heteroallelism
antigen	natural immunity
codominant	phagocytosis
erythroblast	plasma
erythrocyte	Rh series
erythroblastosis fetalis	rhogam
factor VIII	serum

Chapter 8
Prenatal Diagnosis

Although feelings differ among people throughout the world on the use of prenatal diagnosis, it should be recognized that the technique can *save* fetal life as well as provide the means for identifying an abnormal fetus which the parents may elect to have aborted. For example, deliberate intervention in the fetal environment, we learned, was first used to transfuse blood into fetuses with certain blood group incompatibilities, and thus save the lives of the fetuses. That use is rare today because newer methods have been developed to prevent blood group incompatibility between mother and fetus.

**Amniocentesis Is a Major Means
of Performing Prenatal Diagnosis**   The technique most widely used today for prenatal diagnosis is *amniocentesis* (Figure 8-1) which we briefly discussed in Chapter 4. In this technique, a needle is slowly pushed into the pregnant woman's abdomen as precisely as possible so that it enters the amniotic cavity surrounding the fetus. Prior to insertion of the needle the exact location of the fetus, placenta, and amniotic sac is determined by use of ultrasound (Figure 8-2). A beam of very high sound waves echoes off the structures in the mother's abdomen. These sound waves are converted to electric currents, sorted out by computer, and fed into an oscilloscope which shows the internal structures because each tissue has a different density and location within the abdomen.

Once the location of the amniotic sac has been determined and the needle has been inserted, about 20 to 40 ml of a yellowish fluid are removed from the amniotic sac (Figure 8-3). The woman is usually in her 14th to 16th week of pregnancy when this procedure is done. It cannot be performed as accurately or as safely if she is at an earlier stage

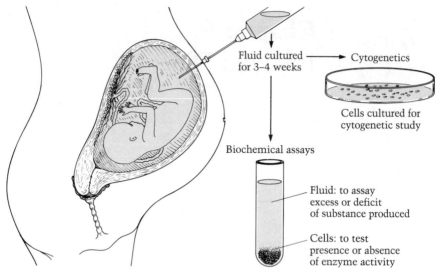

Fluid cultured ——————→ Cytogenetics
for 3–4 weeks

Cells cultured for
cytogenetic study

Biochemical assays

Fluid: to assay
excess or deficit
of substance produced

Cells: to test
presence or absence
of enzyme activity

Figure 8-1
*Amniocentesis is used for the prenatal diagnosis of chromosomal disorders
and genetic defects which have a biochemical disturbance detectable in the
cells or the fluid surrounding them. During the sixteenth week of pregnancy
the woman's abdomen is monitored by ultrasound to determine the proper
site of entry for a needle. The needle is slowly inserted into the amniotic sac
and about 40 ml of amniotic fluid is removed. The fluid is cultured in tissue
culture flasks, and about four weeks later there is sufficient growth to permit
cytological or biochemical assays. Most amniocentesis is done for women at
risk (usually because they are 35 years old or older), yet even for such women,
less than five percent have an abnormality present which leads to an elective
abortion.*

Figure 8-2   Ultrasonography
*The use of ultrasound (sonar)
to locate objects in a fluid is
well-known in submarine
warfare. Its application to
detect internal organs, tumors,
and developing fetuses is a
more recent development. The
ultrasound can reveal soft
tissues not normally detected
by x rays. The sound waves
are not known to be
mutagenic. The use of
ultrasound reduces the risk,
during amniocentesis, that a
needle will be inserted at an
improper location within the
amniotic sac. In this
photograph, the fetus is head
down, facing the camera.*

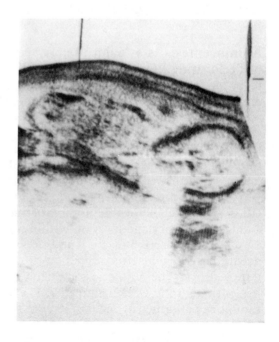

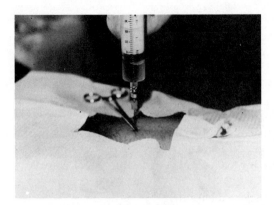

Figure 8-3 Amniotic Fluid
The hypodermic needle is slowly inserted by the physician under the guidance of ultrasound. Fluid from the amnion (about 20 to 40 ml) is very slowly drawn into the syringe. The slow entry and withdrawal reduces the risk of damaging the fetus. (Courtesy of Robert Desnick)

of pregnancy because there is insufficient amniotic fluid at that time. The culturing of the cells takes about four weeks. Thus, if abortion is requested by the parents because of an unfavorable diagnosis, the fetus is usually 18 to 20 weeks old (sometimes as old as 24 weeks). This is considerably later than elective abortions which take place for other than medical reasons, often between the 6th and 12th week of pregnancy. More than 90 percent of nonspontaneous abortions are for nonmedical reasons including illegitimacy, incest, financial hardship, or psychological stress.

An abortion for medical reasons after amniocentesis is usually a late abortion. Early sampling of amniotic fluid is not possible because so little fluid is present in a 9-to-12-week-old amniotic cavity. Furthermore, the risk of damaging the mother or fetus would be higher if amniocentesis were performed earlier in the pregnancy. At present the risk of damage from amniocentesis with ultrasound is very low, but it is by no means a routine procedure like a blood test. It requires a skilled physician and should be done at a hospital.

What the Parents and Physicians
Can Learn from Amniocentesis   For those couples with medical risks to consider, it is not easy to decide whether or not to have children, or to undergo prenatal diagnosis and possible abortion. It is difficult, financially and psychologically, to raise a child with a birth defect. It is even more traumatic for parents who have gone through such an experience once to face that risk again and depend upon luck to provide a normal child the next time. For parents who believe that abortion for medical reasons is not wrong, amniocentesis is an option, for an ever increasing number of disorders can be detected through this technique.

We have seen that the chromosome number of the fetus can be determined from cells cultured from amniotic fluid. It is also possible, by newer techniques, to detect abnormalities of the structure of the chromosomes. This procedure, called *banding analysis,* uses certain fluorescent dyes and staining techniques to reveal a number of bands along the length of the chromosomes. The two most widely used techniques employ quinacrine, a fluorescent dye requiring the use of an ultraviolet microscope, and giemsa, a dye that can be used with standard light microscopes. There are now about 2000 bands that can be recognized for the 23 chromosome pairs in a human being. By use of the banding technique it is even possible to detect where the breaks occurred which produced a translocation or other chromosome rearrangement. Deletions of genetic material in a chromosome may change the size of a band of a chromosome or result in its absence. Sometimes specific disorders, such as retinoblastoma, can be assigned to a specific region of a chromosome by this technique and the gene for the defect is then mapped. Eventually

Table 8-1 Prenatal Diagnosis of Selected Disorders

More than 75 disorders can be diagnosed prenatally. In cases of biochemical disorders whose enzymatic defects are known, the same defects are usually present in the fetal fibroblasts of the amniotic fluid. All chromosomal defects can be detected through prenatal diagnosis, but only women at risk for such disorders are advised to elect amniocentesis. Gross skeletal defects can be detected by ultrasonography (or by x rays). Neural tube defects, certain failures of closure of the abdomen, and nonfunctional kidneys can be detected by an excess or deficit of α-fetoprotein. For certain anatomical malformations (of the limbs, face, and genitals) fetoscopy (fiber optics) can be used to inspect the fetus.

Syndrome or Condition	Technique Used	Product or Feature Assayed
Tay-Sachs*	Amniocentesis	Hexosaminidase A
Hurler*	"	α-1-Iduronidase
Fabry*	"	Ceramidetrihexosidase
Morquio*	"	Galactosamine-6-sulfate sulfatase
Maple syrup urine	"	Branched chain keto acid decarboxylase
Lesch-Nyhan	"	Hypoxanthine-guanine phosphoribosyl transferase
Trisomies 13, 18, 21	"	$2N + 1 = 47$
Klinefelter	"	XXY
Turner	"	XO
Familial multiple malformations	"	Translocation or inversion
Achondroplasia	Ultrasonography	Skeletal abnormalities
Dwarfism with limb defects	Ultrasonography	Skeletal abnormalities
Spina bifida*	Amniocentesis	Excess α-fetoprotein
Anencephaly*	"	Excess α-fetoprotein
Sickle cell anemia*	"	Recombinant DNA fragments of hemoglobin A or S
Agenesis of the kidneys	"	Deficit of α-fetoprotein

*Disorders that are presently predictable through screening of at-risk individuals before they attempt to become parents.

hundreds of genes may be assigned to specific regions of the chromosomes by this technique.

In addition to detecting chromosome defects with virtually 100 percent accuracy, amniocentesis permits a biochemical study of certain inherited disorders (Table 8-1). More than 50 types of birth defects can be studied by testing the cell-free amniotic fluid or the fetal cells in culture. This includes defects of a single gene, such as Tay-Sachs syndrome, which, as we have seen, involves a progressive degeneration of the nervous system, causing blindness, seizures, paralysis, and death at an early age (about 2 to 4 years of age). Many of these inherited diseases are caused by the failure of a gene to produce a particular functional enzyme. Thus some important function within the cell cannot be carried out.

Rather rare and little-known diseases, like the Pompe syndrome, mannosidosis, phenylketonuria, and many others, are detectable by testing for enzyme function. The process is a slow one, often taking scientists several years to identify the specific biochemical process that is causing a disorder. The assay or test to distinguish clearly normal from mutant cells also requires considerable research before it can be used for individuals. There are many laboratories throughout the world studying biochemical defects, and each year a few more rare diseases are added to the list of prenatally detectable birth defects.

Neural Tube Defects Can Also
Be Detected Through Amniocentesis
Sometimes amniocentesis can reveal birth defects caused by an unlucky combination of several genes acting on a single character trait. Such birth defects have a low probability of being transmitted, and the risks of another child having the defect are only about one to five percent. Included in the group are defects such as *spina bifida* and *anencephaly*, deformities of the embryo's developing nervous system (the neural tube). In spina bifida the skin on the back of the fetus may remain open and the spinal cord may be exposed with an imperfectly rolled-up neural tube. Such a fetus leaks many of its circulating proteins into the amniotic sac. Similarly, if the developing neural tube fails to bulge out and form a brain, an open cavity may be present, resulting in severe deformities of the face and head. This produces a lethal condition, anencephaly. Children with spina bifida may live, but they run risks of infection leading to spinal meningitis. Often their neurological defect is so severe that they remain paralyzed for life. Surgery can prevent infections of the spinal cord, but it cannot repair nerve tissue that was malformed, or introduce new nerve tissue.

The substance released by neural tube defects is α-fetoprotein. A mother who is at risk for such a birth can have her amniotic fluid assayed; if the α-fetoprotein is present, she can then decide whether or not to abort the fetus. The α-fetoprotein is also released in certain other rare birth defects, such as gastroschisis, where the muscles and skin of the abdomen fail to close and the viscera are exposed.

**Amniocentesis Can Be Used
to Determine the Sex
of the Fetus for Medical Reasons**   Although some defects are not
directly detectable by enzyme or protein assays, amniocentesis may be
helpful by determining the sex of the fetus. For example, for X-linked
gene defects such as hemophilia and muscular dystrophy, the heterozygous pregnant mother may consider the odds too high (50 percent) that
a son will be affected and she may decide to abort it if amniocentesis
shows that the fetus is male. Such couples might choose to wait for a
girl, who would have to be either homozygous normal, or a carrier for
the disorder, like the mother herself.

**Most Women Who Elect Amniocentesis
Learn That Their Fetus Is Normal**   Most women who have
amniocentesis learn that their child will not have the disease that the
parents are at risk of transmitting. For those parents, the early knowledge
that no abnormality was detected is welcome news and relieves them of
months of anxiety. Anxiety itself may have physiological consequences
on the fetus or the mother during pregnancy. It can also have psychological effects on the woman's relationship with her husband and children.
Even in cases where the prenatal diagnosis determines an abnormality,
the parents who elect not to abort the fetus will have more time to
prepare for the medical and social consequences of their decision.

**Why Amniocentesis Cannot Presently
Be Used to Detect Most Birth Defects**   Unfortunately, most birth
defects are not detectable by amniocentesis. There are about 2800 defects
that have been catalogued as genetic abnormalities, but less than five
percent of all genetic disorders in humans can presently be assigned a
specific biochemical cause. Unless some molecular level of analysis is
available, a birth defect has little chance of being detected by amniocentesis. In the case of X-linked defects, this technique can be used to identify a male fetus, but even here, a pregnant woman would certainly prefer
a test that could tell her whether or not the son would be born normal,
rather than to abort all sons.

**Amniocentesis Is Not Advised
Unless the Parents Face Some Risk**   At present, there are several
situations that should encourage a woman to consider amniocentesis:

1. She had a previous child with a chromosomal defect.
2. Either she or her husband is known to have a chromosome abnormality (e.g., XXX female, balanced translocation).
3. She and her husband are known carriers for a biochemically detectable disorder (e.g., Tay-Sachs disease).

4. She is over 35 years of age (some hospitals say 32 years).
5. She is a known or probable carrier of an X-linked disorder with severe medical problems (e.g., muscular dystrophy, hemophilia).
6. A close relative on either side of the family has had a child with a neural tube defect or other abnormal opening which permits fetal fluid to leak into the amnion.
7. The mother has had two or more spontaneous abortions.
8. The husband and wife are related, as in cousin marriages (Figure 8-4). They should be checked for biochemically detectable ethnic disorders.

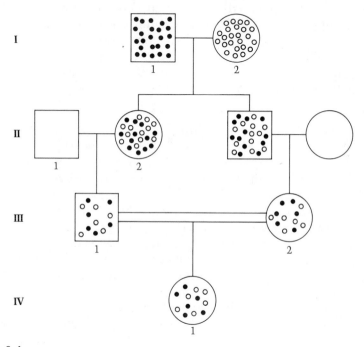

Figure 8-4

First cousin marriages are occasionally sanctioned, and large families may result, as in the case of Eleanor and Franklin Roosevelt who had five children, and Emma and Charles Darwin who had eight children. There are cultural incest taboos against first cousin marriages and genetic reasons for discouraging them. Note that a child of first cousin parents receives 25 percent of the genes present in either of the related great-grandparents. If there was a serious recessive mutation present in one of the great-grandparents, the chances would be 1 in 64 that it would become homozygous in the great-grandchild. This arises from the 1 in 4 probability that III-1 or III-2 has received the mutant gene from I-1. If both have it, then IV-1 has a 1 in 4 chance of being homozygous, and the total odds for homozygosity of the disorder would be $\frac{1}{4} \times \frac{1}{4} \times \frac{1}{4} = \frac{1}{64}$. Rare disorders like the Hurler syndrome, mannosidosis, albinism, or phenylketonuria occur at much higher frequencies in the offspring of first cousin marriages than among children of unrelated parents.

New Technology May Be Used
in the Future for Prenatal Diagnosis   Technology such as fiber optics
may become so refined that *in utero* visual inspection of the fetus (*fetoscopy*) may be made in order to detect gross abnormalities somewhere between the third and fifth months of pregnancy. Severely deformed fetuses, such as those who will become achondroplastic dwarfs, limbless fetuses, or fetuses with severe facial and cranial defects (hydrocephaly), may someday be detectable. Those parents at risk for such defects might elect prenatal diagnosis.

It is also possible that computer-enhanced ultrasonics will become so advanced that clear, detailed images of the fetus and its internal organs may be displayed on the oscilloscope. Such a breakthrough in technology would safely permit routine ultrasonic examination of all pregnant women, and subsequent referral to specialists if some gross abnormality was detected. It would also help parents in deciding whether to go ahead with the pregnancy or to have the fetus aborted, by informing them of the nature and magnitude of the defect.

Genetic Screening Can Sometimes
Detect Couples Who Are at Risk   A heterozygous autosomal mutation usually causes a reduction in the activity of a particular enzyme in the cells. If such a reduction in the activity of the enzyme does occur, the *substrate* or metabolic substance acted on by a specific enzyme may not be metabolized as effectively as it would be in a homozygous normal control. Tay-Sachs carriers are detected by tests that show the activity of the enzyme hexosaminidase A, which is 50 percent less effective in heterozygous cells. When both parents are shown to be carriers of a disease by the process of *genetic screening*, their fetus can be monitored by amniocentesis. Unfortunately, the enzymes causing most metabolic disorders are not known yet. Even when they are known, the carrier cells may be as effective as the homozygous normal cells in clearing a substrate, making an assay impossible. Also, even though some traits may be screened genetically, amniocentesis may not be possible, or may require altogether different techniques. For instance, sickle cell anemia is easily screened in adult blood, but fetal red blood cells cannot be used for prenatal diagnosis. Instead, more complex biochemical techniques, using the DNA for the hemoglobin genes have to be used to identify the sickle cell homozygote.

Amniocentesis Does Not Have to Decrease
the Frequency of a Gene in a Population   It may seem that the increased use of amniocentesis for diagnosing ever larger numbers of birth defects will be *eugenic;* that is, the genes causing these defects will gradually be replaced in the population by human choice. Actually, however, the increased use of amniocentesis should maintain the mutant gene in the population (Figure 8-5). The reason for this unexpected con-

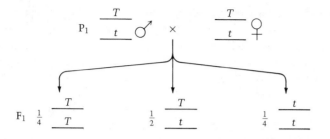

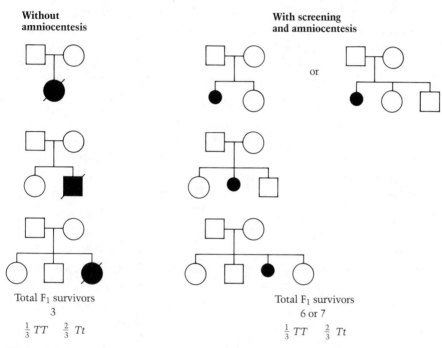

Figure 8-5

Tay-Sachs screening programs identify carriers whose level of hexosaminidase A activity is less than that of the homozygous normal individuals. Such carriers show no ill effects; their normal allele clears their lysosomes of the lipids normally digested in the nerve cells. However, when two Tay-Sachs carriers marry, and are unaware that they are carriers, they have a 25 percent chance of having a child with the homozygous disease. The financial, medical, and psychological consequences of watching such a child die may act as a deterrent to further reproduction. Such couples, in the recent past, usually stopped having children after their Tay-Sachs child was diagnosed.

Screening makes a couple aware of the risk of a Tay-Sachs child, and amniocentesis provides a prenatal diagnosis for homozygosity of the mutant gene. For those willing to elect abortion, a planned family of two or three children is possible. As the comparison shows, more children may result from families with monitored pregnancies than from families for whom these techniques were not an option. Since each normal child with Tay-Sachs carrier parents has a 67 percent chance of being heterozygous, the Tay-Sachs gene may be maintained in the population if the family size is the same as that of the general public or of collateral relatives who do not harbor the Tay-Sachs gene.

clusion comes from a knowledge of genetics and a study of family planning habits. If a couple has a child with a serious, chronic, lethal, or semi-lethal condition like Tay-Sachs syndrome, Hurler syndrome, or cystic fibrosis, they may be so traumatized by the experience that they stop having children for fear that they will have another child with the disease (a 25 percent probability for each of these three disorders). However, if they, through a screening program which detects adult carriers, learn that they are both carriers for Tay-Sachs syndrome, they may decide to go ahead and have more children and to elect amniocentesis when a pregnancy takes place. The fetus, if Tay-Sachs, will probably be aborted *and the parents will try again* until they have a child without Tay-Sachs (a 75 percent probability). But, Mendel's law of segregation predicts that two of the three non-Tay-Sachs children will be carriers like the parents. Thus the mutant gene is still being carried, although it is not expressed in those offspring.

If we contrast the parents at risk who cannot or will not use amniocentesis to detect a potential birth defect with those for whom this technique is an acceptable option, we see a curious fact about the spread of the gene. More children will be born to the couples using amniocentesis (thus maintaining the gene for Tay-Sachs) than to those not using amniocentesis. Therefore, in the long run, amniocentesis may create an unexpected problem, even for those who find abortion for medical reasons an acceptable practice. Although it prevents children with a birth defect from being born, it may also increase the frequency with which such decisions will have to be made. Each generation will have a greater number of potential children with defects than did the previous generation, and the use of amniocentesis will rise correspondingly.

We Are Usually Reluctant to Think of the Long-Range Consequences of Passing on Harmful Genes

There is a way out of this paradox. The most acceptable solution, given our contemporary values, would be for those at risk (who are not opposed to medical abortion) to undergo amniocentesis and abort fetuses shown to be defective, and then to limit their normal (carrier) offspring to a number below the national average. Thus, if Tay-Sachs carriers had one or two normal (carrier) children, and families not at risk for this disorder had two or three children on the average, the gene frequency for this disorder in the population would diminish. It is important to reflect on the reasons for making such a decision. The decision to limit the family size would be made not just for personal family reasons (the new carriers will be as healthy as the carrier parents) but for concern over the population's well-being.

In recent years some environmentalists have made the public aware that pollution is everyone's business and that we cannot, as individuals, abuse the environment without someone else, now or in the future, being disadvantaged by our bad habits. The extension of this philosophy to genetics implies that we unwittingly pollute the *gene pool* (the total of all genes in our breeding population) when we think only of our own family. In raising the issue we must ask ourselves if we have any responsibility to our species as a whole. In most countries reproduction has always been the exclusive right of the individual couples. Thus, considerable debate and careful study of the consequences of our family habits will be required to change what has been essentially an indifference to the gene pool.

Questions to Answer

1. Why is ultrasound used routinely as part of amniocentesis?
2. Can amniocentesis be used to save the life of a fetus? Describe such a situation.
3. Why is amniocentesis usually a second rather than a first trimester procedure?
4. What is the significance of chromosome banding techniques?
5. What is a neural tube defect? How are such defects usually detected by amniocentesis?
6. What categories of fetal illness are detectable by amniocentesis?
7. What criteria are used in counseling a woman to undergo amniocentesis?
8. How does fetoscopy differ from amniocentesis?
9. Under what circumstances would amniocentesis be (a) eugenic, (b) dysgenic, and (c) without effect on gene frequency in the population?
10. Do you as an individual have any responsibility to the gene pool of all humanity? If so, what is that responsibility? If not, why is this of no concern to you?

Terms to Master

α-fetoprotein
amniotic fluid
chromosome banding
dysgenics

eugenics
fetoscopy
neural tube defect
ultrasound

Chapter 9
Genetic Counseling

In 1940 Sheldon Reed at the University of Minnesota coined the term *genetic counseling* and set down some guidelines for its use. First, genetic counseling was separated from all past eugenic movements. While improvement of humanity may have been an idealist's hope, too often that "improvement" was based on naiveté, racism, spurious elitism, or just plain malice. In 1940 the world was becoming aware of the abuses of the Nazi "race hygiene" programs, and the worst features of Nazi eugenics were properly regarded with horror by the rest of humanity.

Genetic Counseling Provides
Information to Referred Clients
Genetic counseling is intended to educate an individual or family about the hereditary or nonhereditary nature of a given trait. The counselor is requested not to give an opinion on the advisability of having or not having a child. That decision has to be the parents'. The counselor does not hold back information on the long-range economic or psychological effects of having a child with a defect, or the medical care available. Everything a parent wants to know about the disorder and its diagnosis, its outcome, the risk of recurrence, its impact on the family, its cost, and its potential presence in collateral relatives, is given to the family. The family may then use this knowledge in deciding whether or not to plan additional pregnancies.

In 1940 there were fewer than 100 documented genetic disorders and the available medical intervention for treatment, early diagnosis, or prevention was virtually nonexistent. Parents were limited to gambling against Mendelian odds.

From the mid-1950s to the present, however, many dramatic changes occurred in human genetics. Chromosomal disorders from nondisjunction became recognizable; numerous disorders due to biochemical defects were uncovered; prenatal diagnosis became possible; treatment for a few disorders became a reality; chromosome banding techniques extended the number of aberrations detectable in a fetus; many complex traits involving the activities of several genes were able to be diagnosed before birth; and fetoscopy became possible as a means of inspecting the fetus for visible defects. While the basic philosophy of genetic counseling has not changed—that is, the parents make the decisions regarding reproduction—the available options have increased, along with the difficult moral decisions for both parents and society.

The Genetic Counseling Session
Usually Takes Place in a Hospital
Most genetic counseling is done at a medical center in a major university or a municipal hospital. This permits better diagnosis because the medical staff members at these centers are more likely to have seen rare genetic disorders than are physicians in private practice. The large hospital is also more likely to have specialists in human genetics (including genetic counselors) who are familiar with both the individual and general features of each trait. The counselor may be an MD or PhD in medical genetics or an MS in genetic counseling. The staff at such medical centers may also be able to devote more time to each case than would a physician in private practice. A thorough study of a family's history and the education of the family about the nature of a genetic disorder may require several hours of work. A practicing physician with a large clientele may not have enough time to spend with one family. Also, even if the physician does have the time, his or her services are probably prohibitively expensive.

A request for genetic counseling most often occurs when parents learn from their physician that they have a child with a birth defect. The physician may suspect a genetic defect and refer the parents to a specialist or genetic counselor.

Why an Accurate Diagnosis Is
Essential for Genetic Counseling
The first task for the parents who have been told that their newborn child has a birth defect is to get an accurate diagnosis. If the child is a dwarf, what type of defect is involved? Does the child have an autosomal dominant achondroplastic dwarfism? Is the child one of several types of pituitary dwarfs or midgets, due usually to autosomal recessives? Does the child have thanatophoric dwarfism, similar superficially to achondroplastic dwarfism, but inherited as an autosomal recessive? There are also several types of dwarfism caused by endocrine defects (other than pituitary). Turner females are small and have specific defects because of a chromosomal disorder (the XO condition). A metabolic defect causing failure to thrive can slow the development of the infant and stunt its growth. These varied causes of dwarf-

ism demonstrate the need for an accurate medical diagnosis of the child, including studies of clinical features, bone x rays, and metabolic studies of blood serum and urine.

An accurate diagnosis cannot usually be done by correspondence or telephone consultation. There is also much more involved in making an accurate diagnosis than merely seeing the child. For example, even if a child's features resemble Hurler syndrome features, this may not be the proper diagnosis. There are more than a dozen disorders with Hurleroid features (pug nose, shallow nasal bridge, thick lips, coarse skin texture, and mental retardation), each with its own problems of management and life expectancy. Some are autosomal recessive; others are X-linked; some arise from chromosomal aberrations.

The Counselor Must Be Sensitive
to the Needs of the Clients   After the child's disorder is accurately diagnosed, counseling can begin (Table 9-1). First the parents are informed about the prospects for their child and the problems they will

Table 9-1 Steps in Genetic Counseling

Genetic counseling does more than provide worried couples with an estimate of the risks involved in their having children. As the table demonstrates, accurate diagnosis is essential before any counseling can take place. Several hours are required to obtain the pedigree, determine the risks to members of the family, and inform them of their available options. The counselor needs to be tactful, skilled in teaching, sensitive to the multiple needs of the persons being counseled, and very knowledgeable about the disorder involved. The counselor does not tell the family members what they should do, but is obliged to be fully informative and accurate so that they can make informed choices.

Basis for genetic counseling
  Parents at risk
  Affected child or relative

Diagnosis by primary physician or specialist

Responsibilities of the genetic counselor
  Determines pedigree
  Provides risk figures for family members
  Informs family of available options
  Gives notes and sends follow-up letter with detailed counseling information
  Gives oral quiz to check understanding

Options available (if desired)
  Prenatal diagnosis (when technically possible)
  Family planning
  Sterilization
  Artificial insemination, adoption, or other reproductive option

Follow-up
  Counselor refers family for psychological help, if needed
  Counselor refers family for financial help, if needed
  Counselor schedules additional meetings with medical geneticist or primary physician for medical management of future problems
  Counselor checks family for special needs (travel to obtain proper services, special education programs, special physical needs such as orthopedic shoes, etc.)

have to face based on the severity of the disease. For example, the severity of cystic fibrosis varies, but many parents of such children will probably experience the anguish of seeing their child hospitalized numerous times for pneumonia, emphysema, pancreatitis, and heart failure, and death will be likely when the child reaches mid-adolescence or early adulthood. The parents will also learn how much time they will probably spend dislodging accumulated mucus from the child's lungs. At the same time, the counselor will take the family history and explain to the parents the genetics involved and what the risk figures mean.

Sometimes a team approach is used. A specialist MD does the diagnosis and counsels about medical management. A geneticist discusses the risks involved in a subsequent pregnancy and the options available for prenatal diagnosis. A medical social worker may discuss the economic and psychological problems facing the family.

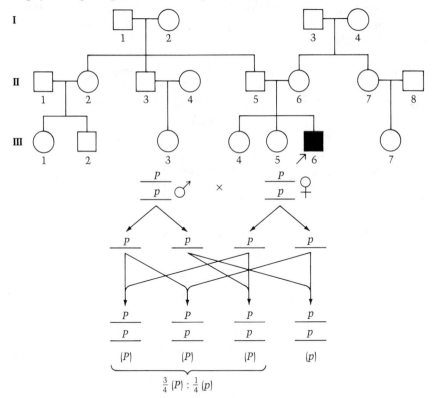

Figure 9-1
The typical autosomal recessive child has parents who are both heterozygous for the mutant gene and who have a 25 percent chance of producing an affected child. In the Pompe syndrome, the typical pedigree shows no prior family history. The genotypes involved show the genetic basis for the children of parents II-5 and II-6. Of the three children only III-6 can be assigned the genotype pp. Both III-4 and III-5 have the normal allele P but there is no way to tell by pedigree analysis if they are heterozygous or homozygous for P. A biochemical test for acid maltase activity in their cells could establish that genotype.

Genetic Counseling Involves Communicative Skills Appropriate to the Family Being Counseled

For a family whose history reveals an autosomal recessive, autosomal dominant, or X-linked recessive trait, the genetics are straightforward and usually easy to follow (Figures 9-1 through 9-3). However, there are bound to be different levels of awareness among families. A college-educated parent who majored in biology is more likely to grasp the Mendelian genetics than a less educated person. The counselor needs to take this into consideration when using analogies and technical terms to explain the genetics. The parents should be asked questions following the counseling to make sure they understood what was presented. It is a good idea, during counseling, to write down terms, rough pedigrees, and key ideas as the conversation develops and to give the parents those notes when they leave. It also helps to send a follow-up letter, summarizing the key points, so that the parents have a reference source when they explain to other members of the family what problems to expect.

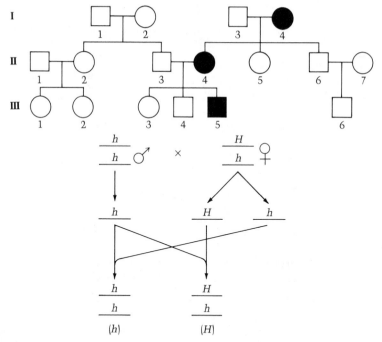

Figure 9-2

Huntington disease is an autosomal dominant trait that is usually passed on by a parent with the disease. Since the symptoms do not usually show up until completion of the reproductive years, the gene is often transmitted to the next generation. Note that in human dominant traits the person with the trait is usually heterozygous. It is very unusual for two heterozygotes with a rare dominant disorder to marry each other and even more unlikely that a child of such a match (a homozygous mutant) would ever be born or live to reproduce. The genotypes shown here illustrate what may happen if II-3 and II-4 have children.

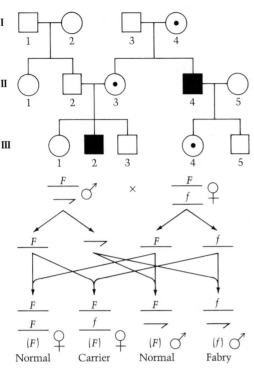

Figure 9-3
In the case of X-linked recessive traits, one or more males in a line of descent will show the mutant disorder. In a disease like Fabry syndrome, the males live to reproduce and thus generate obligate carriers (the circles with a central dot). The genotypes of the parents, II-2 and II-3, demonstrate the passage of the mutant gene to the hemizygous son, III-2.

During counseling, amniocentesis, if available for a particular disorder (see Chapter 8), should be presented as an option. Here the counselor has to be sensitive to the parents' values. If the parents have moral reservations about abortion they will probably not wish to use amniocentesis, or they may decide to use it for a different reason—a positive diagnosis can give parents several months of lead time to prepare for the financial and social problems they will have to face, or a welcome negative diagnosis would set their minds at ease.

Genetic Counseling Sometimes Involves Complex Genetics

Occasionally (contrary to genetic expectation) a family history reveals no helpful information. This is especially true for autosomal dominant disorders (Figure 9-4). In such sporadic cases (e.g., retinoblastoma, achondroplasia, and neurofibromatosis) the recurrence risk is not an insignificant 1 in 20,000, but as much as 50 percent because one of the parents may be a mosaic whose gonads contain a substantial amount of the mutant cells, but whose somatic tissues are normal. From literature based on studies of hundreds of families, a genetic counselor can provide a realistic risk figure for recurrence. This is why the practicing physician may be at a disadvantage. There are so many unusual and complex features of a genetic disorder (mosaicism, allelic differences, environmental and genetic background differences) that only a person working regularly with human genetics and birth defects can possibly be familiar with them.

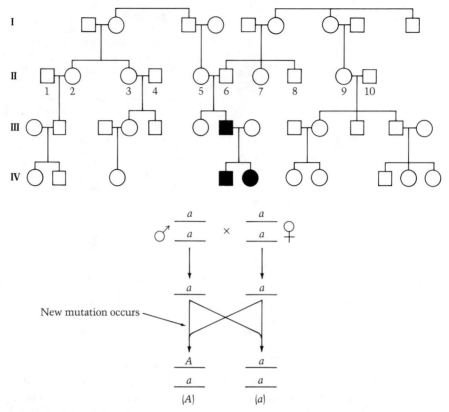

Figure 9-4
Achondroplasia is an autosomal dominant mutation which arises as a new mutation about once in 100,000 births. The typical pedigree shows a sudden appearance of the dominant trait from two normal parents. The new mutant individual, being heterozygous, then has a 50 percent chance of transmitting the disease. In the majority of cases where new mutations of a gene arise, the alteration takes place during the formation of a sperm or an egg, and thus only one offspring from the parents is affected. Such cases, where no prior history for the mutation exists and only one child is affected, are called sporadic mutations. Note that the genotypes of the parents, II-5 and II-6, are assumed to be homozygous normal, and the new mutation occurs during the meiotic process as a sporadic event.

Retinoblastoma Illustrates
the Complexity of Counseling   A good example of such a complex dominant disorder is retinoblastoma. In this disease a cancer of the eye becomes noticeable within a year or two after birth. One or more tumors develop in the retina of one or both eyes. A seriously affected eye is surgically removed, but if the eye is in an early stage of cancer, the tumor can be killed with x rays, laser beams, or liquid nitrogen (cryosurgery). About 80 to 90 percent of children with this cancer are now curable. If the tumor has developed to an advanced stage the cancer cells travel along the optic nerve and may enter the brain. Such metastases are usu-

ally fatal. For this reason chemotherapy or x rays are used when multiple or extensive tumors are found in the eye.

Parents usually recognize that something is wrong with their baby either because they find the child does not follow a hand or object passed in front of it, or because the child's eye seems to have a peculiar sheen or cloudy reflection of light, or because the eye coordination is poor, with one of the eyes crossed (strabismus) or wandering.

If the tumor is destroyed the child may have both health and sight restored. Such individuals may go on to marry and reproduce. Persons with bilateral retinoblastoma (having had the cancer in both eyes) will have a 50 percent probability of transmitting the mutant gene to their children. This is the Mendelian ratio characteristic of a heterozygous dominant mutant when crossed with a homozygous nonmutant individual.

As seen in Figure 9-5 the new mutation causing retinoblastoma may arise after one or more mitotic cell divisions have taken place from a homozygous normal zygote, producing a mutant cell line and a normal cell line. The mosaic embryo may result in a normal person whose eyes are unaffected because all the retinal cells of the embryo are derived from the normal cell line. The mosaic individual would be unaware that there were mutant cells present in his or her gonads. As shown in the illustration, such a phenotypically normal person may produce affected children.

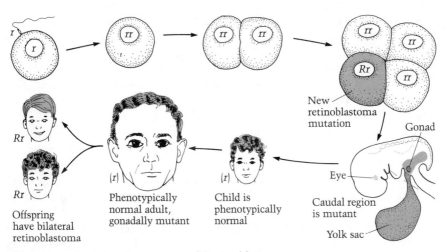

Figure 9-5 Cryptic Mosaicism and Retinoblastoma
When dominant mutations occur after fertilization, as in the four-celled cleavage stage shown here, the new dominant mutation (in this case, to retinoblastoma) is incorporated into the yolk sac and gonad of the embryo but not into the head region. The yolk sac of this male embryo is the source of the primordial germ cells which proliferate in the testes and form the future sperm. This child is a phenotypically normal male. Since he is a cryptic mosaic, if he marries as an adult he will be at risk for having more than one child with bilateral retinoblastoma, a cancer of the eyes in children.

There are many other times in the life cycle when the new mutation may arise. Taking all factors into consideration, including clinical features such as severity or numbers of tumors, whether the tumors appeared in the child's first year or some later year, and family history, the genetic counselor attempts to evaluate the risks for the person requesting the counseling.

Some, But Not All, Parents Are Psychologically Ready for Genetic Counseling

We will consider two true cases studied by the author. The first case involves a typical autosomal recessive disease, Hurler syndrome. This disease can be diagnosed biochemically because certain substances, dermatan sulfate and heparan sulfate, are present in the urine and in several body tissues. These substances are partial degradations of mucopolysaccharides, and the child's symptoms are produced from the engorgement of tissues with these products. The Hurler child lacks an enzyme needed for breaking down these complex molecules into simpler ones. There is no treatment available, and the child will die in a few years, probably from respiratory infections.

The Hurler baby in this case is the first child for a teenage mother who is not married. She is a carrier of the mutant gene. Her boyfriend is disturbed to learn that he is also a carrier of the disorder, and he does not wish to marry his girlfriend because of the risk of recurrence. The mother feels that she is being punished by God for having an affair. She is a high school dropout and she does not understand, emotionally or intellectually, that the mutant gene for the Hurler syndrome had been present in her from birth, and had nothing to do with her lifestyle. Although she knows the child is seriously ill, she has not been told that the condition will result in the child's death at an early age (about five to ten years). When she is better prepared, psychologically, she will be counseled about her child's prognosis. Here an advisor must call in a social worker and a psychologist to help the young woman cope with her guilt problems and her financial needs. At the present time, genetics is of little concern to her.

The psychologist or social worker will be informed by the genetic counselor that a 25 percent risk of recurrence exists if she and her boyfriend are reconciled and have another pregnancy. The referral to a specialist (in this case to deal with the mother's guilt) permits a person more skilled than the genetic counselor in such matters to work with her on this problem. The team effort of physician, genetic counselor, and psychologist will work out a strategy to help the mother. If she does go back to her boyfriend she will need advice on family planning, amniocentesis, and the genetics of Hurler syndrome.

The second case involves the third child for some parents who are being counseled. Their third baby, a boy, has several birth anomalies including a congenital heart defect, a cleft palate, an asymmetrical head, and now a failure to thrive. At six months his length and weight are far below normal. He is not very responsive and looks thin and grey. (A combination of defects need not be given the name of a new syndrome. More often the baby is referred to as having multiple birth defects.) In this case, the father is normal, except for a hypospadias (an abnormal placement of the opening of his penile urethra), which does not affect his reproductive capacity. The mother and two daughters are normal. The father's sister is married, and has had two first-trimester spontaneous abortions. She also has a daughter with Down syndrome.

Although an initial cytological check is normal (2N = 46) for the baby boy, a repeat test, using chromosome banding techniques, reveals a partial trisomy-15 (Figure 9-6). The father's blood is tested and a 15–21 translocation is found. Note that all the pieces involved in the translocation are present, so that the chromosome number and size are normal. The girl with Down syndrome received a partial trisomy-21. The abortions were probably responses to severe monosomy-15 or trisomy-15 combinations.

This is probably the most complex situation to try to explain to nongeneticists. The parents have to learn about chromosomes and translocations in order to understand their child's defect and the risks involved for additional children. They need to trace the paths of normal and abnormal chromosome combinations at meiosis. Then they have to tie the chromosomal situation to the development of the fetus.

This family, although not college-educated, understood what was happening. The father accepted the fact that he was a balanced translocation carrier without guilt. The parents considered themselves lucky to have had two normal children. The father elected to have a vasectomy and requested that his daughters have their chromosomes checked so that if either of them is a carrier, the family will be able to give advice about amniocentesis when the children are older. Also, the parents contacted the father's sister to let her know that, by electing amniocentesis, she could have a child without Down syndrome or the partial trisomy-15. In this family, abortion for medical reasons posed no moral dilemma. They were presented with the facts, risks, and options and made decisions based on their value system.

The Need for Genetic Knowledge
Is Presently Greater Than
Its Availability

Although genetic counseling is very helpful in discussing risks for chromosomal disorders and single gene mutations, it is less effective for polygenic traits (cleft lip or palate, congenital heart disease, spina bifida, and mental retardation) where the genetics may be too complex to be adequately worked out. By averaging all families to arrive at a risk figure, some useful information becomes available; but some of the genotypic combinations may have higher risks than others. Until more is known about the developmental or molecular basis of such birth defects, this is the best method available for obtaining information. That is small comfort to the diabetic wondering if her child

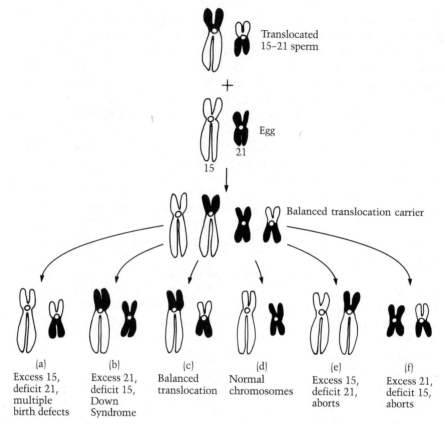

(a) Excess 15, deficit 21, multiple birth defects

(b) Excess 21, deficit 15, Down Syndrome

(c) Balanced translocation

(d) Normal chromosomes

(e) Excess 15, deficit 21, aborts

(f) Excess 21, deficit 15, aborts

Figure 9-6

Mr. P. is the father of two normal daughters and a six-month-old boy with multiple defects. Mr. P's sister has had two spontaneous abortions and a daughter with Down syndrome. Mr. P's karyotype reveals his diploid number to be 46, but he is heterozygous for a translocation of chromosomes 15 and 21. At meiosis, the distribution of Mr. P's chromosomes could have any one of the following six results:

(a) the situation for the abnormal boy
(b) the Down syndrome found in Mr. P's niece
(c) the same balanced condition found in Mr. P.
(d) a perfectly normal 15 and 21
(e) an aborted fetus having too little 21 and too much 15
(f) an aborted fetus having too little 15 accompanied by an excess of 21

Note that in this family the parents and all the zygotic possibilities have the same chromosome number, $2N = 46$. Most translocations leading to Down syndrome, however, involve a different exchange with chromosome 21. In these cases the long arm of 21 joins the long arm of 13, 14, or 15. The two small arms (all heterochromatic) form a chromosome which usually gets lost. The normal-looking parent is $2N - 1 = 45$, but that parent has the full euchromatic genes of 21 and 13, 14, or 15. When the 15-21 (or other appropriate) combination moves to one pole with a normal 21, the resulting gamete leads to Down syndrome. Such translocations are called **Robertsonian** translocations, and they produce infants with Down syndrome who are $2N = 46$.

will have diabetes, or the normal son of a schizophrenic mother wondering if his children will have a high risk of mental illness.

Of the many thousands of single gene defects possible in humans, only about 2800 have been identified phenotypically. Less than ten percent of these have had their molecular bases identified. There are probably immense numbers of combinations leading to polygenic defects, but virtually nothing is known about most of them genetically. Only in the case of neural tube defects (and a few nonclosures of the abdomen) is there an opportunity to use amniocentesis, because the elevated α-fetoprotein level indicates the presence of a problem.

In a typical year at a large referral hospital some 30 percent of pediatric admissions involve birth defects or genetic disorders. Table 9-2 shows how these are distributed. Unfortunately the vast majority of these families would not have been helped by amniocentesis even if this were a morally acceptable option for them, because we lack a knowledge of the biochemical bases of most of these disorders. The exact genetics of many of these conditions is also unknown, and these families have to rely on imprecise risk figures. Despite the limitations of today's knowledge, however, the rate of progress in human genetics is rapid and the quality and accuracy of genetic counseling will continue to improve.

What Influences Parents in Deciding Whether or Not to Exercise Their Reproductive Options?

Parents who are counseled may retain that knowledge for many years, as studies at the Dight Institute of Human Genetics in Minneapolis have shown. The parents' decisions will vary according to their value systems, but, in general, parents who face a chronic, serious illness in a child, such as muscular dystrophy, hemophilia, cystic fibrosis, Tay-Sachs, or familial Down syndrome, will have fewer children than the average. After the birth of an affected child they will often cease reproducing or elect amniocentesis if it is available. If the disease is low-risk (about 5 percent) and moderately or only temporarily impairing, such as a congenital heart defect, the family size will be less dramatically affected. No change in family size has been shown to occur for milder defects, such as cleft lip or albinism, or ones for which there is good medical treatment, such as diabetes, and club foot. If the disorder is lethal at birth or prior to birth (absence of kidneys, anencephaly) the parents are likely to try again. If the impairment is grossly abnormal and lingers on for several months or years (profoundly retarded hydrocephalic, grotesque facial dysmorphia), the parents would be far less likely to have additional children. In short, it is not the severity of the disorder itself that acts as an inhibitor to reproduction, but the prospect of caring for such a child.

Who Should See a Counselor?

Any parent of a child who has a birth defect or genetic disorder would be well-advised to see a counselor. Since

Table 9-2 Birth Defects and Genetic Disorders Seen in One Year at a Major Medical Center

Type of Disorder	Conditions	Number of Patients
Classification of disorders of the admitted patients		
Chromosomal (6 conditions)	Down, Turner, Edward, "G", 4p-, XO/XY mosaic	20
Single gene defects (128 conditions, 270 patients)	5 or more cases of each of 9 disorders	100
	2–4 cases of each of 36 disorders	87
	1 case of each	83
Heart defects (polygenic) (12 conditions)	Tetralogy of Fallot, transposition of great vessels, atrial septal defect, ventricular septal defect, coarctation of aorta, single ventricle, atrial-ventricular canal, Eisenmenger's syndrome, etc.	92
Other nonchromosomal congenital defects (28 conditions)	Hydrocephalus, meningomyelocele, cleft palate, tracheal-esophageal fistula, omphalocele, imperforate anus, urinary tract anomaly, etc.	103
Ten most frequently encountered defects		
	Cystic fibrosis	52
	Tetralogy of Fallot	24
	Multiple congenital anomalies	19
	Ventricular congenital anomalies	15
	Hydrocephalus	13
	Meningomyelocele	13
	Atrial septal defect	12
	Transposition of great vessels	11
	Cleft lip and palate	11
	Down syndrome	10

Source: The data were obtained from the University of Minnesota Hospital, Dight Institute for Human Genetics, Genetics Rounds weekly list, compiled by Dr. R. Desnick and staff.

*This list of the ten most common defects varies according to the hospital surveyed. In a black community, sickle cell anemia would be on this list, and cystic fibrosis would not because it is primarily seen in Caucasians. If the hospital specialized in a particular disease (e.g., eye diseases or cancer) the results would also be different.

preventative medicine can today be extended to the fetus, prospective parents who are willing to exercise some of the medical options available, could benefit from genetic counseling. Counseling is advisable for:

1. Any person who has a birth defect or genetic disorder.
2. Any sibling of an individual with such a defect.
3. Persons who marry late or plan to have children at an age of 35 or older.

4. Any individual who has a close relative with a disorder (parent, grandparent, aunt, uncle, cousin).
5. Any individual in a hazardous occupation involving exposure to excessive or chronic radiation or chemicals.
6. Any person in a family with a high genetic load (lots of surgery or medical care for various abnormalities).
7. Females who have been or remain on large doses of medications (e.g., dilantin for epilepsy) which can affect a developing fetus and cause birth defects.
8. Persons contemplating a consanguineous marriage (e.g., first or second cousins).
9. Parents who have had one or more spontaneous abortions.
10. Individuals concerned about ethnic disorders (e.g., Tay-Sachs, sickle cell anemia), especially if their spouse is of the same ancestry.

Questions to Answer

1. What restrictions did Sheldon Reed put on the role of the genetic counselor?
2. Why is the practicing physician usually not involved in genetic counseling?
3. How does genetic counseling differ from eugenics?
4. Describe a typical genetic counseling session.
5. What information would you try to convey to a mother who has an illegitimate child with Hurler syndrome and who believes she was being punished for her sin?
6. What information would you give to (a) the 20-year-old unmarried brother of a woman with Huntington disease; (b) the sister of a boy who died of Tay-Sachs disease who says she will marry?
7. What portion of known human genetic defects can be diagnosed prenatally today?
8. Should physicians be sued for failing to inform parents that a medical condition present in the family is hereditary or that prenatal diagnosis is available? Defend your viewpoint.

Terms to Master

cryptic mosaic
genetic counselor

risk recurrence
translocation

Part 4
Multiple
Gene Effects

No practice in the history of genetics has been more harmful or false than the assignment of single genes to single character traits. While it is true that single genes can and do sometimes produce dramatic illnesses and deformities, affecting single traits or causing a cascade of effects, it is rare, if not impossible, for a socially significant behavioral characteristic to be determined by a single gene. Humans vary in their musical, artistic, athletic, and intellectual aptitudes. We do not know how many of these abilities are attributable to genetic, and how many to environmental, factors, but we do know that the genetic component in every case involves more than just a single gene.

With even more certainty we can reject a single-gene basis for poverty (which almost always results from cultural or economic factors), mental illness (most, but not all, psychoses and probably all neuroses), criminality, and mental retardation (especially noninstitutionalized individuals without brain damage).

Throughout history, the misapplication of genetics to human affairs has been prevalent. Not only the fallacy that single genes are responsible for single character traits, but also the belief that inferred (rather than demonstrated) multiple genes are responsible for social character (polygenic inheritance), have played major roles in the history of genetics. These two misconceptions have frequently been used to justify racism (white supremacy, inferiority of Jews, inferiority of Eastern and Southern European immigrants, intellectual deficit of Afro-Americans) and spurious elitism (the innate intellectual superiority of the rich, the privileged, the professional, and the highly educated classes).

In these chapters we will learn how Mendel's second law of independent assortment provides a basis for the astonishing variation among individuals. Some human traits, such as skin color, clearly involve poly-

144

genic (multiple factor) inheritance. For instance, the quantity of melanin which pigments the skin involves two or more participating pairs of genes. Other traits, such as cleft lip, spina bifida, and congenital heart defects, run in families, but with a low incidence, which leads geneticists to conclude that such traits frequently involve both genetic and environmental factors.

Because society has so often been poorly served by those who invoke genetic determinism with inadequate proof, hidden assumptions, and disguised biases, it is important that we learn how to distinguish legitimate experimental demonstrations of polygenic and multiple factor inheritance from those that rely on analogy, indirect approaches, and questionable data.

A good guiding principle to follow is never to identify a social trait or cultural difference as substantially determined by genes unless we can provide accurate and complete genetic evidence to prove the thesis. If we cannot do so, then we must ask ourselves if we are not merely expressing personal biases or beliefs. This does not mean that such traits are necessarily exclusively determined by environment. It means that our present knowledge of the genetics of human behavioral traits is not yet complete enough for conclusions to be drawn.

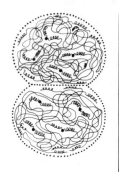

Chapter 10
Complex Traits and
Multiple Factor Inheritance

Mendel's law of segregation readily accounts for the duality of our hereditary traits. Each parent contributes one allele for a specific inherited trait. The dominance of one allele can temporarily obscure the contribution of the other parent, but the presence of the two contributions is revealed in crosses between heterozygous individuals and homozygous individuals showing the recessive trait. These are called *test crosses* when used in fruit flies, mice, corn, and other experimental organisms. The ratio of one dominant to one recessive is obtained whenever the heterozygote expressing the dominant trait is crossed with the homozygote bearing and expressing the recessive trait.

While many human disorders are single gene defects, some involve the participation of several genes. Such traits are called *polygenic* or *multiple factor* traits. Frequently the expression of polygenic traits is determined or influenced by environmental conditions.

Polygenic or Multifactorial Traits
in Humans Are Difficult to Demonstrate
The evidence for polygenic inheritance in humans is indirect. This is not the case with fruit flies or other laboratory species which geneticists can analyze. The successful analysis of polygenic traits in experimental organisms demonstrates that these genes are present on different chromosomes, each can be mapped or localized to a specific place on a chromosome, and, in some instances, a chromosome may harbor two or more such genes. This analysis is done experimentally, using techniques derived from Mendel's laws as well as some additional techniques that permit chromosome mapping and were developed by Thomas Hunt Morgan and his students in

1910–1915. No such thorough analysis of a quantitative or polygenic trait in humans is presently possible.

Mendel's Second Law Governs
the Distribution of Most Nonallelic Traits

When two genes, not allelic to one another, are located on separate chromosomes, they will be distributed during meiosis in a predictable way, giving rise to Mendel's law of independent assortment. In this pattern of inheritance an individual who is heterozygous for two pairs of genes, *Aa Bb*, produces four types of haploid sex cells. Consider a person who is a carrier, *Aa*, for the recessive, mild, biochemical defect, alkaptonuria, which produces a dark-colored urine. This person may also be heterozygous for a second trait, brown eyes, *Bb*. Independent assortment of the *Aa Bb* double heterozygote during reduction division will yield gametes in a ratio of 1 *AB* : 1 *Ab* : 1 *aB* : 1 *ab*.

Mendel's use of two pairs of alleles, one for pea color, the other for pea texture, is illustrated in Figure 10-1. The checkerboard or Punnett square permits a quick identification of the 16 ways in which four sperm (pollen in peas) encounter four eggs (ovules in peas).

Mendel's Second Law Is Seldom
Illustrated in Human Pedigrees

When humans marry it is rare for two doubly heterozygous individuals for traits like alkaptonuria and eye color to select each other as mates unless it is a consanguineous marriage (e.g., a marriage between first cousins, who have a large number of familial traits in common) or a marriage between two individuals from a small, isolated community founded by a few initial families. The chances are also extremely rare for a test cross situation in which the mate of this double heterozygote is a double homozygote, *aa bb*, with the harmless black urine of alkaptonuria and blue eyes.

To complicate matters, even if two such individuals were to select each other, or a test cross like the one described could be set up, the couple would probably have only two children, and not several dozen. Therefore, we would need several such marriages and we would need to pool the data from all of them in order to work out Mendel's second law of independent assortment. Fortunately for us, the laws were worked out with peas, fruit flies, mice, chicken, maize, and many other organisms which are easier to work with, and we can feel confident that the same pattern of inheritance extends to humans.

The Test Cross for Nonallelic Traits
Demonstrates Independent Assortment

If it were possible to carry out the test cross in humans for these two traits, the results would yield a ratio of 1 (*AB*) : 1 (*Ab*) : 1 (*aB*) : 1 (*ab*). The independent segregation of

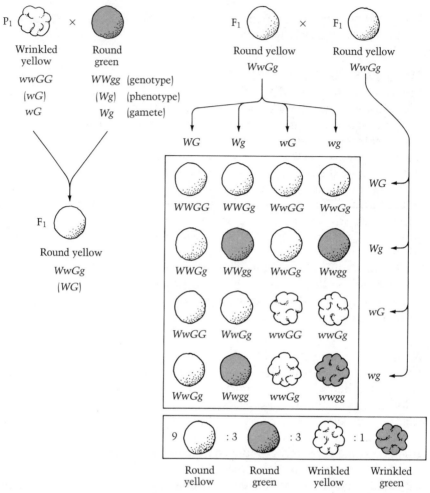

Figure 10-1   Independent Assortment in Plants
Mendel's law of independent assortment can be illustrated by a cross of peas, combining two traits, the color of the seed coat (yellow or green) and the texture of the mature peas (round or wrinkled). When parental wrinkled-yellow plants are crossed with round-green plants, the F_1 are uniformly round-yellow, showing that yellow is dominant to green and round is dominant to wrinkled.

When the hybrid F_1 plant pollinates itself, an F_2 population emerges showing the 16 ways zygotes form when four types of pollen encounter the same four types of ovules. The ratio, determined from the Punnett square, is $\frac{9}{16}$ round-yellow:$\frac{3}{16}$ round-green:$\frac{3}{16}$ wrinkled-yellow:$\frac{1}{16}$ wrinkled-green.

each pair of alleles in the heterozygote would result in four gametic combinations (shown in Figure 10-2 for Mendel's peas). Since each of these gametes would encounter the double recessive during the test cross fertilization, the test cross would show us the make-up of the four gametes. We would read these four phenotypes as one-fourth normal urine and brown eyes; one-fourth normal urine and blue eyes; one-fourth black urine and brown eyes; and one-fourth black urine and blue eyes.

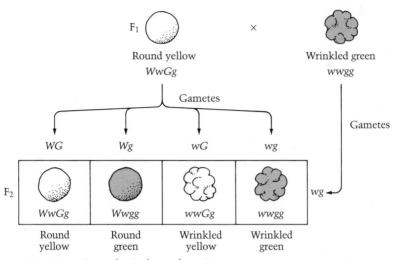

Figure 10-2   Test Cross for Independent Assortment
Geneticists find it easier to test for the heterozygosity of traits by using a test cross. In this cross the parents of the F_1 were homozygous for either of two traits, pea color (green or yellow) or pea texture (wrinkled or round). The F_1 shows the dominance of yellow over green and round over wrinkled, because they are heterozygous for both of these traits. The F_1 hybrid can be test crossed to the double recessive, ww gg. This guarantees that the phenotypes of the F_2 progeny reflect the genotypes of the gametes from the doubly hybrid F_1 plant. Since the random assortment of the two heterozygous pairs of genes yields four gametic genotypes, the diploids formed in the test cross correspond to those same four combinations of traits in the ratio of $\frac{1}{4}$ round-yellow:$\frac{1}{4}$ round-green:$\frac{1}{4}$ wrinkled-yellow:$\frac{1}{4}$ wrinkled-green.

The equivalent results of an $F_1 \times F_1$ cross are far more complex because four different gametic types exist for the sperm and four for the eggs, resulting in 16 diploid genotypes. These will form a 9:3:3:1 ratio with $\frac{1}{16}$ in the double recessive category (*ab*) and $\frac{9}{16}$ in the double dominant category (*AB*). There will be two classes which will express a dominant and a recessive, (*Ab*) and (*aB*). Each of these classes will account for $\frac{3}{16}$ of the conceptions. About 1905, geneticist R. C. Punnett devised a simple grid for determining the zygotic combinations from the gametic genotypes. The Punnett square, which is shown in Figure 10-1, is often used in laboratory studies of species with large numbers of progeny, especially plants and insects.

The independent assortment, or segregation, of two or more pairs of traits, each carried on a different chromosome pair, is explained by the pairing of homologous chromosomes and their alignment during reductional division (see Figure 10-3). A paternally derived chromosome (from the father's sperm) forms a side-by-side pairing with a maternally derived chromosome (from the mother's egg). During alignment (metaphase I), this pair of homologs may be oriented in a cell so that the maternal chromosome faces the top of the cell and the homologous paternal chromosome, underneath it, faces the bottom of the cell. A second pair of homologs is equally likely to have the maternal chromosome facing the

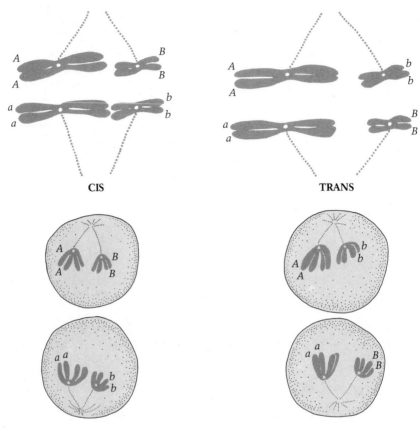

Figure 10-3
The key to independent assortment is the random alignment of paternal and maternal chromosomes. If the alleles A and B are brought in by the sperm, and a and b are present in the egg, the adult formed by this union is heterozygous, Aa Bb. When that hybrid individual undergoes meiosis, the homologous chromosomes carrying genes for these two traits will align randomly. Half the alignments will show the cis-alignment, with a and b facing one pole, and A and B the opposite pole. The other half will show a trans-alignment, with a and B going to one pole, and A and b going to the opposite pole. The gametic ratio for the double hybrid is thus: $\frac{1}{4}$ ab:$\frac{1}{4}$ AB:$\frac{1}{4}$ Ab:$\frac{1}{4}$ aB.

top of the cell. Since this cis-alignment (*AB* facing the top and *ab* facing the bottom) is just as probable as the trans-alignment (*Ab* facing the top and *aB* facing the bottom), there are four equally likely combinations, resulting in the 1:1:1:1 ratio that the test cross demonstrates.

It Is Unlikely That One of Your Gametes Has the Same Genotype As the One Your Father or Mother Contributed to You   During the formation of your own reproductive cells, with 23 pairs of chromosomes in each, the probability of your father's set of 23 chromosomes ending up in one

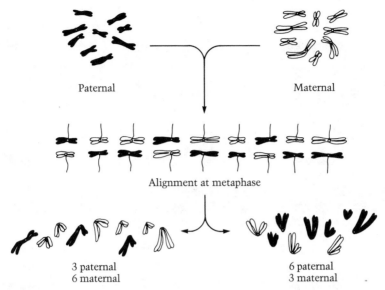

Paternal   Maternal

Alignment at metaphase

3 paternal   6 paternal
6 maternal   3 maternal

Figure 10-4
In the reductional division of meiosis the alignment of paternal and maternal chromosomes is random. Since the alignment of the chromosomes is in two planes, each plane will contain some paternal and some maternal chromosomes. The diagram shows one possible distribution of mixed paternal and maternal chromosomes for nine pairs of homologs. In a human germ cell there would be 23 pairs, and thus, in theory, $(\frac{1}{2})^{23}$, or 1 in about 8.4 million, would be the odds of having a sperm or egg with all paternal or all maternal chromosomes. In reality the odds are even greater, because there is an additional shuffling of paternal and maternal genes within each pair of homologous chromosomes, called crossing over. For these reasons it is virtually impossible for you to produce a gamete with exactly the same genetic composition as either of your parents.

cell and your mother's in another cell is only one in 2^{23}, or about one in eight million (see Figure 10-4). Clearly the paternal and maternal genes are shuffled during meiosis so that the next generation will have only a remote possibility of having an identical genotype to any individual of the previous generation.

Another Process, Crossing Over, Shuffles the Maternal and Paternal Genes Between Homologs

Another process, called *crossing over*, makes these odds even higher than one in eight million. Virtually no sperm or egg ever carries a genotype identical to that of one of the parents, or for that matter, any other individual who ever lived. This is because crossing over takes place, which assures that a paternal and a maternal chromosome within a pair of homologs, will exchange pieces, resulting in sequences of genes derived from both parents ending up in each chromosome (Figure 10-5).

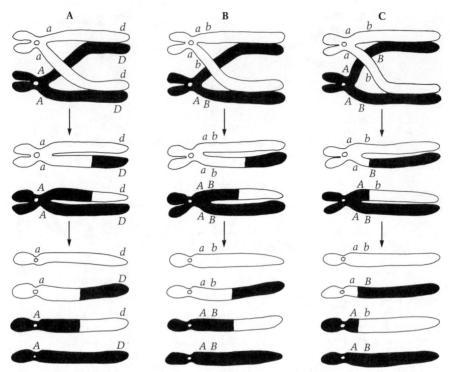

Figure 10-5 Cytological and Genetic Crossing Over
When a pair of homologs is in the reductional division phase of meiosis the chromatids become twisted around each other. A physical exchange (cytological crossover) occurs, and pieces of the paternal and maternal chromatids are joined together. This leads to a genetic crossover if two nonallelic genes are separated during this process. In situation A the genes a and d are far apart on the chromosome. Almost any cytological crossover results in the physical exchange of maternal and paternal chromatids carrying these genes. In situation B, however, the genes a and b are close together. The cytological exchange does not occur between them; thus a and b remain together. When, in situation C, the physical exchange occurs between a and b, a genetic crossover occurs with recombinant chromatids Ab and aB entering into the gametes of the adult undergoing crossing over.

Crossing over occurs during the prophase (after homologous pairing) in the first meiotic division. The number of crossovers within a pair of homologs is related to the length of the chromosomes. Crossing over is less likely to occur near the tips of chromosomes, near centromeres, or near other crossovers. Also, a single crossover is more likely than a double or triple crossover.

Crossing Over Can Be Used to Make a Map of the Genes on a Chromosome
If two nonallelic gene pairs are on the same pair of homologous chromosomes, a crossover occurring at any point *between* them will result in recombination. If the two different

Figure 10-6
*This snapshot of the Drosophila Group at Columbia University shows
Thomas Hunt Morgan and his students enjoying a welcome-back party for
A. H. Sturtevant, who had just been released from the U.S. Army after the
armistice of World War I. The guest of honor is* Pithecanthropus erectus, *an
ape-man dressed in Sturtevant's army fatigues. H. J. Muller sits to Morgan's
left, Sturtevant smokes a cigar and C. B. Bridges reads to Pithecanthropus.
The Drosophila Group helped to establish the chromosome theory of heredity
and introduced concepts such as crossing over, mapping, nondisjunction,
chromosome rearrangements, multiple allelism, and gene mutation. (Courtesy
Muller Archives)*

genes are far apart, near the opposite ends of the chromosome, crossing
over between them is very likely to occur (they are said to be loosely
linked). If the two genes are very close to one another, a crossover between them would be unlikely (they are said to be tightly linked). T. H.
Morgan and his student A. H. Sturtevant (a 19-year-old sophomore at the
time) conceived of the idea of mapping the recombinant genes from this
knowledge (Figure 10-6). As a result, the sequence of genes on a chromosome and their relative distance from one another can be mapped
(Figure 10-7), producing a *linkage map.*

Deletions of a Piece of a Chromosome
Can Also Be Used to Map Genes
Fairly detailed linkage maps
have been constructed for the hundreds of mutant traits that have been
found in species such as fruit flies, mice, and other experimental organisms. Human chromosome maps can also be constructed from breeding
analysis, but it is a difficult and slow process, since the desired recombination depends on the marriages that occur.

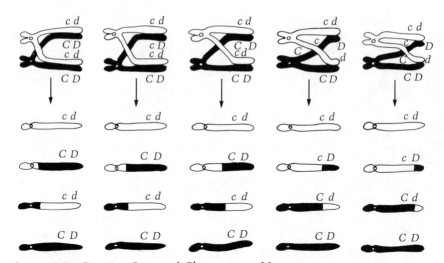

Figure 10-7   Crossing Over and Chromosome Maps
*The frequency of crossing over can be used to map genes on chromosomes.
This is shown here for the genes c and d. Five diploid germ cells undergoing
reductional division show cytological crossovers occurring at different places
along the length of the chromosome. Only in the two cases on the right does
the cytological exchange land between c and d. This yields four recombinant
chromatids (cD or Cd). The frequency of crossovers is thus 4 in 20, or 20
percent, which is then converted to map units (1 percent crossing over is 1
map unit). The 20 map-unit distance for genes c and d can then be indicated
on a line.*

More recently some additional processes have been used for mapping
human chromosomes in detail. One of these is *deletion mapping.* A defect like retinoblastoma may sometimes occur in a child who has other
defects that are caused by the deletion of a segment of the long arm of
chromosome 13 (the q is the long arm and the p is the short arm of a
chromosome). The 13q deletion can result in retinoblastoma as well.
Retinoblastoma usually arises because of gene mutation, but in those
rarer cases where the 13q segment is deleted, the effect is the same.

The 13q deletions of several different families have been compared
and the region that is missing from all of them has been used to define
the locus for retinoblastoma. Similar mapping has been done with other
forms of abnormal chromosomes associated with retinoblastoma and other
dominant disorders.

Fusing Cells of Different Species
Permits Assignment of Genes to Chromosomes   A third method of

locating genes on human chromosomes uses *somatic cell hybridization*
(Figure 10-8). For example, a human cell can be fused with a mouse cell
and then allowed to multiply by mitotic division. During this process
most of the human chromosomes are lost. In a few cases a cell may
contain a single human chromosome in a mouse nucleus. The human

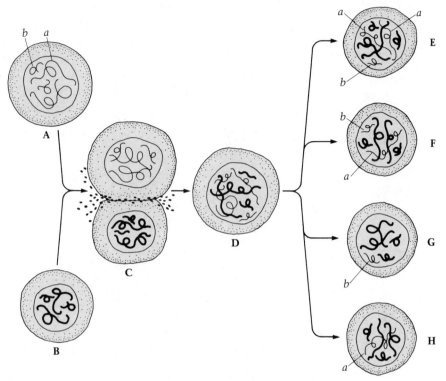

Figure 10-8   Somatic Cell Hybridization

If a human cell (A) and a mouse cell (B) are brought together in the presence of the sendai virus (C) they fuse along their cell membranes, and their cytoplasms commingle. At mitosis the chromosomes become mixed and the nucleus (D) of each daughter cell contains both the human and the mouse chromosome sets. Usually the mouse chromosomes persist, but most of the human chromosomes become lost through nondisjunction (E, F), leaving some (G, H) that may have only a single human chromosome. If a particular biochemical function is found in E, F, and G, but not H, then the small chromosome (b) contains that function because (b) is present in E, F, and G, but not H. If the function is present in E, F, and H, but not G, then the function is on chromosome (a) because E, F, and H have (a) but G does not. It is much easier to map genes once they have been assigned to a specific chromosome.

proteins or the products of the activities of the human proteins from this chromosome may be detected by using such a colony of hybrid cells (called a cell line). The location of the individual genes on this chromosome may be identified by using x rays to fragment the human chromosome. If a specific human protein is present in a hybrid cell containing a broken piece of a chromosome, it may be concluded that the missing part of the chromosome does not contain the gene that makes the protein. By obtaining different fragments that produce a given protein and looking at the segments that are common to all of them, the section that carries a specific gene can be mapped.

Some Traits Do Not Appear
to Follow Mendel's Laws

When Morgan began his work with fruit flies about 1909, he encountered a few cases where segregation did not occur according to Mendelian laws. He put these aside and concentrated on the ones that did segregate as predicted by Mendel. However, one of Morgan's students, H. J. Muller, worked on some of the complex traits. These traits were variable in expression, could not be rendered homozygous, and could be intensified or diminished in their expression through

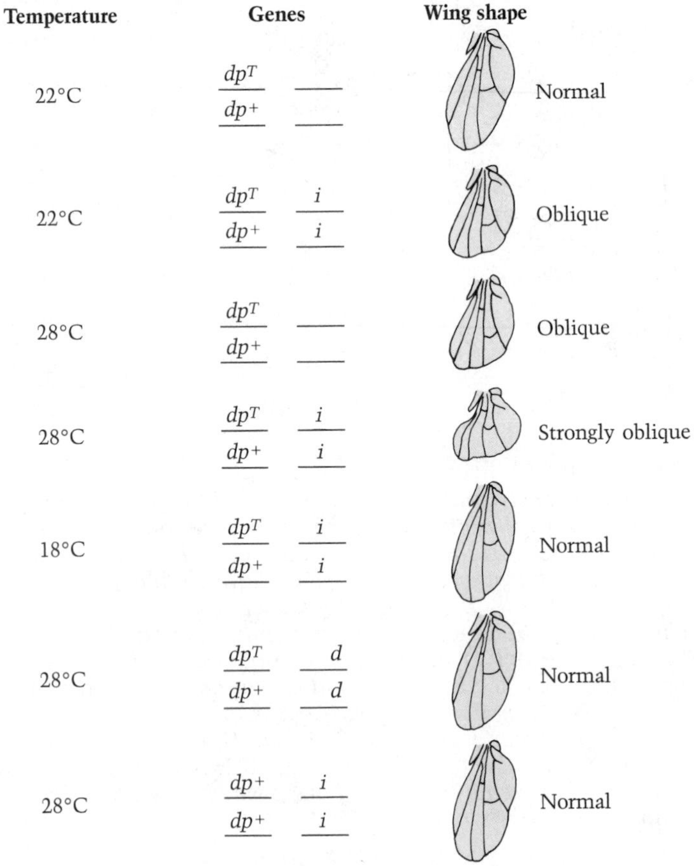

Temperature	Genes		Wing shape	
22°C	$\dfrac{dp^T}{dp^+}$	$\dfrac{\rule{1.5em}{0.4pt}}{\rule{1.5em}{0.4pt}}$		Normal
22°C	$\dfrac{dp^T}{dp^+}$	$\dfrac{i}{i}$		Oblique
28°C	$\dfrac{dp^T}{dp^+}$	$\dfrac{\rule{1.5em}{0.4pt}}{\rule{1.5em}{0.4pt}}$		Oblique
28°C	$\dfrac{dp^T}{dp^+}$	$\dfrac{i}{i}$		Strongly oblique
18°C	$\dfrac{dp^T}{dp^+}$	$\dfrac{i}{i}$		Normal
28°C	$\dfrac{dp^T}{dp^+}$	$\dfrac{d}{d}$		Normal
28°C	$\dfrac{dp^+}{dp^+}$	$\dfrac{i}{i}$		Normal

Figure 10-9 Truncate Wings: An Inconstant Trait

The wing of the fruit-fly is oval-shaped. A mutant gene arose which caused it to be snipped or obliquely truncated in shape. The number of flies showing truncation varied with each generation, as did the degree of truncation. A chief gene for truncation must be present, but expression of the chief gene also requires higher temperature or one or more genetic modifiers. Neither the higher temperature nor the genetic modifiers can express truncation if the chief gene is absent. In this analysis all the environmental and genetic components could be isolated or combined at will. Complex human traits are not subject to such genetic analysis and thus their interpretation remains controversial. In the figure, dp^T = truncate, dp^+ = normal, i = intensifier, and d = diminisher.

constant selection. Some geneticists speculated that the gene for such an inconstant trait varied or was unstable. Such traits seemed to be more like those human traits such as longevity, intelligence, height, weight, talent, insanity, and general health, which were variable, inconstant, and apparently non-Mendelian traits because of their wide range of expression and uncertain inheritance in families and populations.

Muller's analysis of one of these traits, truncate wings, took eight years (Figure 10-9). He worked out a careful genetic analysis that revealed several factors in the expression of the truncate trait. He found that truncate wings were obliquely shaped, varying from a mildly expressed, snub-like wing tip to a severely blistered, reduced wing missing half of its distal tissue.

A Chief Gene May Not Be Expressed
Without the Help of Modifiers
In order for a fly to express truncation, a mutant gene on the second chromosome must be present. Muller called this the *chief gene* for truncate wings. Truncate wings cannot be made homozygous because it is a recessive lethal. At the same time, in the heterozygous state, it can act as a dominant, producing the truncate wing phenotype. At most, 50 percent of the progeny of a truncate fly mated to a nontruncate fly will be truncate. In practice there were usually fewer than 50 percent. Furthermore, the progeny that were truncate varied from mild to extreme in their expression of the trait.

Muller found that there were *genetic modifiers* present in the stock which could intensify or diminish the expression of truncation in a heterozygote. In fact, the heterozygote lacking intensifying modifiers does not show truncation. Muller used breeding analysis to locate the chromosomes carrying each of these modifiers. He then used the techniques of Morgan and Sturtevant to map each modifier on its chromosome.

If the gene for truncation is absent, the modifiers cannot alter the wing shape. By combining (through breeding) the chromosomes carrying intensifying modifiers and the chromosome carrying the gene for the truncate trait, Muller could predict in advance the percentages of truncate and nontruncate flies and the percentages of extreme and slight expressions of truncation.

Sometimes Environmental Factors Can
Modify the Expression of a Chief Gene
In addition Muller showed that high temperature (28° C or 80° F) intensifies truncate expression and low temperature (18° C or 65° F) diminishes its expression. Not only is a complex trait dependent on a polygenic inheritance, but it is also subject to environmental modification for its expression.

Muller's analysis was important for several reasons. He proved that the variability of a hereditary trait is not a reflection of an unstable or fluctuating gene. He proved that the genetic and environmental components of inconstant and complex traits can be identified by difficult and tedious genetic analysis. Finally, he drew an important inference. If evolution involves the selection of minor differences of character traits, then the recombination of paternal and maternal genes by meiotic reductional division gives an immense variety of combinations of modifiers and chief genes for many character traits. Diversity would be a characteristic of each new generation, with some polygenic combinations surviving and spreading more rapidly than others. Yet, even such successful combinations would be shuffled by meiosis, and many of the progeny would lose the right combinations for successful competition in life. Change would be gradual, arising out of the diversity in character expression.

Inbreeding and Selection Can Be Used
to Analyze a Complex Trait

A different experimental approach to complex traits was developed by W. Johannsen in Denmark. Johannsen used size as a character trait in the common garden bean, *Phaseolus.* If a plot of bean plants is freely pollinated by bees, the harvest

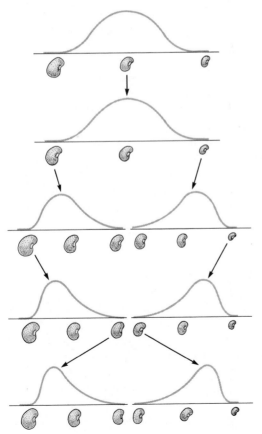

Figure 10-10
When beans are outbred by cross-pollinating different plants in a field, a full range of bean sizes results. These form a normal distribution curve. By selecting the smallest (or largest) bean in a population and inbreeding it (by self-fertilization), Johannsen shifted the shape of the curve. Continued selection and inbreeding over several generations establishes a pure line. In the pure line the shape of the curve remains constant regardless of the size of the bean harvested. Johannsen used the term genotype to designate the genetic limits imposed on the shape of the curve in a pure line. The term phenotype designated the appearance, because a medium-sized bean from an outbred population would look the same as the largest bean in the pure line inbred for small beans, or the smallest bean in the pure line inbred for large beans.

of beans will show a range of sizes from small to large. This represents an outbred strain of beans.

Johannsen selected the smallest bean and planted it, and when it reached a flowering stage he deliberately self-fertilized it and then protected it against visitation by bees. This inbred strain of beans produced a range of bean sizes that varied from small to medium (Figure 10-10). If a curve was constructed for the quantities of each size obtained for the inbred and outbred strains, the two strains were clearly different. The outbred curve resembled a normal bell-shaped curve, and the inbred curve was skewed toward the smaller bean sizes, but there was still a considerable quantity of medium-sized beans.

Repeated Selection and Inbreeding
Eventually Produce a Pure Line   Johannsen repeated his selection, using the smallest bean of the second generation to produce a third generation, inbred like its predecessor. While the curve was similar to the previous generation's, it had shifted somewhat, showing fewer medium-sized beans. Johannsen kept selecting for ten or more generations, hoping to produce a strain of all small beans of a uniform size. After five or six generations there was no further shift in the average bean size. Using beans from the tenth (or later) generations, he then selected and bred the largest bean from his inbred small-bean strain. Would this reverse the effect of the selection in the previous generations? No, the curve was identical to that of the smallest bean from the same generation.

Johannsen concluded that selection and continual inbreeding can lead to homozygosity. This establishes a strain that repeats its curve; it is a *pure line*. However, even this pure line varies. The variation in a pure line is not due to genetic differences because selection of any bean from it will not change the shape of the curve for the pure line. Regardless of which bean size you choose, they will all give identical curves. The pure line variation is environmental. The homozygous genotype provides the limits for the expression of bean size. Environmental enrichment in a pure line for small size can cause medium size, but it cannot produce a very large-sized bean, because its genotype restricts further growth.

Note that Johannsen's analysis of bean size distinguishes between *genotype* (the inferred or known genetic constitution of an individual) and *phenotype* (the observed trait or characteristic expression of an individual). In Johannsen's analysis it was possible to select a pure line for small size (ranging from small to medium), a pure line for large size (ranging from medium to large), and an outbred line (ranging from small to large). The largest bean of the small-size line, the smallest bean of the large-size line, and a medium bean of the outbred line would have the *same phenotype*. This would all be medium-sized beans. Yet each would produce progeny with a different mean bean size, reflecting the fact that each has a *different genotype*.

Independent Assortment and Crossing Over
Were Used to Demonstrate
the Chromosome Theory of Heredity

When fruit flies were studied thoroughly by T. H. Morgan and his students between 1910 and 1915, a remarkable parallel was revealed between their chromosomes and the genetic analysis of the mutant genes that were found in that five-year period. The diploid chromosome number of *Drosophila melanogaster* was eight, consisting of two rod-shaped, two dot-shaped, and two pairs of V-shaped chromosomes. The female and male differed only in the rod-shaped chromosome, which was present as a pair in the female. The male had one rod and one J-shaped chromosome. The rod was called an X, and the male's J-shaped chromosome was called a Y. The female had a pair of X chromosomes and three pairs of autosomes. The male had an X and a Y plus the same three pairs of autosomes. The length of the X was about half the length of the V-shaped chromosomes. The dot-shaped chromosomes were very small, each less than ten percent as long as a rod.

When mutant genes were mapped, all of the sex-linked traits (such as white eyes) formed a rod-shaped map about 70 map units in length. The other genes (autosomal, non-sex-linked) were assigned by Mendel's law of independent assortment, mostly to one of two groups. Each of the two groups formed a map about 110 map units in length. There was also a third group to which only one or two genes had been assigned. Morgan's student, Muller, argued that a parallel existed between the chromosomes seen under the microscope and the inferred maps of these chromosomes determined by genetic crosses of the many mutant genes in fruit flies (Figure 10-11). If the X chromosome represented the sex-linked map (called group I), the two very large groups (II and III) represented the V-shaped autosomes, and the small group (IV) represented the dots, then the chromosome theory was established by the complete correspondence between the genetic maps and the chromosome number, size, and function.

Nondisjunction Offered Another,
More Dramatic Proof
of the Chromosome Theory

C. B. Bridges, also in Morgan's laboratory, analyzed some unusual exceptions to the expected offspring in crosses involving the X-linked inheritance of eye color. He realized that the genetic event could be interpreted by assuming a change had taken place in the chromosome number at fertilization. He had discovered nondisjunction and predicted that the unusual flies of his cross had $2N-1=7$ or $2N+1=9$, instead of the normal $2N=8$. This prediction was startling to Morgan and many of the other geneticists. Bridges published his analysis with the title "Nondisjunction as proof of the chromosome theory of heredity." Figure 10-12 illustrates the cytological and genetic consequences of nondisjunction in the white eye cross. Note that, in fruit flies, unlike humans, the XXY is a fertile female (in humans it

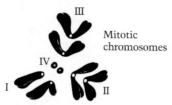

Mitotic
chromosomes

Group I		Group II		Group III		Group IV	
0.0	yellow	0.0	aristaless	0.0	roughoid	0.0	bent
1.5	white	2.0	asteroid			0.5	shaven
3.0	notch	11.0	echinoid			0.7	eyeless
6.9	bifid	13.0	dumpy				
7.5	ruby	15.5	clot				
13.7	crossveinless			20.0	divergent		
20.0	cut			26.0	sepia		
27.5	tan	31.0	dachsus				
33.0	vermilion			40.2	tilt		
36.1	miniature	43.5	black				
43.0	sable			44.0	scarlet		
44.4	garnet	54.5	purple				
54.2	smallwing	57.5	cinnabar	50.0	curled		
56.5	forked						
57.0	bar			58.5	spineless		
59.6	beadex	67.0	vestigial	58.7	bithorax		
				62.0	stripe		
65.0	cleft	75.5	curved	70.7	ebony		
70.0	bobbed	83.5	fringed	75.7	cardinal		
		90.0	humpy	91.1	rough		
				93.8	beaded		
		100.5	plexus	100.7	claret		
		105.0	brown				
				106.2	minute-g		
		107.0	speck				

Figure 10-11   Proof of the Chromosome Theory by Correlation
The work of Morgan and his students provided two demonstrations of the chromosome theory of heredity. Cytological study showed a 2N chromosome of 8, consisting of one pair of sex chromosomes and three pairs of autosomes. Morgan's discovery of crossing over and Sturtevant's suggestion that the frequency of crossing over can actually be used to map genes, enabled the Drosophila workers to localize the dozens of new mutations that were found. They found that the mutations fell into four groups. One was a group of sex-linked traits (group I) which showed linkage (the traits did not assort independently of one another). This permitted them to map these genes in a linear sequence, the total distance from tip to tip of the genetic map being 60 map units. Groups II, III, and IV all showed independent assortment when any member of one linkage group was crossed to any member of the other two linkage groups. Within linkage group IV, however, only one or two mutants were known to exist, and their distance was only a fraction of 1 map unit. Muller argued that the function (X-linked or autosomal), length, and linkage relations formed a near-perfect match with the size, number, and function of the four pairs of chromosomes found in Drosophila. This was clearly a demonstration of the underlying chromosomal basis of the genetic results obtained by breeding analysis.

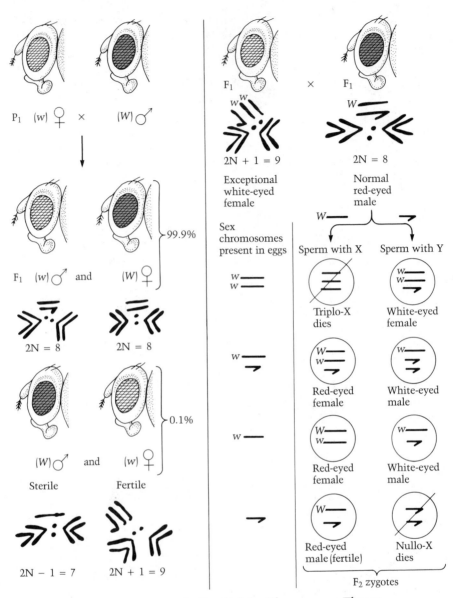

Figure 10-12 Nondisjunction As Proof of the Chromosome Theory
When C. B. Bridges analyzed nondisjunction and associated it with the white eye in inheritance patterns, he believed he had a powerful demonstration that genes were specifically associated with chromosomes. To Morgan and most other geneticists, nondisjunction was convincing because the sterility and unexpected red eye color of the exceptional F_1 males produced by P_1 homozygous white females was associated with the XO condition of those F_1 males. The return of fertility in the F_2 red-eyed males from the unexpected F_1 white-eyed females (also derived from the P_1 white-eyed females) was accompanied by the exact number and type of chromosomes which Bridges had predicted. The "exceptions to prove the rule" extended Muller's correlation of maps and chromosomes and provided an experimental demonstration in which cytological results could be inferred from the genetic analysis of unusual flies.

is a Klinefelter male) and the XO is a sterile male (in humans it is a Turner female). In humans, the Y is a male-determining factor. In fruit flies the Y only guarantees fertility to a male. It is the number of X chromosomes per set of autosomes that determines fruit fly sexuality (one X for males, two for females in diploid cells).

The Analysis of Most Complex Traits in Humans Is Difficult or Presently Impossible

The analysis of truncate wings which Muller carried out is clear and relatively uncomplicated. He identified the chief gene and genetic modifiers for his variable trait. Such an analysis is presently impossible for humans, however. To do such genetic analysis, the geneticist must mate a particular individual with another genetically suitable individual and obtain large numbers of progeny. Fruit fly progeny (but not human offspring) can be inbred without any qualms about incest. A pair of flies can produce 100 or more progeny. Genetic stocks can then be constructed, tagging chromosomes with easy-to-follow marker genes, and these can then be used to track down elusive invisible modifiers. Humans, on the other hand, cannot produce more than one or two dozen children, even if they want to do so to help out an investigator. Human beings cannot be manipulated to suit a geneticist's convenience. Biologically and morally, the scientist's investigation of human genetics is limited.

Human traits of social significance such as intelligence, personality, longevity, or overall health are almost always polygenic or environmental. The direct proof of any genetic component for such traits is difficult, and for most character traits such proof does not exist. By analogy or extrapolation from genetic studies on mice, fruit flies, and other laboratory organisms, certain human traits are thought to be polygenic and to have environmental components resulting in their expression and variability. These traits vary in their social significance, and thus controversy over the interpretation of some of these traits is inevitable.

Human intelligence is probably the most controversial. Whether or not height and weight are polygenic traits rarely becomes a topic of public debate. Traits such as insanity (the schizophrenias, manic depression, character pathologies) are frequently assigned a polygenic component, and such interpretations are often challenged and debated in the professional journals. Less research is available on the heritability of talent, life span, and overall health. Considerable doubt exists about polygenic components for ambiguous traits such as criminality, laziness, poverty, licentious behavior, or lewdness. However, at one time, all of these were naively believed to be heavily influenced by our genotypes.

As we shall see, a polygenic basis for skin color is genetically valid, with the environment playing a modest role at best in the determination of this trait. Despite the clear genetic basis for skin color, there are social values, prejudices, and fallacies that are imposed on our biological understanding of race. We shall also see how difficult it is to obtain solid evidence for the heritability of intelligence and how advocates of environmental and biological determinism have used their social philosophies to alter public policy regarding this trait.

Other traits have recently been proposed as having possible polygenic bases. Some behavioral biology scholars have demonstrated that traits such as aggression, gregariousness, cooperativeness, territoriality, and loyalty can be selected in mammals, birds, or fish. They also assume that these same traits in humans are subject to an evolution through the natural selection of genotypes affecting the central nervous system. Other scholars in this field, and many geneticists, have serious doubts about the heritability of these behavioral traits in humans.

Each of these controversial traits merits our attention. We need to distinguish what is known from what needs to be known in order to judge the merits of the conflicting interpretations.

Questions to Answer

1. How do polygenic traits differ from monogenic ones?
2. What is a test cross?
3. What is the meiotic basis of independent assortment? Of Mendelian segregation?
4. Why would it be difficult in your own family to show whether two traits assorted independently or not?
5. What is the value of a chromosome deletion for mapping individual genes?
6. What is meant by a "chief gene"? What is its relation to variable traits?
7. How did Johannsen demonstrate the genotypic constancy of a pure line?
8. Why can't human quantitative traits be considered examples of chief genes and modifiers, or of polygenic inheritance?

Terms to Master

chief gene	modifier
cis-alignment	multiple factor inheritance
crossing over	polygenic inheritance
deletion map	Punnett square
environmentalism •	pure line
genetic determinism	recombination
inbreeding	somatic cell hybridization
independent assortment	test cross
linkage map	trans-alignment

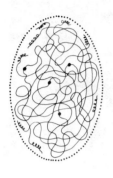

Chapter 11
Skin Color Inheritance
As a Quantitative Trait

Human skin varies in color—seasonally from environmental factors and permanently for genetic reasons. Since skin color is often the most striking feature we observe when we meet a person for the first time, it is the most commonly used basis for classifying humans into races.

Skin color is almost entirely a consequence of melanin synthesis and its distribution in the epidermal or outer layer of skin. To a much lesser degree, the density of capillaries may affect skin color. This latter effect is seen especially in the ruddy complexions of some very light-skinned individuals.

Skin Color Is Primarily
a Function of Melanin Synthesis Melanin consists of small granules (yellow to black in color) which are synthesized from the amino acid tyrosine by the enzyme tyrosinase. Caucasians produce very small amounts of melanin, but blacks have an abundant quantity distributed throughout the epidermis.

The synthesis of melanin occurs in specialized cells, *melanocytes*, which arise from the embryonic junction of the neural tube (the *neural crest* cells). In the human embryo this occurs at about the eleventh week and continues until the end of the fourth month, when these melanocyte precursors from the neural crest are embedded throughout the epidermis. They are also located in hair bulbs and the surrounding follicles of the hair bulbs. In Orientals and Negroids there is a dermal deposit of melanocytes in the sacral region (the base of the spine) producing a characteristic blue spot (Figure 11-1). Such dermally-located melanocytes do not usually occur in Caucasians.

Figure 11-1 The Sacral Spot
In Orientals and persons of black ancestry there is a deposition of melanin in the dermal layer of the skin overlying the base of the spinal column. This forms a bluish, bruise-like, triangular pattern called the sacral spot. As the infant ages the spot fades. It may also extend to other parts of the body, as on this child's buttocks, left shoulder, and back. (Courtesy of Dr. A. R. Rhodes)

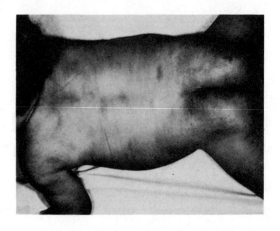

A melanocyte synthesizes melanin in a specialized cytoplasmic organelle called the melanosome (Figure 11-2). The melanin granules are then distributed to projecting filaments or dendrites. In the surrounding epidermis there are cells called *keratinocytes* which nip off branches of the dendrites. The ingested melanin granules (mature melanosomes) are then frequently gathered into a cap over the nucleus of the cell (Figure 11-3). This cap faces the outside, rather than the dermal side, of the cell and provides the nucleus with protection from the sun's ultraviolet radiation. Ultraviolet radiation causes DNA alterations, especially the rupture of adjacent thymine molecules which link abnormally into dimers, preventing normal gene function and raising the probability of chromosome breakage.

Melanocytes Differ in Size, Structure, and Distribution in the Body

The difference between black skin and white skin is not due to the number of melanocytes present (Figure 11-4). There is little difference in the distribution or density of melano-

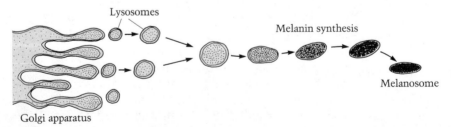

Figure 11-2 The Production of Melanosomes
In the melanocyte, lysosome-like buds leave the Golgi membrane network. These small spheres synthesize melanin and elongate into lentil-shaped melanosomes. Each melanosome, when mature, contains densely-packed melanin granules. The size, number, and location of melanosomes, as well as the size of the melanin granules, determine the skin color.

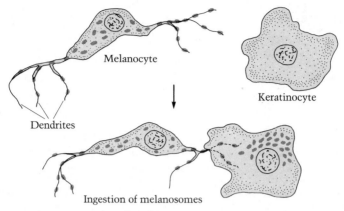

Figure 11-3 The Deposition of Melanin
The melanocyte synthesizes melanosomes, which are then circulated to thin, finger-like projections called dendrites. Surrounding keratinocytes, mostly in the epidermis or outer layer of skin, engulf the dendrites in a process called phagocytosis. The melanosomes ingested by the keratinocytes are assembled to form a protective cap around the nucleus, facing the light. This cap shields the chromosomes from radiation damage induced by the sun's ultraviolet light.

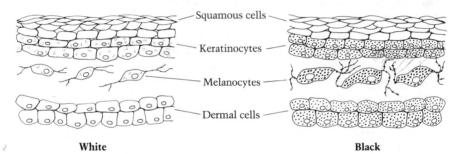

Figure 11-4 Skin Color Physiology
The major difference between the skins of white and black persons resides in the activity of the melanocytes. In whites the melanocytes produce fewer melanosomes that go to the epidermis. In blacks they produce larger and more numerous melanosomes, which are distributed to both epidermal and dermal layers of the skin. Whites and blacks have the same number of melanocytes in their skin, but the larger melanocytes in black skin have more dendrites and contain more melanosomes. The outermost layer of cells become flattened and are called squamous cells. As they die they flake off.

cytes among the different races. What does differ strikingly is the function of the melanocyte. The cell body in the melanocyte of a black-skinned person is larger and has many more dendrites emanating from it.

Within an individual there is considerable variation in melanocyte number. In an adult white male the penis has 2380 melanocytes/mm^2, about the same as the face (2310/mm^2). Lighter areas of the body have fewer (1120/mm^2 in the finger, 1000/mm^2 in the thighs, and 800/mm^2

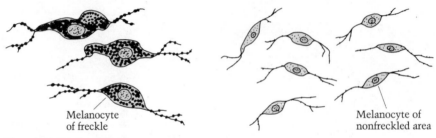

Melanocyte
of freckle

Melanocyte of
nonfreckled area

Figure 11-5   Freckles

Freckles are most often associated with red-haired individuals, but they can appear in the skins of other persons as well. The freckle is a localized region where the melanocytes are larger and more active (as in the skins of brown and black individuals), but not more numerous. The nonfreckled areas contain the smaller melanocytes of white skin.

in the abdomen). Tanning will not alter those numbers, but it will increase the function of the cells, resulting in more melanin deposition.

The role of melanocyte function is most strikingly seen in freckled individuals who have *fewer* melanocytes in the freckled areas than in the surrounding nonfreckled epidermis. The melanocytes in the freckled areas are large and thickly dendritic; the melanocytes in the nonfreckled areas are minute and have few dendrites (Figure 11-5).

The number, size, and shape of melanin granules produced by a melanocyte are genetically controlled. So too is the rate of transfer of melanin from melanocyte dendrites to epidermal keratinocytes. Finally, there are genetic controls on the distribution and concentration of the melanin granules in the keratinocytes and in the layers of the epidermis that harbor them.

Genes can even control melanin synthesis itself, as in certain forms of albinism. In experimental systems using mammals or birds, skin color mutations have been found for the differentiation of the neural crest cells which migrate to the epidermis. There are mutations controlling the distribution and migration routes of the neural crest cells producing banding, spotting, and other patterns. In mice some 40 genes (with a total of 70 alleles) which control melanization have been isolated.

Skin Color Is
a Quantitative Trait   From long-standing observation of interracial crossing, we know that skin color is not a simple monogenic trait. The children of a white and black couple are intermediate in color, and their descendants show a range of skin colors. Two or more genes must be involved because we know the skin color is inherited and not acquired.

We have seen in Chapter 10 that some variable traits can be analyzed into a chief gene with modifiers, or into a large indeterminate number of genetic factors whose distribution produces a normal curve. A third mechanism is also possible. Heribert Nilsson-Ehle studied a quantitative trait—seed coat coloring—in cereal grains such as wheat, rye, and oats.

Some grains have a dark dull-red color and others have a light blond color. In a P_1 cross of red and blond, Nilsson-Ehle obtained F_1 of a uniform pink color. These F_1 plants were inbred and produced F_2 with a distribution suggesting independent assortment of two pairs of genes (Figure 11-6). There were five categories of color—red, dark pink, pink, light pink, and blond, in a ratio of 1:4:6:4:1.

Nilsson-Ehle proposed that two genes, A and B, are on nonhomologous chromosomes. Each of these contributes equally to the red color. Their alleles, a and b, do not contribute to red color. Thus the homozygote $AA\ BB$ is full red and the homozygote $aa\ bb$ is blond. The F_1 of a P_1 cross between $AA\ BB$ and $aa\ bb$ is doubly hybrid, $Aa\ Bb$. Since the F_1 has two factors contributing to redness, instead of four or none, it is intermediate or pink in color. The $F_1 \times F_1$ cross results in 16 zygotic combinations, $\frac{1}{16}$ being $AA\ BB$ or red with four contributing factors, $\frac{4}{16}$ having three red-producing factors, $\frac{6}{16}$ having two, $\frac{4}{16}$ having one, and $\frac{1}{16}$ being $aa\ bb$ with no red-producing factors.

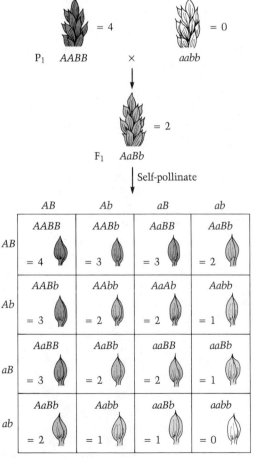

Figure 11-6   Quantitative Inheritance

H. Nilsson-Ehle demonstrated a polygenic mechanism for a quantitative trait, the color of seed coats in cereal grains. From crosses of red-coated strains with blond-coated strains, he obtained F_1 pink strains. $F_1 \times F_1$ crosses gave a distribution of F_2 plants with five categories of seed coats: red, dark pink, pink, light pink, and blond. The ratio of these categories, 1:4:6:4:1, suggested two pairs of alleles assorting independently. The Punnett square illustrates the way the 16 zygotes lead to this distribution of phenotypes. Note that the trait does not involve dominance. Pink is neither red nor blond, but an intermediate color reflecting the quantity of red and blond present.

F_2: $\frac{1}{16} = (4)$ $\frac{4}{16} = (3)$ $\frac{6}{16} = (2)$ $\frac{4}{16} = (1)$ $\frac{1}{16} = (0)$

Later Nilsson-Ehle found strains with three or four pairs of factors whose F_2 distribution of color shades resembled the normal curve distributions of Johannsen's beans. Unlike the bean size trait, cereal coat color had no environmental component affecting the variations in shades. Nilsson-Ehle's cereal coat color is an example of polygenic inheritance for a quantitative trait whose contributing factors can be determined from the fixed number of phenotypes produced among the progeny.

Davenport Established the First Genetic Model of Human Skin Color

The first attempt to study the heredity of human skin color was made by C. B. Davenport in 1913. He went to Jamaica to do his studies because it was one of the few places where marriages between whites and blacks (a racial mixing called miscegenation) were legal. Most states in the United States did not permit mixed marriages in those days. Davenport used a portion of the body normally shielded from the sun, the inner thigh, to measure skin color quantitatively. He also used a color wheel consisting of movable sectors of black, red, yellow, and white disks and rotated the sectored disks by twisting a wooden rod serving as an axis until he found a combination whose composite color matched the skin of the thigh of the individual he examined.

From a study of several dozen racially mixed families (white and black) Davenport worked out a model based on two chief genes for melanin synthesis. These were both autosomal and they assorted independently and were most likely on different chromosomes. Davenport's study was valuable because it destroyed several myths about interracial marriage.

In Davenport's model (which was based on Nilsson-Ehle's study of cereal coat color) two genes, A and B, produce large amounts of melanin. Their alleles, a and b, produce small amounts of melanin. A West African black is homozygous for both of the heavy melanin-producing genes. A Northern European is homozygous for the a and b alleles, which produce modest quantities of melanin. Davenport selected black and white parents who best matched the two extremes. Genotypically the P_1 cross would be $AA\ BB \times aa\ bb$ (Figure 11-7). Note that the black parent has four heavy melanizing genes and the white parent has none. Their children will all have a uniformly brown skin color, intermediate between the white and the black skin of the parents. The F_1 skin color is heterozygous for both melanizing genes. We can represent it as $Aa\ Bb$. Note that the F_1 has two of the heavy melanizing factors.

Some Brown-Skinned Parents Have Children Lighter or Darker Than Themselves

When Davenport studied pairs of brown-skinned individuals who had a black and a white parent, he found that the grandchildren of these interracial crosses varied in skin color

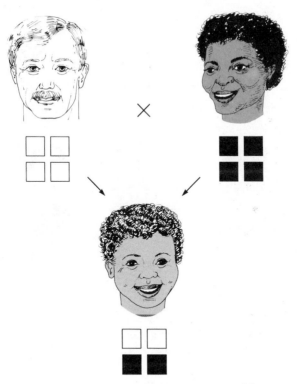

Figure 11-7   Skin Color Inheritance: The Davenport Model
The Davenport model attributes human skin color to two chief genes, each of which contributes equally to melanin production. The black square represents one such melanin-producing gene. In white individuals all four of the alleles produce a minimum amount of melanin (enough, however, to protect the skin from skin cancer in most temperate and polar regions). The white and black parents in a mixed marriage have children who are of an intermediate brown skin color. Such heterozygotes have two of the four alleles that produce large amounts of melanin. Since the effects of the four alleles are additive, the skin color has half the pigment found in the black parent.

from white to black (Figure 11-8). We can see readily that each of the F_1 heterozygotes produces four types of haploid gametes: *AB; Ab; aB;* and *ab*. The Punnett square prediction for the 16 possible genotypic combinations can be interpreted phenotypically by the number of melanizing factors present, using zero, one, two, three, or four for each genotype. Individuals with zero are white, with four are black, with two are brown, with one are light brown, and with three are dark brown. This gives us a ratio of 1 white:4 light brown:6 brown:4 dark brown:1 black among the 16 possible genotypes for the children of such parents. While the "white" child may have skin as light as a European's, that child would be *socially* (not biologically) classified as black.

Other traits of the African ancestry may assort independently of skin color, most of these being polygenic traits, such as facial features. Traits like hair color and texture would involve fewer genes than the shape of one's mouth or nose. If an individual with white skin did not have par-

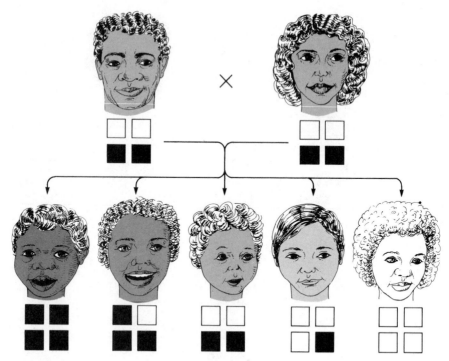

Figure 11-8   Skin Color Variability in Black Families
*Two Afro-American individuals, each a product of an intermarriage between a
white person and a black person, are heterozygous for the two chief genes for
melanin production. Independent assortment provides four gametic
possibilities: ¼ contain both of the full melanin-producing alleles; ½ contain
one of each type of alleles; and ¼ contain two of the alleles producing very
little melanin. This leads to 16 possible zygotic unions in a ratio of
$\frac{1}{16}$ black:$\frac{4}{16}$ dark brown:$\frac{6}{16}$ intermediate brown:$\frac{4}{16}$ light brown:$\frac{1}{16}$ white.
Note that the skin color factors will assort independently from hair texture,
hair color, and facial features.*

ticularly prominent facial features characteristic of African ancestry, that
individual could "pass for white" in a white community and thus pass
some of the genes from his or her African ancestors into the white breeding pool. As we shall see, interracial matings provide a *genetic bridge*
between races, and that is why we remain one species.

Note that brown children vary in their genotypes although their skin
color is phenotypically identical. Some may be heterozygous at both loci,
just as their parents were. Others may be *AA bb* or *aa BB*.

Brown-Skinned Parents May Produce
Children Whose Color Is Like Their Own
or Children of a Range of Colors
When matings occur
between brown-skinned individuals whose white ancestors are many
generations removed (as in Alex Haley's *Roots*) there is no way phenotypically that we can tell the genotype of the chief genes for melanin

production. In such brown × brown matings there may be all brown children (if the parents are *AA bb* × *aa BB* or of an identical homozygous genotype, e.g., *AA bb* × *AA bb*) or there may be a range (if the parents are of diverse genotypes, such as *AA bb* × *Aa Bb* or *Aa Bb* × *Aa Bb*).

Two White Parents Cannot Have a Child Considerably Darker Than Either Parent

Davenport's analysis proved that a person who passes for white cannot give rise to black children (i.e., children with dark skins) if married to a white person (i.e., a person with no African ancestry). The origin of the fallacy that a person with a distant African ancestor can produce a "throwback" to ancestral times probably came from the observation that two light brown individuals (e.g., *Aa bb* × *aa Bb*) can produce a child who is darker than either parent (the *Aa Bb* genotype is an intermediate brown). In addition, by independent assortment, such a child may have more pronounced African ancestral facial features than either parent.

Clearly, however, this cannot happen when a white person of black ancestry marries a white person with no black ancestry. In all such cases the child cannot be darker than the parent with the African ancestry because the other parent is homozygous *aa bb*, and thus contributes no heavy melanin-producing genes. To a modest degree, however, the child could be darker if other genetic modifiers are present in the white parent and if these modifiers are lacking in the parent of black ancestry. A person of Mediterranean ancestry, for example, could harbor such genetic modifiers, but a person of Northern European ancestry would be less likely to carry them. Even in cases like these it is genetically impossible for the skin color and facial features of the child to correspond to those of a West African.

Skin Color Function May Be Adaptive to the External Environment

Numerous theories have been proposed for the adaptive functions of skin color. Perhaps all of these models are involved to some extent, rather than just one of them. That an adaptive change is related to geographical location is supported by *Gloger's rule*. Gloger noted that the fur color of mammals is black in humid tropics, brown in the desert, and white or light-colored in the arctic. Skin color in humans is correspondingly distributed.

Fur color and skin color within a species are not necessarily the same. All chimpanzees have dark furs but their skin colors vary from white to black. Gloger's rule does not necessarily imply that coloration is for protective purposes (camouflage); nocturnal animals in both the tropics and the arctic obey Gloger's rule.

Thermoregulation, the fact that heat is reflected by light surfaces and absorbed by dark surfaces, is not a conclusive explanation of skin color variation either, since it would be disadvantageous in the arctic to reflect needed heat and equally disadvantageous in the tropics to absorb additional heat. The maintenance of a constant body temperature actually involves many factors other than skin or fur color.

Another theory of skin color function is the cold adaptation theory. People with dark skins are more susceptible to frostbite than those with light skin. Studies done during the Korean War, where the winter was severe, revealed higher frequencies of frostbite among black U.S. soldiers than among white U.S. soldiers. Whether this was due to the skin color itself or to other factors associated with being black is not known.

Skin Color May Be Associated with Vitamin D Synthesis

Since human beings have sparse, thin hair over most of their bodies, in contrast with the thick pelts of other primates, melanin serves an important function in humans by regulating the amount of ultraviolet that enters the epidermis. The ultraviolet stimulates synthesis of vitamin D. In northern latitudes when winter sunshine is at a minimum, insufficient vitamin D production can lead to rickets, an abnormality of bone and calcium metabolism. This is particularly true in children who are kept indoors and bundled up with clothes to keep them warm. Light-skinned individuals have less melanin to filter out ultraviolet light and so are able to synthesize the needed vitamin D.

Advocates of the vitamin D theory of skin color believe that light skin followed Gloger's rule as an anti-rachitic (rickets-limiting) measure, the darker children dying more frequently than the lighter ones in glacial areas as human populations moved north from Africa and the Middle East into Europe. Since vitamin D in excess amounts is toxic, selection would work in the opposite direction where the days are long and the sun's rays are most intense. Thus, in the tropics, heavy melanization would prevent overproduction of vitamin D.

Skin Color Can Protect Individuals from Skin Cancers

Another theory of skin color function involves sunburn and skin cancer. It is well-known that ultraviolet radiation is absorbed by DNA and leads to gene mutation and chromosome aberrations. Skin cancer is more prevalent in California and Florida than in the New England states. It is also more frequent among farmers and outdoor professional athletes than among those who work indoors. In the far northern and southern latitudes melanin is not as essential as it is near the equator; heavy melanization is essential to prevent cancers in the exposed skin of individuals living in equatorial regions.

Spending a lot of time in the summer sunlight can be a painful experience for a light-skinned individual. The sunburned skin turns red, is painful to the touch, and blisters. This reaction of the skin cells to excessive ultraviolet radiation is a disadvantage in equatorial regions where summer-like sunlight is present all year round, especially in open plains.

Highly melanized skin (brown to black) is advantageous in such areas. In a forested area, however, where the sunlight is scattered and absorbed by leaves, populations are not selected for dark skin color by this mechanism.

Albinism Is a Disorder
of Melanin Synthesis or Deposition
Occasionally an individual is born with striking, milk-white skin and hair. These individuals are albinos. Albinism has been recognized since ancient times, and it appears in all populations of human beings. It is also a trait found in many vertebrates, including amphibians, reptiles, birds, and mammals.

There are three forms of human albinism. The familiar milk-white albinos with white hair and light-sensitive eyes are *oculocutaneous albinos* lacking the enzyme tyrosinase. Such albinos have a normal number of melanocytes and melanosomes, but no melanin granules are present in them. A second type of oculocutaneous albinos have functional tyrosinase but their melanosomes contain less than the normal amount of melanin for their ethnic or racial type. They lack melanin pigment in their retinas as children and frequently experience squinting eyes and photophobia just as do the tyrosinase-negative albinos. Their skin has an unusual number of scattered dark spots (nevi) which, when exposed to the sun, produce a freckled tan. Their hair color varies from off-white to light brown. Both types of oculocutaneous albinos are autosomal recessives and both are controlled by an independently assorting gene. The third albinic form is X-linked and only affects the eyes. Such *ocular albinos* have normal skin and hair pigmentation but their eyes lack retinal melanin (although they are not as deficient in it as tyrosinase-negative albinos).

All three forms of albinos have abnormal neural anatomies involving the visual pathway of the brain, resulting in a quivering of the eyes (nystagmus) and also in defective depth perception. They require sunglasses in the daytime. Without them their eyes would tear copiously and smart from the intense glare.

Albinism Can Be an Ethnic Disorder
in Some Populations
Among the Cuna Indians of the San Blas Islands off the west coast of Panana, albinos are common and are called moon-eyed people because they see best during the moonlight and feel more comfortable working in the dark. As is true of other albinos, they suffer a high incidence of skin cancers and malignant melanomas (colorless, however) from exposure to ultraviolet light.

Albinism is rare in most parts of the world, the incidence of tyrosinase-negative births being about 1 in 35,000. The tyrosinase-positive type is more frequently seen in blacks (1 in 14,000) than in whites (1 in 60,000).

Although albinism is restricted to defects of melanocyte function, there is a spotted abnormality, piebaldism, which is associated with a skin color mutation. This is an autosomal dominant defect of the migration of melanocytes from the neural crest. Melanocytes are absent in the portions of the body that are depigmented. Most piebald individuals have a triangular white blaze or patch of hair on their heads. The melanocytes are missing in both the skin and hairbulbs of that region. Piebalding of the skin in blacks is more pronounced than it is in whites because of the contrast between pigmented and nonpigmented areas. This may account for the erroneous belief that this is an ethnic disorder of blacks. Piebaldism is more easily seen in Caucasians by the white forelock of the hair.

Questions to Answer

1. What are the major components of skin color?
2. What is a melanocyte? How does its pigment metabolism lead to skin color?
3. Do blacks and whites have the same number of melanocytes? Discuss the significance of your answer.
4. What are the assumptions of the Davenport model of black skin color?
5. Why is it virtually impossible for a white person married to a person of black ancestry to have children darker than the partner with black ancestors?
6. Why does independent assortment between two brown-skinned parents (each having a white parent and a black parent) not result in a 9:3:3:1 ratio?
7. What is the main difference between a biological description of skin color phenotypes and a social definition of a black or white person?
8. What is Gloger's rule?
9. How is skin color related to (a) thermoregulation, (b) vitamin D synthesis, and (c) skin color?
10. Discuss the three forms of human albinism.

Terms to Master

albinism
genetic bridge
Gloger's rule
keratinocyte
melanin
melanocyte

melanosome
miscegenation
quantitative trait
sacral spot
tyrosinase

Chapter 12
Miscegenation and the Emergence of New Races

Although Davenport's model is useful for identifying the two major genes involved in the melanization of human skin, it is inadequate because skin colors form a continuous spectrum rather than five discrete shades. To account for these more subtle gradations in color, Curt Stern in 1953 added three more genes, for a five-gene model of melanization. Stern's model generates 11 categories of skin color, from the West African black skin to the Northern European white skin (Figure 12-1).

If we were to simplify the inheritance by attributing equal quantitative contributions to each of Stern's genes, we then would find, as in Davenport's original cross of black × white Jamaicans, that the intermediate F_1 color was brown. Those F_1 individuals would have five melanin-producing genes as compared to ten in the black parent and none in the white parent. Such F_1 brown individuals, in turn, would have children segregating five heterozygous pairs of melanin-producing genes. Those with one or two factors would be tan; three or four factors, light brown; six or seven factors, dark brown; and eight or nine factors, very dark brown. The uncommon child with no factors at all would have skin that looked like the white Northern European skin, and the equally rare child with ten factors would have skin that resembled the black West African skin.

It should be emphasized that Mendel's law of independent assortment would predict that $\frac{1}{16}$ of the children from a two-gene model would be these rare white or black children, and $\frac{1}{1024}$ from the five-gene model. Since the white or tan children (those with zero, one, or two factors) from Stern's model could pass for whites if their facial features and hair texture were not characteristically like their African ancestry, the total of very light-skinned individuals would be $\frac{32}{1024}$ or 1 in 32, a frequency that is much closer to the incidence actually found in Davenport's study.

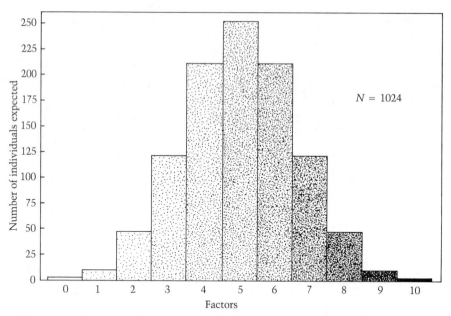

Figure 12-1

Stern proposed a five-gene model for skin color inheritance because there are more than five skin colors among the progeny of biracial hybrid parents. The frequency of extreme skin colors (white or black) in such matings provides a rough idea of the number of genes involved. In the five-gene model there are 11 categories of skin color, and the probability of an F$_2$ child from brown-skinned biracial parents being as black or as white as the original P$_1$ grandparent is only 1 in 1024. The larger the number of genes participating in the determination of skin color the more uniform the miscegenated population will appear, with very few individuals having noticeably dark or light skins. Stern's use of five genes is, however, only a model and the precise number of genes involved in the synthesis and distribution of melanin in human skin is not known.

Miscegenation Was Officially Discouraged
But Illegally Practiced in North America   When the northeast coast of North America was successfully settled, beginning with the Pilgrims in the early 17th century, entire families left their English homes and sought freedom in the New World. Their religious and cultural traditions made them adopt a policy of sexual and cultural isolation from the American Indians. Further south, another group of British settlers arrived somewhat earlier in Jamestown, Virginia, to help establish a new colony for the Crown. Although the Jamestown colony had fewer families and more soldiers than the Plymouth settlement, cultural isolation prevailed there also despite the occasional accounts of miscegenation such as the love of Pocahontas for John Smith and her subsequent marriage to John Rolfe.

African slaves were introduced into the colonies gradually during the 17th century, as domestics and field hands, particularly in the hotter,

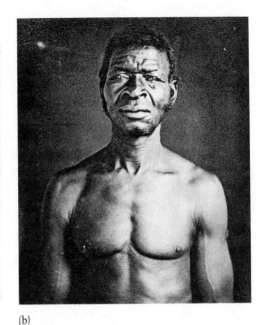

(a)   (b)

Figure 12-2
The slave trade in North America flourished from 1730 to 1810, the period when most of the African slaves were imported. Of all the slaves brought to the New World, the United States imported 36 percent. The last legally sanctioned slavery was abolished in Puerto Rico in 1873, Cuba in 1886, and Brazil in 1888. (a) A handbill for a slave auction. (b) A daguerreotype of a slave taken in 1850. [(a) American Antiquarian Society. (b) Peabody Museum, Harvard University.]

more humid, climates of the South (Figure 12-2). Since the African slaves had few legal rights, it was not uncommon for black females to be raped by their white owners. The children of such one-way miscegenation were culturally identified as black and raised as slaves. Because of this the black slaves who were imported from West Africa had limited opportunity to maintain a biological homogeneity of their gene pool. Their specific native African differences were gradually altered by the addition of genes from their white owners. As members of the same species, all races of humans can freely interbreed, and there began to be evidence of race mixture in North America in the form of brown-skinned individuals.

A Small Percentage of Miscegenation over Several Generations Alters the Phenotypes of Afro-Americans

To some extent, especially after the emancipation of slaves, those individuals of African ancestry whose skin color was as light as a European's could abandon their cultural heritage and claim that they were white. Thus genes from the African ancestry (other than those for melanization of the skin) could enter the white gene pool.

During the 10 to 12 generations that Afro-Americans have resided in North America, their phenotypes have become considerably lighter than their West African forebears. About 20 to 30 percent of the genes in Afro-Americans are derived from white matings over the past three centuries. In any given generation the percentage of mixture is low—about two to three percent—but as the generations increase in number the process of skin lightening increases. Although slavery has been abolished since the end of the Civil War, such white × black matings have continued at this low rate, either through extramarital liaisons or, since the mid-20th century, through marriages (formerly forbidden by state laws which were struck down in the 1950s as unconstitutional by the Supreme Court).

The result of the North American historical and cultural experience has been the relative isolation of the genes of the American Indians from mixtures with the white or black gene pools. This was accomplished by the westward movement of Indians as white settlers took over their land; the massive devastation of Indian populations upon first contact with whites (especially from smallpox, measles, and tuberculosis); and the confinement of Indians to reservations rather than their integration into households as slaves. Some of the eastern tribes, however, did have significant mixtures of genes with either the blacks or the whites with whom they engaged in trade.

Culture, Not Biology, Defines
a Person As Black or White   The black and white racial mixture in North America was highly significant, especially for the black population. Because North American values defined race in two exclusive categories—white and nonwhite—all those not exclusively white were forced to be part of the black culture, regardless of the ease with which some of them could pass for white.

The Latin American Pattern
of Miscegenation Differed
from the North American Pattern   When the Portuguese colonized Brazil and the Spaniards conquered the Aztec and Incan civilizations, the ships bearing Europeans carried almost exclusively male conquistadors. Few Spanish or Portuguese women were permitted by their families to participate in such hazardous ventures. The consequence of this first encounter of white males with South and Central American Indians was the creation of a new hybrid class of children, *mestizos*.

At the same time, the effects of those 15th and 16th century encounters between whites and Indians decimated the Indian populations. There were about 25 million Indians in South America when Columbus discovered the western hemisphere, but by 1600 disease had reduced the population to 1.7 million (Table 12-1).

Table 12-1   The Demographic Disaster: American Indian Population
in Central Mexico

The discovery of the New World was a catastrophe for its inhabitants in Central and
South America. Portuguese and Spanish explorers and conquerors introduced smallpox,
typhus, measles, and influenza. Black slaves were exported from Africa by royal license
of the Spanish Crown in 1518. The Africans introduced malaria, yellow fever, and
trachoma (a parasite causing blindness). Many Indians were slaughtered in battle and by
other acts of violence, but these numbers were relatively small compared to the numbers
who were devastated by disease. Smallpox, influenza, and measles also devastated North
American Indians and greatly reduced their numbers, especially along the east coast from
Florida to Maine.

Year	Population
1519	25,200,000
1532	16,800,000
1548	6,300,000
1568	2,650,000
1580	1,900,000
1595	1,375,000
1605	1,075,000

Source: From M. Morner, 1967.

Black slaves, who had long been imported from West Africa to Portugal, were brought in by their Portuguese owners shortly after the settlement of the new colony began. These slaves spoke Portuguese, and
were regarded as more friendly and approachable by many of the op-
pressed Indians who sought relief from their white conquerors. This contact led to black × Indian matings and the production of *zambo* offspring.
Many Portuguese males also satisfied their sexual needs with black domestic slaves, fathering *mulatto* children. The terms mestizo, zambo,
and mulatto are not technical terms and are rarely used now because
they have overtones of racism. They are used here only because they
were of historical significance for several centuries in establishing legal
and cultural rights in Latin America.

Throughout the 16th and early 17th centuries, the Portuguese conquerors obtained and cultivated their own land, fathered mestizo children, and then petitioned the king for permission to bequeath their hold-
ings to their children. Not enough Portuguese women were permitted to
abandon their homes in Portugal and risk a dangerous voyage and an
even more hazardous life in Brazil. To encourage settlement and development of the colony, these mestizos were legitimized, as were their
parents' marriages. Distinctions were then drawn between the legitimized mestizos (sometimes called creoles) and the illegitimate mestizos.

A Three-Way Miscegenation Occurred
in Many Latin American Countries
As the generations passed, a
three-way miscegenation took place with mestizo × mulatto, zambo ×

Table 12-2   The Pigmentocracy of Spanish America (circa 1520–1650)

The three-way pattern of racial mixture in Spanish America led to new racial types whose legal and social status varied. Although Indians legally were considered second only to the Spaniards in terms of their rights of property and employment, they were the least accepted socially because they did not speak Spanish and because the Spaniards wished to minimize their power and destroy their religious beliefs. This pigmentocracy began to collapse in the 17th century. It was impossible to maintain strict, well-defined castes based on skin color and other features. The three-way miscegenation was obliterating the individual Old World characteristics and forming a Latin American type whose ancestry was blurred. A similar hierarchy existed for the Portuguese.

Hierarchy by Law	Hierarchy by Social Status
1. Spaniards	1. Spaniards (born in Spain or having two Spanish parents)
2. Indians	2. Creoles (legitimate Spanish × Indian offspring)
3. Mestizos	3. Mestizos (illegitimate Spanish × Indian offspring)
4. Mulattos, zambos, and free blacks	4. Mulattos, zambos, and free blacks
5. Slaves	5. Slaves
	6. Indians

Source: Adapted from Morner, 1967.

mestizo, and mulatto × zambo crosses, as well as matings of those three hybrid forms with black, Indian, and white members of the population. At first the Portuguese created a "pigmentocracy" with the Portuguese in a favored position, politically and economically, followed by the mestizo, mulatto, zambo, black, and Indian populations (Table 12-2). The Indians were of lowest social status because they were ineffective slaves whose susceptibility to infectious diseases and the ease with which they could run away in their own country made them economic liabilities. As the three-way miscegenation progressed, however, distinctions between groups were hard to maintain by visible appearance, and those who came from socially lower groups in the pigmentocracy, who could pass for mestizo or Portuguese, readily did so.

New Races Arise
from the Miscegenation
of Previous Races   The consequence of a population in which Portuguese men were plentiful and Portuguese women were scarce was the rapid creation of a new racial type, the Latin American, bearing part Indian and black ancestry mixed with varying amounts of Portuguese ancestry. Since the prevailing social system favored the Portuguese, before and after Brazilian independence from Portugal, the Brazilian model defined persons as white if they had any white ancestry at all.

The Brazilians today are still highly heterogeneous, with the darker-skinned populations in the northern portions where the plantations were extensive, and the lighter-skinned populations, of more recent European immigration, in the southern portions. German, Russian, Asian, and Italian immigrants have settled in Brazil (mostly as families) but have not miscegenated to the same extent as the original Portuguese conquerors did. Later Portuguese immigrants, arriving as families, also kept a relative genetic isolation from the highly miscegenated populations they found there. With such a large portion of the Brazilian population having mixed ancestry, and a cultural tradition that defines such individuals as white, these various racial types have an excellent chance of eventually merging into a new race.

It Is Difficult to Define
the Term Race Without Bias
The term race is difficult to define because, to geneticists, it is a dynamic state of a population, not a fixed collection of traits. Human gene frequencies vary in each generation in response to natural selection, migration, changes in cultural habits, and social opportunities. All of these influence the mixing of populations and the shifting proportions of genes in any geographical region. The technical definition of a race is a population whose physical traits, gene frequencies, and chromosome arrangements are demonstrably different from those of another population of the same species. The number of races varies according to the number of differences an anthropologist or geneticist sees.

For most people these technical distinctions are less important for defining race than is skin color, which is the most widely used criterion for classifying the white, black, Oriental, and American Indian races. Since the technical distinctions are mostly for the convenience of the scholars and researchers, race is a relative term, not an absolute one. For example, Asian Indians in the United States are considered Caucasian but in South Africa they are considered Colored. Our popular classifications reflect cultural biases. Race is not a meaningless term if the criteria for classification are provided. Those who object to prejudice based on racial classifications often choose synonyms for them, such as ethnic groups, variant populations, genetic strains, breeds, or people (as in the phrase "Jewish people" when both a cultural tradition and a common ancestry are emphasized).

A Stable Equilibrium May Arise
from Prolonged Miscegenation
Human populations have never been isolated for periods long enough to generate new species. In order for that to happen many thousands of generations would have to elapse to build up differences in chromosome structure and new gene functions. Less than one percent of the genes between any two human populations differ from one another, yet from those genetic differences races formed. The human races today are considerably different from those of the pre-Columbian world. Colonizations in Asia, the Pacific Islands, the western

hemisphere, and Africa are creating new racial types through miscegenations on a large scale which are particularly pronounced and distinctive in Hawaii, the Philippines, South Africa, Brazil, and the Caribbean Islands.

In the United States the two-way miscegenation between whites and blacks is continuing at a modest rate of one to three percent. If this rate is maintained then the gradual mixing of blacks and whites will eventually reach a genetic equilibrium, forming a new race. This equilibrium could be achieved more rapidly if all social barriers against miscegenation disappeared. It could be delayed and partially prevented if a strong social

Table 12-3   Equilibrium by Miscegenation: North America

Assuming that approximately ten percent of the people who arrived in North America since it was first colonized were African, the gradual race mixing that has taken place can be projected to an equilibrium. If one or two percent of the genes in the Afro-American population are derived each generation from Caucasians, then that rate will reach equilibrium about 200 to 300 years from now. Assuming no major changes in immigration or in the values affecting the rate of miscegenation, the Davenport model (two pairs of chief genes for melanin) shows an increase in tan or light brown skin color and a dramatic decrease in dark or black skin. The Stern model (five pairs of chief genes for melanin production) indicates that white-skinned (Anglo-Saxon type) individuals would be a minority; most Americans would look like the Mediterranean populations. Very few people would have brown skins, and virtually none would have dark brown or black skins. Social values accepting miscegenation would hasten the process. Strong ethnic or racial marriage patterns which exclude miscegenation would delay it.

Skin Color Genes	Phenotype	Percentage
Davenport Model		
0	white	65.6
1	light brown	29.2
2	brown	4.9
3	dark brown	0.4
4	black	0.01
Stern Model		
0	white	43.4
1	tan (Latin)	37.8
2	light brown	14.8
3		3.4
4	brown	0.5
5		0.05
6	dark brown	4×10^{-3}
7	very dark brown	2×10^{-4}
8		6×10^{-6}
9	black	1×10^{-7}
10		1×10^{-9}

Source: Adapted from Stern, *Principles of Human Genetics* (San Francisco: Freeman, 1967).

pressure developed which led to a rejection of interracial matings (Table 12-3).

Using the Davenport two-gene model, the melanin distribution in the United States at equilibrium (about 300 years from now) would show 66 percent of the population with skin colors as light as the original Pilgrims and 0.01 percent as black as the West Africans who were originally brought over as slaves. Using the Stern model, 48 percent would have the original Pilgrim skin color and only three in a billion would look as black as their original black West African ancestors.

The Same Data, Interpreted
by People with Different Values,
Could Lead to Opposite Conclusions
Indeed, using Stern's five-gene model, the typical American skin color would look like today's Mediterranean (Spanish, Italian, Greek). Skin colors darker than that would constitute less than 0.5 percent of all the population. This redistribution of pigment-producing genes could be interpreted in opposite ways by the values of many contemporary Brazilians or white citizens of the United States. The Brazilian, who defines a white as a person with any white ancestry, could claim that the blacks have disappeared from the United States. More than 99 percent of the United States would be white, from their point of view, using either Stern's or Davenport's model. A person in the United States, however, where any individual with a black ancestor is considered black, would look at the same data and come to the opposite conclusion. Stern's model would make the whites a minority, with 52 percent of the population having some of their genes for skin color derived from a black ancestor. Even in Davenport's model the black population would be roughly 34 percent of the total population.

The models assume that the rate of miscegenation will not be less than three percent per generation, that the mean number of children reaching maturity is the same for whites and blacks, and that values favoring or discouraging miscegenation will not change. Any departure from these assumptions will, of course, alter the number of generations needed to reach an equilibrium.

Human Races Are Not Fixed
and Will Continue to Change
Human races have been evolving and will continue to do so in the future. As travel becomes available to a nation's inhabitants, diversity in marriages and matings will increase. Americans whose ancestors came to America in the 17th or 18th century frequently find mixtures of English, Irish, German, French, Dutch, Polish, Italian, Greek, or Russian in the family pedigree. They may no longer have the strong ethnic identity of those Americans whose ancestors came to this country only one or two generations ago (Figure 12-3).

Each major war brings with it a wave of hybrid children, as in the GI occupations of Germany, Italy, Japan, Korea, and Vietnam. Most of

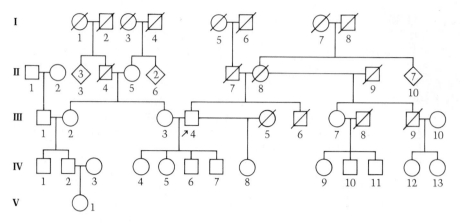

Pedigree member	Ancestral or national identification
I-1, 2	American of Dutch, English ancestry (in U.S. since 1760)
I-3, 4	American of German, English ancestry (in U.S. since 1760)
I-5	French, Swedish (European)
I-6	Swedish (European)
I-7, 8	Polish (European, Ashkenazic)
II-1	English (in U.S. for several generations)
II-2	Mexican, Spanish ancestry
II-3, 4	American of Dutch, English ancestry
II-5, 6	American of German, English ancestry
II-7	Swedish (European)
II-8	Polish (born in U.S., Ashkenazic)
II-9	Russian (European, Ashkenazic)
II-10	Polish (born in U.S., Ashkenazic)
III-1	American-Mexican (born in U.S.)
III-2	American, mostly White Anglo-Saxon Protestant
III-3	American, mostly White Anglo-Saxon Protestant
III-4	American, mixed
III-5	Russian-American, Ashkenazic (born in U.S.)

Pedigree member	Ancestral or national identification
III-6	American, mixed
III-7	American, Ashkenazic
III-8	American, White Anglo-Saxon Protestant, English, Scottish ancestry
III-9	American, Ashkenazic
III-10	American, Russian Ashkenazic
IV-1	American, mixed
IV-2	American, mixed
IV-3	Afro-American, mixed
IV-4	American, mixed
IV-5	American, mixed
IV-6	American, mixed
IV-7	American, mixed
IV-8	American, mixed
IV-9	American, mixed
IV-10	American, mixed
IV-11	American, mixed
IV-12	American, Ashkenazic
IV-13	American, Ashkenazic
V-1	Afro-American, mixed

Figure 12-3 The "Melting Pot" in the United States

Ethnic and religious identifications often become obliterated by intermarriage. As an example, in the author's immediate family, at least eight ethnic components are present. Religious identifications are equally diverse with Huguenot, Baptist, Lutheran, Roman Catholic, Jewish, and Unitarian representation, as well as numerous other Protestant sects. The author is the propositus (III-4), and ancestral components are described in the accompanying list.

these children will eventually marry, thus introducing new genes into the United States gene pool or into the gene pool of their homeland if they were not brought back by their fathers.

In the far distant future, the current human races may disappear. A tawny race, similar to Polynesians, may arise which would reflect the mixture of white, black, and Oriental populations. This is likely to occur with continued developments in rapid intercontinental travel, if literacy becomes universal, and if national values become secondary to individual or international values.

From a strictly genetic perspective, race mixtures can be advantageous by reducing the amount of homozygosity of harmful genes or by breaking up disadvantageous polygenic combinations through genetic recombination. The genes are still there, but outbreeding makes it difficult for them to come together. However, it can also be a disadvantage if certain traits are highly valued within a race (small height among pygmies; dark or light skin; straight or curly hair; thin or full facial features). Those traditional values, of course, rapidly erode when isolated populations encounter different traits associated with new people whose ideas, wealth, and power may make them attractive as potential spouses.

Questions to Answer

1. How does Stern's model of skin color differ from Davenport's?
2. What historical factors initially determined miscegenation patterns of whites, blacks, and Indians in North America?
3. What historical factors initially determined miscegenation patterns of whites, blacks, and Indians in South America?
4. What percentage of genes from whites have been introduced into North American blacks? How did this happen?
5. What percentage of genes from blacks have been introduced into whites in North America? How did this happen?
6. What was the major reason for the rapid conquest of the Indians by their Latin conquerors?
7. How did Brazil attempt to establish a pigmentocracy and why did it fail?
8. Are the skin colors of humanity today the same (in percentage) as in 1492? Discuss your position.
9. How would the Brazilian and North American white interpret the equilibrium results of white and black skin color phenotype frequency? What would be the basis for their interpretations?
10. Is ethnic identity possible for more than a few generations in the United States or is the "melting pot" theory more realistic? Defend your point of view.

Terms to Master

genetic equilibrium	pigmentocracy
melting pot	race
mestizo	zambo
mulatto	

Chapter 13
The Early Eugenics Movement: A Misuse of Genetics

Eugenics has had virtually no success as a social program to improve the heredity of humanity or to prevent our heredity from deteriorating. Two eugenics programs arose independently of each other in the last half of the 19th century. The English movement, founded by Sir Francis Galton, had as its goal the improvement of humanity by encouraging the talented to bear more children. The American movement was initiated by several social reformers who erroneously identified certain related families as bearers of socially undesirable traits.

Galton's Contributions
to Genetics Were Modest
Francis Galton (1822–1911) was Charles Darwin's cousin (Figure 13-1). He entered medical school at the age of 15 but left after two years and quickly finished up his bachelor's degree when his father died leaving him independently wealthy. Galton funded his own expeditions to Africa and Asia and published articles on anthropology and geography. He believed measurement was the most objective approach to the study of human behavior and heredity. He invented the correlation coefficient and was one of the pioneer founders of experimental psychology. Galton introduced finger printing for human identification. He showed, experimentally, that hereditary traits are not transmitted by the blood (although we noted earlier that our language still has vestiges of that belief, as when we refer to royalty as "blue bloods"). Galton's contributions to heredity were uneven. He did show that for quantitative traits, such as adult human height, there is a "regression to the mean" (Figure 13-2). That is, children of tall parents, while taller than average, tend to be shorter than their parents. The reverse is true for children of short parents. However, Galton was incorrect in his

Figure 13-1
Francis Galton (1822–1911)
was a precocious child who
later became a pioneer in
anthropology, psychology,
meteorology, and statistics.
Galton's belief that traits such
as genius are inherited became
the basis for a moral
philosophy of eugenics, or
differential breeding to
promote the spread of
desirable traits. (The Granger
Collection)

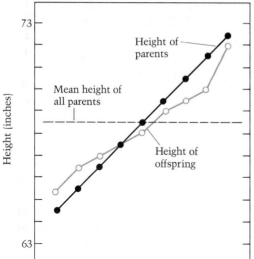

Figure 13-2   Galton's Law of
Regression
Galton showed that the
extremes of a quantitative
trait tended to revert toward
the average for that trait. Thus
tall men had sons shorter than
themselves (but nevertheless
taller than average), and short
men had taller sons (but who
were also smaller than
average). This regression to the
mean or "law of mediocrity"
reflected a blending which
was characteristic of traits
with many genetic or
environmental components. A
similar regression to the mean
is found for body weight and
human intelligence.

assumption that each past ancestor contributed proportionately to the
appearance of an individual. Nevertheless, his mathematical approach to
heredity eventually led to population genetics.

Galton Claimed That
Genius Was Inherited   In 1869 Galton wrote the book *Hereditary
Genius* in which he attempted to prove the heritability of eminence. He
used individuals who were cited in biographical dictionaries or whose
obituaries exceeded a certain length in the *London Times* for a given

year. He excluded royalty (since they did not earn their fame or titles by merit) and emphasized writers, jurists, militarists, church figures, scientists, physicians, explorers, and others whose contributions to culture, Galton believed, constituted eminence. Galton calculated that one eminent individual was born out of every 4000 people. By following the pedigrees of these individuals, Galton concluded that there was about a 20 percent chance that an eminent individual had an eminent parent or an eminent child. As a control, Galton used the children invited to live with bishops and cardinals, who had, of course, an adopted status, since they were usually orphans of a relative or friend or children whose parents could no longer care for them. These children, he claimed, had the environmental advantages of good homes, wealth, culture, refined conversation, and many of the qualities that are found in the homes of the eminent. Furthermore, bishops and cardinals are elected on merit and thus are as eminent as those in secular professions. Galton claimed that few of these children achieved eminence, in or out of the church, and from this he concluded that genius, the underlying talent leading to eminence, was inherited.

There are many defects in Galton's assumptions and experimental approach but his arguments were persuasive to his contemporaries. He concluded that those who were eminent had a moral duty to have larger families so that the society could benefit from the talents their children would use in their careers. Galton coined the term eugenics and used moral persuasion in an effort to better humanity.

In the United States a Corrupted Heredity Was Assigned to the Jukes, Kallikaks, and Ishmaelites

In contrast to Galton's positive eugenics program in England, the American program could best be characterized as *negative eugenics*. The objective of the movement was to prevent deterioration of healthy American stock.

The movement began as a reaction to the failure of certain programs promoted by social reformers. When inmates of lunatic asylums were freed from their chains and treated with some compassion as ill rather than as demonically possessed, it was noted that the insanity did not disappear. Likewise, poverty did not cease to exist when communities provided welfare funds to help the unemployed while they sought jobs. Crime did not disappear when rehabilitation programs were tried out in some prisons. Experimental utopian farms, based on the beliefs that property is communal and an extended family is more harmonious than an individual marriage, frequently failed due to the boredom, jealousy, laziness, or dissension among some of the participating members.

These social environmental approaches to problems were often inadequate in conception and economic support and there were many social circumstances that made such token efforts ineffective. When these social approaches that seemed so rational to the early reformers failed to do away with crime, poverty, retardation, insanity, and poor health, the inevitable question was this: What was the basis for the persistence of these problems in American society?

In a study of prison inmates in upper New York State about 1875, Richard Dugdale noted that many of the prisoners were related. He traced the family for several generations and concluded that the many branches of this family were all traceable to an immigrant from Holland who came to New York City in 1720 and then moved upstate to settle. Since a major portion of the prisoners, past and present, were members of this family, and many other members of the family were paupers and illiterates, Dugdale concluded that these individuals shared a defective environment. Later interpreters of Dugdale's book believed that the social problems of these individuals stemmed from a hereditary weakness which limited their moral judgment, intellect, and resistance to disease. These interpreters founded the *hereditarian philosophy*, a belief that behavioral traits have innate origins. Dugdale was an environmentalist but his study was taken over by hereditarians. Dugdale's name for the family was based on a slang word—jukes—referring to the habit of chickens of dropping eggs wherever they happened to find a convenient spot and fouling their nests and themselves. The Jukes family became the prototype for other kindreds—the Kallikaks, the tribe of Ishmael, and the Nams—who represented one or more socially undesirable traits.

Compulsory Sterilization Laws
Were Passed in 30 States
Those who accepted the hereditarian conclusions and research sought legislative reforms to contain these families before they outbred, and thus further corrupted the stock of a healthy America with their allegedly defective heredity. Between 1906 and 1930 some 30 states passed eugenic sterilization laws. These laws were used to sterilize, by vasectomy in males and tubal ligation in females, individuals who were insane (with a family history of insanity), "feeble-minded" (mentally retarded), or "habitually" criminal. Some 30,000 Americans were involuntarily sterilized before these laws were ruled unconstitutional and repealed in most of these states.

Xenophobia Was Added to
the American Eugenics Movement
The Jukes, Kallikaks, and similar kindreds had one feature in common. They were white Anglo-Saxon Protestant families differing from their neighbors primarily in social status. However, their status as scapegoats for social problems changed after World War I for several reasons. Compulsory education, a period of full employment, and the training many of them received in the army eliminated much of the poverty, illiteracy, and crime associated with them. An even more important factor was the change in immigration patterns in the closing decade of the 19th century. Most immigrants were from

Eastern and Southern Europe. Their languages were Russian, Polish, Italian, Yiddish, and Greek, and their culture was strange to their neighbors who were predominantly Anglo-Saxon settlers and descendants. The poor showing made by such immigrants on English language intelligence tests and the reported high failure rate of their children in public schools alarmed supporters of the American eugenics movement. The corruption of American stock was no longer attributed to the Jukes or Kallikaks but to the aliens—the "outcasts" and "failures" who fled to America in hopes of finding wealth and success in the New World. Their critics pointed to the ghettos, the crimes in their impoverished streets, and their slow process of accepting and adjusting to American ways. A fear of foreigners is called *xenophobia* and many Americans believed that continued immigration would change the United States in unacceptable ways.

During the early 1920s supporters of the American eugenics movement, led by lobbyists from the Eugenics Record Office of Cold Spring Harbor, New York, successfully marshalled their evidence that the immigrants were dangerous to American cultural standards. In 1921 and 1924 restrictive immigration laws were passed to establish limited quotas for various nations based on the census of 1890, a time when few immigrants from Eastern and Southern Europe had arrived in America. Although the chief contributors to these restrictive immigration laws were not geneticists, their views on the heritability of social failure were not challenged by geneticists active in research. There were many geneticists who did not share the bias against foreigners, but they did not speak out against it, believing that polemics, social controversy, and the application of pure science to humanity were unbecoming to the scientist. Thus, by default, the spurious views of prejudiced eugenists were accepted enthusiastically by the majority of Americans who felt that "bad heredity," not societal neglect, was the major cause of the many failures within a democracy.

In the 1930s There Were Many Criticisms and Failures of the Eugenics Movement

Positive eugenics did not work in Great Britain because raising large families is time-consuming and expensive. Raising a large family may also diminish substantially the contributions a talented individual can make since that person's energies are channeled into raising children instead of professional work. Throughout the first quarter of the 20th century, the eminent (as defined by Galton) had an average or less than average family size in England.

The American eugenics movement was set back by the rise of Nazism in Europe, the growth of the Ku Klux Klan in the United States, and the Great Depression (1929–1939). The German program of eugenics was based on the American model but was applied by the Nazis on a mass scale to Jews and Gypsies, regardless of individual merit. It was also used to sterilize those suffering from a variety of congenital illnesses (Figure 13-3). Many leaders of the American eugenists applauded Hitler's

Die Spinne

Manch Opfer blieb im Netze hangen / Von Schmeichentönen eingefangen
Zerreißt das Netz der Heuchelei / Ihr macht die deutsche Jugend frei

Figure 13-3

Adolf Hitler's anti-Semitism became an established tenet of the Nazi Party. The Nazis modified the eugenic sterilization laws of the American eugenics movement and applied them on a large scale throughout Germany. By extension to races which were deemed inferior, the eugenic measures led to the persecution of Jews and eventually to their mass extermination in special concentration camps set up in Eastern Europe. This propaganda poster depicts the theme of the Jew as an enemy of humankind. The German phrase states, "Many victims remain hung in the net, caught by tones of flattery; tear the web of hypocrisy, make German youth free." (From Die Stürmer, *Julius Streicher, no. 26, 1934.)*

first efforts to enforce rigorously a program of "racial hygiene," and rejected Jewish protests as paranoid or hysterical overreaction. Most Americans, however, were turned against Nazi eugenics because they interpreted its use in Germany as political rather than medical.

The white supremacy, anti-Catholicism, and anti-Semitism of the Ku Klux Klan were regarded with suspicion by educated Americans, and the resemblance of many of the Klan's beliefs to those of the eugenics movement discredited eugenics as a valid program.

The Depression was undoubtedly the most influential social factor in destroying the American eugenics movement. With one-third of the nation impoverished by unemployment or low wages, the American people were no longer willing to accept their plight as a defect of their heredity.

In 1932 the third and last International Congress of Eugenics met in New York City. While many eugenists were still looking for defective families and ways to encourage differential reproduction, some of the speakers criticized the eugenics movement. Geneticist H. J. Muller denounced the movement for its spurious elitism, racism, and sexism. Although he believed in the worth of eugenics, Muller rejected any program that equated worth with wealth and unearned privilege.

Several Fallacies Destroyed
the Early Eugenics Movements   Although studies through the 1920s supported the view that college-educated people had smaller family sizes than did unskilled laborers, the mean family size had changed by the early 1930s, and this assumption could no longer be made. Margaret Sanger and other pioneers established the birth control movement, permitting family planning in the more developed nations. This greatly reduced family size in the late 1920s, for both the rich and the poor. In some countries (Sweden and Holland) the highly educated had even more children than did the laborers.

Today there is virtually no difference in family size between college-educated and blue-collar people. Factors other than social standing determine mean family size, including family income, lifestyle, and peer pressure. The educated and gifted are not the only people capable of family planning.

Although the American eugenics movement assumed that families such as the Jukes and Kallikaks were rare and readily identifiable through analysis of prison records, IQ tests, or welfare rolls, that assumption was false. These families represented only a trivial fraction of all prisoners, most of whom were unrelated. Similarly the vast numbers of so-called paupers were not related and their poverty disappeared when work was abundant. The validity of IQ tests for detecting such families was seriously challenged by the results of the World War I army alpha tests. Some 47 percent of the white inductees were "feeble-minded" according to the standards previously applied to Jukes and Kallikaks. Few Americans were willing to believe that nearly half of their population had defective heredity or should be involuntarily sterilized.

Underlying the fallacy of the hereditary defectiveness of a small number of families was the pre-Mendelian notion that heredity was directly passed on from parent to child. However, as we now know, the majority of defective mutations in humans are not expressed in the offspring and thus, for example, the children of a "feeble-minded" parent are likely not to be retarded at all. This was proven when compulsory education became widely adopted. Furthermore severely retarded individuals more frequently arise from normal parents because heterozygotes far outnumber homozygotes for the few conditions which actually arise from single-gene determinants of mental function.

New Knowledge of Genetics Weakened
the Claims of Early Eugenists

During the 1920s population genetics developed rapidly. The "feeble-minded" actually had a diverse number of environmental, single-gene, and polygenic origins. Even if negative eugenics had been rigorously applied to retarded individuals, along with compulsory sterilization, a substantial reduction in retardation would have taken dozens or hundreds of generations because these cases represented only a small fraction of the new cases that emerged each generation from non-feeble-minded parents.

A thorough understanding of genetics led to a questioning of the basic assumptions of the American eugenics movement. Perhaps the most naive and poorly researched assumption was the belief that social traits were primarily determined at conception rather than by upbringing and family, community, and national environment.

Although eugenics, as it was conceived in the first three decades of the 20th century, was flawed with mistaken views and deserved to fail, certain issues raised by eugenists remain with us and should be faced by our own and future generations. Foremost among these issues is the course of human evolution in developed nations where the traditional checks on population have been replaced or reversed to some extent by public health practices and modern medicine. The implications of these changes in human survival will be discussed in our chapter on genetic load.

Questions to Answer

1. What was Galton's position on eugenics?
2. How did Galton define eminence? What objective standard did he use to classify it? What objection would you raise to his standard?
3. Why did Galton conclude that eminence (genius) was an inherited trait? What criticism of his evidence would you make?
4. How did the United States eugenics movement differ in its origins from Galton's movement?
5. What were the accomplishments of the Eugenics Record Office? What assumptions were made in rating the worth of ethnic groups? What arguments would you use to refute those assumptions?
6. How did Nazism utilize eugenic views?
7. Why did the eugenics movement fade into oblivion by 1940?

8. What genetic assumptions are used by advocates of compulsory sterilization? What are the strengths and weaknesses of these assumptions?

Terms to Master

negative eugenics   regression to the mean
positive eugenics   xenophobia

Chapter 14
Intelligence Testing and the IQ Controversy

Although Galton was the first to measure the incidence of eminence and to study it as a possible inherited trait, he never developed a test for individual intellectual ability. Various mental tests were developed by other psychologists in the late 19th century, including tests for perception of touch, reaction time, and memory. None of these, however, measured intelligence. Alfred Binet, a French psychologist, believed that general intelligence was a complex of many mental functions. He interviewed parents and teachers in Paris to determine the skills children acquired at different stages of their development and then he chose items and tasks that he believed were part of the experience of all children living in Paris.

A Test for Mental Age Was Converted
to an Intelligence Quotient

The test Binet designed in 1905 was used by the Parisian school system to identify those who were most likely to benefit from higher education. Binet defined the *mental age* for a given task as the age at which 75 percent of the children could perform it successfully. The total of all successfully performed tasks of various mental ages could then be determined for an individual child of any chronological age. The better that child did, compared with others of the same chronological age, the higher was that child's intelligence.

In 1916, a German psychologist, W. Stern, devised the *intelligence quotient*, or IQ, by dividing the mental age by the chronological age (later psychologists multiplied the IQ by 100 to remove the decimal and provide a score of from 0 to 200). The IQ test was quickly adopted in the United States by H. H. Goddard, for use in evaluating and classifying those who were then called feeble-minded, and by L. Terman for the

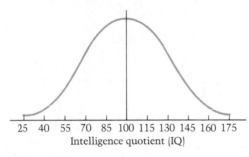

Range	Descriptive classifications
86–115	Normal, average
71–85	Slow learner, borderline, moron
56–70	Mildly subnormal, mildly
	retarded, feebleminded
41–55	Retarded
26–40	Severely retarded, imbecile
0–25	Profoundly retarded, idiot
116–130	Bright
131–145	Gifted
146–200	Genius

Figure 14-1

The IQ curve is based on half the population being above and half below a mean IQ of 100. Questions of varying difficulty are selected so that a large tested population (e.g., a rural area, an ethnic group, a poverty area) can be used to establish a normal curve distribution. Departures from the mean IQ may arise when smaller or restricted groups are tested. Whether these differences are predominantly genetic or environmental in origin is highly controversial and unresolved at this time.

The 15-point partitions on either side of the mean have been used to define several categories of mental ability. The terminology, past and present, has often implied inborn potentials or unalterable capacities for learning. IQ scores are not as readily accepted in public schools or among parents as they were in the past, and separate tests for verbal, mathematical, and logical skills are more widely used today to evaluate academic interest and aptitude.

educational ranking of children in the public schools, for genetic studies of genius, and for differentiating the intelligence of various groups (people of different sexes, races, occupations, and geographical locations).

The IQ test (and a modified form of it, the alpha test, used by the U.S. Army in World War I) was soon used for tens of millions of children and adults. In its ideal form, where the subjects are numerous (as would be the entire population of public school children in a state), the resulting distribution of IQs forms a bell-shaped normal curve (Figure 14-1). This arises because the selection of items originally used for the test was matched against this desired distribution, and refinements were made until the normal distribution was achieved.

The IQ Curve Classifies People into Named Categories

Ranging from Idiocy to Genius   Since the IQ curve is continuous, certain arbitrary intervals were defined for separating the various categories of intelligence. These can be convenience intervals of ten points, or statistical intervals of about 17 points in either direction. An IQ of 100 is used to define the mean of a large population. The interval or range for normal individuals on the convenience scale is thus 91–110, or 84–117 on the statistical scale. Sometimes a third interval scale is used, with intervals of 15 points. The normal range using this scale is 86–115.

If the 15-point scale is used, individuals who are two or more intervals below the mean are classified as subnormal. Three categories of subnormality can then be defined. Individuals in the 71–85 range are considered slow learners, the range from 56–70 consists of mildly sub-normal individuals; and those from 0–55 are severely subnormal individuals.

The Terms for IQ Categories
Have Changed Several Times
in Response to Public Criticism   The choice of terms for IQ categories has varied with the social attitudes in school and society. Before 1940 individuals in the 71–85 group were called morons, those in the 56–70 IQ group were called feeble-minded, and those in the 0–55 group were known as idiots and imbeciles (an arbitrary IQ score of about 25 defined the upper limit of idiocy). Today the mentally defective are referred to as educationally subnormal. Those formerly labeled as idiots and imbeciles have been renamed for educational purposes as severely and profoundly retarded (or just retarded when referring to either or both groups). Those in the 56–70 IQ group are called mildly retarded, slow learners, or exceptional children.

On the other side of the mean, little change has taken place in the terminology. Those with IQs of 116–130 are called bright, and those in the 131–145 group, gifted. There has been no agreement on the starting point of genius; it has been variously started at 145, 160, and 175, depending on the number of intervals above normal that it is arbitrarily assumed to be. The categories of bright and gifted individuals would also vary with a shift of the starting point for genius.

Advocates of IQ Testing Favor
a Genetic Basis for Intelligence   The IQ was believed by some of its early proponents to represent general intelligence, a difficult-to-define collection of skills including rapidity of solving tasks, effectiveness of recall, mathematical and logical skills, verbal fluency, vocabulary range, and three-dimensional perception. No one of these skills by itself measured intelligence, but general intelligence was assumed to underlie all of these skills. A person's IQ score was also believed to represent an innate ability that was relatively stable, rarely increasing or decreasing by more than ten points. The IQ was thought to measure not only academic potential, but the potential for success or achievement in any occupation. General intelligence was believed to be as essential for ordinary trades or occupations as it was for the professions.

As knowledge of Mendelian genetics was applied to quantitative traits in plants and animals, such as size or color inheritance, many psychologists drew an analogy between the normal curve for IQ and the normal curve for quantitative traits. They believed that general intelligence was a consequence of a large number of genes, some augmenting and others diminishing the mental capacities that determine general intelligence. Such a model accounted for Galton's well-known regression to the mean which stated that quantitative traits such as height tended to the mean.

According to this law of regression, more intelligent parents would have somewhat less intelligent children (most of whom would still be brighter than average). Similarly, less intelligent parents would have children who were brighter than the parents (but usually not quite at the normal level). In both instances a rare assortment of these genes could result in a gifted person having a retarded child or a retarded parent having a gifted child. Since such cases were encountered often enough by psychologists, the *polygenic model of intelligence* became the favored model of human intelligence and IQ scores. As we shall see, this is not the only model that can account for these results.

Uses of IQ Tests Have Not Always
Been Beneficial for Children

For more than fifty years IQ tests have been used to separate slow learners from normal and brighter students in elementary and secondary schools. There is nothing inherently wrong with this practice, as long as the slower children are then given the extra help they need. Unfortunately, however, many school districts have historically given little attention to the slow learners. Instead of providing special small classes and skilled teachers trained in learning disabilities, many school districts have often shunted the slower students into unstimulating programs and neglected their problems. Similarly, gifted students have rarely been given enriched programs to stimulate their enthusiasm for learning. Only in large cities and in some affluent districts were there special classes or schools for these gifted children.

In a democracy, it is difficult for school districts to treat brighter or slower children any differently from other students. Separating the bright students is often considered undemocratic because it may foster a snobbish elitism; separating the poor learners is often considered stigmatizing. Thus there is a tension between those who wish to set aside students according to their abilities and those who do not feel that this would be beneficial to some or all of the students.

Another unfortunate result of IQ testing in the public schools is labeling. A child identified as having an IQ of 140 is more apt to be encouraged by the teachers than is an average or slower child. The child with an IQ of 85 is likely to be considered dull and unresponsive, and treated as such. If a teacher, through prior reading of IQ scores, expects a child to do poorly, this expectation may very well slow the student's progress in the class. The teacher may unconsciously judge the student by this label rather than by the student's achievement.

The IQ Controversy Involves
Both Research Studies and
Social Policies Based on IQ Testing

Both as a research tool and as a practical test used to classify school children, the IQ test has been

Table 14-1 Basic Assumptions in the Nature-Nurture Controversy

The hereditarian believes that most (about 80 percent or more) of what determines intelligence is genetic. The assumed model is polygenic, with dozens or hundreds of genes for cognition being assorted and recombined each generation. This is the assumed basis for Galton's regression to the mean and for Johannsen's variation of bean size by selection. The argument is based on analogies and correlations using twins or family relation studies. Hereditarians assume a significant level of importance for genetic factors in other traits, including personality, longevity, and racial differences in accomplishments.

The environmentalist assumes that the family life, schooling, neighborhood, lifestyle, income, and cultural values to which a child is exposed will be the predominant influences on intelligence, talent, personality, and other socially-significant traits. Like the hereditarian, the environmentalist uses correlations, comparative social studies, and historical analogies. There is no substantial experimental work to conclusively support either position.

Hereditarian	*Environmentalist*
Intelligence is polygenic.	Intelligence is learned.
The slow learner has fewer genes for cognition.	The slow learner comes from a deprived environment.
Longevity requires long-lived ancestors.	Longevity is largely a matter of lifestyle.
Personality is inherited.	Personality is learned from family interactions.
Genius is inborn.	Genius combines opportunity, stimulation, and correct timing.
Races vary in their mean number of genes for cognition.	Races vary in affluence, effects of prejudice, and cultural traditions.

defended and denounced since it was introduced (Table 14-1). Those who denounce it do so for several reasons. Some critics are environmentalists who believe that all individuals who are not brain-damaged acquire an IQ from their family, neighborhood, and culture rather than from some innate collection of genes. Some critics, who are not particularly concerned with the scientific mechanisms or definitions of intelligence, look upon IQ testing as a political instrument that prevents the poor and certain minority groups from having adequate schools and educational services, and therefore they oppose such testing. Other critics accept the IQ test, provided it is not used for making judgments about racial or class differences. Others feel that, because the cultures of the different races and classes vary enormously, the IQ test cannot be culture-free, and so is bound to be unfair to some students.

Supporters of IQ tests claim that adequate controls for environmental versus inherited differences exist in twin studies, adoption studies, and class or status comparisons. They claim that while environments can shift IQs to their upper or lower limits, the range for any given person is not unlimited, but fixed genetically. Supporters of IQ tests believe that good teachers can and will devise individualized teaching methods for the various IQ levels in a classroom if provided with the IQ test results.

Table 14-2   Comparison of World War I Army Alpha Test Scores

The army alpha intelligence test was administered to all inducted men in World War I. The results were used by both hereditarians and environmentalists to criticize each other's position. Whites, in general, fared better than blacks. Black draftees from Northern states, however, had higher scores than white draftees from Southern states. In this latter case, an environmentalist interpretation, not a genetic one, best supported the interpretation of these results. The better-financed Northern states provided better schools and educational opportunities than did the poorly-financed Southern states. Such contradictions to the racial myth of the genetic superiority of whites were ignored, and the advocates of the American eugenics movement continued to promote its program of anti-miscegenation, restrictive immigration, and compulsory sterilization of the "defective." The scores on the army alpha test were not IQ scores; they were the raw scores from a standardized test uncorrected for age differences.

Whites	Median Score	Blacks	Median Score
Mississippi	41.25	Pennsylvania	42.00
Kentucky	41.50	New York	45.02
Arkansas	41.55	Illinois	47.35
Georgia	42.12	Ohio	49.50

Source: Adapted from A. Chase, *The Legacy of Malthus* (New York: Knopf, 1976), p. 228.

They believe that the teachers would be unable to do so without IQ test results, and so all students would be taught at the same pace, given the same homework, and judged by the same standards.

Throughout its 60-year history, the IQ test has remained controversial, and has been used to support both liberal and conservative causes. The army alpha test results on World War I recruits revealed that the mean IQ of Northern blacks was higher than the mean IQ of Southern whites (Table 14-2), a view often cited by liberals. This reflected the differences in state aid to public schooling and cultural attitudes, rather than the innate superiority or inferiority of either of these groups. Other studies which have compared racial differences have consistently shown blacks to score lower than whites when their cultures are superficially similar (Northern whites versus Northern blacks; Southern whites versus Southern blacks). These data are often used by conservatives who object to funding public programs to help those who are disadvantaged.

Accusations of Racism Have Made the IQ a Questionable Concept in Education   The IQ test has received its most severe criticism since the publication in 1968 by Arthur Jensen in the *Harvard Educational Review* of a sizable technical article on the intellectual differences between whites and blacks. Jensen, an educational psychologist, claimed that blacks did as well as whites in verbal skills but not in quantitative or cognitive tasks. He proposed that a different strategy be used in the early school years to promote those skills which black students handle well and to devise better methods for

teaching blacks some of the cognitive skills. Jensen's main conclusion, that a 15-point difference exists between the mean IQs of blacks and whites and that this difference is about 80 percent genetic, created a furor that has not subsided. He also criticized federal programs of compensatory education, calling them ineffective because the preschool teaching (often based on noncognitive approaches) did not increase the IQs or the performance of the children when they later entered the public school system.

That there are mean IQ differences between races and social classes is not the issue. What is being debated is the inferred cause of these differences. Are the children of unskilled workers likely to have careers like their parents or can they become lawyers, engineers, physicians, business executives, or college professors? In practice, social mobility is limited and many of those who are raised in poverty cannot hope to fill the ranks of the college-educated professions. The question is whether this is environmentally or genetically determined.

Environmentalists and Hereditarians Differ in Their Fields of Specialization and Their Interpretations of Intelligence
Curiously most advocates of the hereditarian interpretation of intelligence (with an alleged 80–90 percent heritability of IQ) are not geneticists (Table 14-3). They are largely psychologists, educational specialists, and persons in other fields (e.g., physics) who have followed the literature carefully. Ironically some of the strongest critics of the hereditarian viewpoint are geneticists.

From a genetic viewpoint the case for a major hereditary component of IQ is based on analogy to physical quantitative traits such as skin color and height. In such traits the genetics can be demonstrated directly (as in the skin color crosses of whites with blacks) or indirectly (most instances of average or less-than-average height in middle-class families do not reflect malnutrition, and taller-than-average individuals can often occur in families whose variety and quantity of food are considerably less than in other families. Unlike experimental tests on fruit flies and cereal grains, direct tests on humans for hereditary traits are limited by ethics and luck. The indirect approaches used in human mental tests are flawed by the clearly different environments of compared groups. For example, females originally scored lower on IQ tests than males, but adjustments have been made to compensate for the fact that girls score higher on the verbal parts of the test, and boys generally score higher on the mathematical parts. These adjustments are made by decreasing the number of questions graded in the pertinent parts when scoring a boy's or girl's test. The difference may be environmental, reflecting parental and social attitudes towards sex role identification. For instance, girls are not usually given mechanical, chemical, and scientific toys; boys are not usually given dolls, carriages, and kitchen toys. Or it may be that innate sexual or developmental differences exist for verbal or mathematical reasoning.

Table 14-3   Major Contributors to the Nature-Nurture Controversy

Hereditarians and Environmentalists   Hereditarians believe that inherited factors (nature) play a more important role in determining human traits than do environmental factors (nurture). Galton was a pioneer in the measurement of human abilities and he founded the eugenics movement (1902) to promote the improvement of health, intelligence, and longevity. Bateson, while never active in the eugenics movement and the best experimental geneticist among the hereditarians, was convinced of the genetic inferiority and superiority of certain nationalities and races and expressed his views in his essays (1924). Davenport was a zoologist who took an interest in applied heredity, especially human genetics. His work was sometimes inaccurate and often carelessly designed. He attracted many bigots as coworkers, and they were successful in lobbying for restrictive immigration laws and compulsory sterilization laws (1910–1925). Goddard was a psychologist who studied mentally retarded persons institutionalized in New Jersey. He published an influential, popular book, *The Kallikaks* (1919), in which he claimed that most "feeble-minded persons were products of defective inheritance." Terman published many volumes on the genetic basis of genius (1925–1957). He was a leading advocate of mass-administered IQ tests in the public schools. Jensen revived the argument in 1968 by claiming that blacks have an IQ 15 points lower than whites because they receive fewer genes for cognition than do whites. Herrnstein in 1971 argued that a genetic meritocracy exists for occupation, with heredity playing the major role in determining one's likely profession and one's success in it.

Environmentalists believe that family and cultural factors determine intelligence, occupation, and success in life, except in a few cases where a pathological condition exists. Boas, the anthropologist, debunked those who sought physical evidence of the inferiority of races and nationalities. J. B. Watson was a psychologist who introduced behaviorism. It was Watson who believed that any child, depending on how the child was raised, could become a thief, banker, engineer, or saint. Kamin has exposed the fraud, self-deception, error, and hidden values of some of the major advocates of the hereditarian view. Like Lewontin, who is a population geneticist, Kamin is convinced that work in this field is subjective and reflects the political prejudices of the investigator or of society.

Three geneticists, Muller, Haldane, and Dobzhansky, were among the most highly regarded experimental and theoretical geneticists of this century. They believed that both genetic and environmental components are involved in individual differences. Muller criticized the eugenics movement for its belief that there were genetically determined beneficial and detrimental social traits, but he believed that there were genetic differences in intelligence, talent, health, personality, and longevity. Haldane lost favor with the Communist Party in Great Britain because he too advocated the dual contribution of heredity and environment for these traits. Dobzhansky was opposed to eugenics, but accepted the basic premise of both genetic and environmental influences leading to variations in individuals.

Primarily Hereditarian	*Primarily Environmentalist*	*Not in Either Position*
Francis Galton (P)	Franz Boas (A)	H. J. Muller (G)
William Bateson (G)	John B. Watson (P)	J. B. S. Haldane (G)
C. B. Davenport (G)	Leon Kamin (P)*	T. Dobzhansky (G)
H. H. Goddard (P)	Richard Lewontin (G)*	

Primarily Hereditarian	Primarily Environmentalist	Not in Either Position
Lewis Terman (P)		
Arthur Jensen (P)*		
R. J. Herrnstein (P)*		

Note: A = anthropologist, G = geneticist, and P = psychologist.

*Indicates a scientist active in the controversy in the 1970s; all the others have died.

Table 14-4 Characteristics of Families of Lower
and Higher Socioeconomic Status

Even before a child attends kindergarten, there are profound differences between the lifestyle of the disadvantaged child and that of the child who lives in affluence. The comparison of low economic status and professional circumstances shown here indicates a few of the ways economic circumstances can affect a child even before school begins. Undoubtedly, there are many low-economic-status families who represent exceptions to these generalizations (e.g., poor families that visit museums, play music, or read books from the public libraries), just as there are affluent families who thwart their children's education (e.g., professionals too busy to pay attention to their children, too tired to read at home, or too materialistic to appreciate art, music, or literature). A greater proportion of professional students come from professional homes than from low-economic-status households. Similarly, a greater proportion of low-economic-status jobholders come from low-economic-status families than from professional families. American society does not prevent upward mobility for the talented but the mobility is not random in either direction.

Low-Economic-Status Family	Professional Family
Low income restricts most purchases to necessities.	High income provides many resources: hi-fi, books, music lessons, paintings.
Parents rarely attended college.	Parents college-educated.
Books rarely read at home.	Parents usually read a lot; give books to children.
Toys usually traditional: dolls, guns, sports equipment.	Toys varied, with lots of educational games and intellectually stimulating, supervised activities.
Parents cannot afford to travel.	Children taken on weekend trips or vacations; travel to Europe and to other states is common.
Children stay mostly in neighborhood.	Children often taken to museums, plays, musical concerts.
Neighborhood schools and community may lack stimulation or adequate funding.	Local taxes or political power in affluent neighborhoods provides better teachers and services.
Relatives, friends, and neighbors mostly nonprofessional; academic goals may not be emphasized.	Relatives, friends, and neighbors are mostly professional or managerial; academic goals are strongly emphasized.

Blacks, Hispanics, and other minority cultures often have different exposures to middle-class attitudes and skills than do whites. They have fewer professional models at home and in their neighborhoods than do members of the white middle-class, which is often taken as the standard of average capacity.

The environmental factors that can lead to lower IQ also include nutrition. Children of lower birth weight may have lower IQs. If minority-group mothers do not understand what comprises a balanced diet, do not have funds to buy sufficiently varied foods, or do not have access to markets for such foods, then they risk vitamin, calorie, or protein malnutrition when pregnant. While such mothers will probably not starve, their children may be deprived of fully-developed nervous systems. In identical twin births, however, the mean IQ difference between two twins differing in birth weight is about 5–10 points, which would seem to indicate that other factors besides nutrition contribute to IQ differences. It is also unlikely that an adequate food program in poverty areas would bring the IQ of minority children (who differ by some 15 points) to a normal level with middle-class whites. Clearly nutrition alone cannot raise IQ levels appreciably.

IQ is strongly correlated with other environmental factors such as the number and kind of books at home, family income, and parental education (Table 14-4). Low IQ and high IQ are both clearly related to environmental factors. Hereditarians attribute low IQ to inadequate genetic constitutions, and environmentalists attribute it to the noxious effects of a poor environment. Most geneticists are not satisfied with either position because the observations lack convincing controls to demonstrate either model. In the absence of compelling evidence to support a strong genetic basis for intelligence, it seems wiser for schools and society to assume an environmental basis. Innate genetic models too easily make those groups and individuals with low IQs subject to discrimination, segregation, and neglect.

Questions to Answer

1. What were Binet's objectives in introducing the concept of mental age?
2. (a) What is an intelligence quotient?
   (b) What assumptions are involved?
3. Would IQ be of the same social significance if it were called an academic quotient rather than an intelligence quotient? Discuss your position.
4. How would you reconcile individual differences with democratic rights applicable to all citizens?

5. What are some of the findings of the U.S. Army alpha test from World War I and their relation to racial differences?
6. Why would most geneticists reject both the hereditarian and environmentalist positions on the IQ controversy?

Terms to Master

environmentalist

hereditarian

intelligence quotient

mental age

polygenic model of intelligence

Chapter 15
Genius and Retardation

Persons of unusual talent whose work provokes awe or admiration are sometimes called geniuses. The term itself is difficult to define. Often it is applied to people who evoke fear or hatred, as in cases of infamous dictators, frauds, and financial manipulators who are popularly called evil

(a)   (b)

Figure 15-1

John Stuart Mill and Johann Wolfgang von Goethe were both precocious children. Mill learned Greek at age three and Latin at age eight. His father taught him geometry and algebra at age eight; he taught himself calculus when he was eleven. In his adult life Mill was universally recognized for his contributions to political philosophy. Goethe began to write plays when he was eight years old. He was recognized as a leading poet and writer in Germany while he was still in his early 20s. [(a) National Portrait Gallery, London. (b) Courtesy of the Goethe Institute.]

geniuses. There has been general agreement in considering J. W. von Goethe, W. A. Mozart, and John Stuart Mill as geniuses because they were all precocious and they all made lasting contributions to their fields (Figure 15-1). Charles Darwin, Thomas Alva Edison, and Albert Einstein would not have been considered geniuses by their school teachers (nor by their parents) because their talents emerged in their early adult years (Figure 15-2). Yet we now recognize these men as geniuses. Precocity is not essential for the emergence of extraordinary talent in most endeavors.

Persons of extraordinary accomplishment frequently come from parents and upbringings that are not unusual in accomplishment, affluence, or educational opportunities. Genius can emerge in a member of any race and any level of social standing. It can even occur in an adopted child whose modestly educated parents are puzzled by the musical gifts or the insatiable reading habits of the child they have reared. The frequency of genius depends upon the standard used for its measurement. Based on IQ test scores, an IQ of 145 or higher can be used to define genius if it is defined by three intervals above the mean, but many other studies have used 135 or 140 as the lower end of the genius level. However, IQ scores may be unrelated to genius in artistic endeavors, especially music and art.

As we learned in Chapter 13, Francis Galton, himself a precocious child, defined eminence by using dictionaries of biography and found its frequency to be 1 in 4000. Others have used listings in *Who's Who* (1 in 1500) or specialized categories (Nobel laureates, Pulitzer Prize winners,

      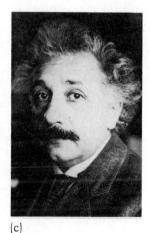

(a)   (b)   (c)

Figure 15-2

In contrast to precocious geniuses, Darwin, Edison, and Einstein were not distinguished as children. Darwin dropped out of both medical school and theology school. Edison, considered disruptive and a poor learner, was withdrawn from school and taught by his own mother. Einstein was considered a misfit, showing little interest in his elementary and high school coursework, and satisfying only the minimal expectations of his teachers. [(a) Brown Brothers. (b) Department of the Interior, National Park Service, Edison National Historic Site. (c) National Archives.]

membership in scholarly organizations such as the Royal Society or the National Academy of Sciences) where selection procedures involve peer judgment and evidence of major contribution to a field. Clearly persons with unusual talents can be recognized, but no agreement on the minimum standard for excellence exists. The highly intelligent show a range of abilities in their differences.

While Galton's attempt to demonstrate the inheritance of eminence, over a century ago, is flawed by the absence of data for the orphaned children raised by eminent church leaders and by the vague standards for eminence that he used, his premise that genius is biologically inherited has been favored by some and denounced by others in the 20th century.

Genius Is Often Attributed to Hereditary Factors
In support of the hereditarian view are the difficult-to-explain instances of genius and talent flourishing among modestly educated, poor, and untalented families, where the gifted children received no special tutoring, stimulation, or enriched upbringing to account for the emergence of unusual achievement. If it is an environmental heritage that accounts for the special gifts of the Darwins and Einsteins, then why do their brothers and sisters and their own children only rarely achieve an eminence or talent comparable in stature? Whatever environmental factors are involved are elusive and difficult to characterize, especially because they can be independent of the wealth, status, or educational level of the gifted child's family (Table 15-1).

Table 15-1 Occupations of Fathers of Eminent Geniuses (1500–1850)

The well-known contributors to western civilization listed here were mostly children of middle-class or professional households. A few had more modest backgrounds. Prior to the mid-19th century, it was difficult for a genius to rise from an uneducated, impoverished, or lower-class household. Universal education, introduced in the past century, has permitted many more eminent geniuses to come from modest circumstances, although the majority still come from middle-class households.

Those of you who do not know what major field or contribution is associated with each of these individuals can turn to Table 15-5 for the answers. The individuals listed here span some 350 years, from 1500–1850. The father's occupation is emphasized because married women were rarely permitted to have independent incomes, professional education, or activities not approved by their husbands (or by their fathers before they were married).

Genius	Father's Occupation	Genius	Father's Occupation
1. John Stuart Mill	Writer, historian, corporation executive	4. Robert E. Lee	Governor of Virginia
2. Robert Fulton	Farmer, minor public official	5. Nathaniel Hawthorne	Sea captain
3. Hans Christian Andersen	Shoemaker	6. Hans Christian Oersted	Apothecary

Genius	Father's Occupation	Genius	Father's Occupation
7. George Eliot (Mary Ann Evans)	Carpenter, surveyor	29. Lord Byron (George Gordon)	Nobility, no profession
8. Charlotte Brontë	Weaver, clergyman	30. Daniel Webster	Farmer, officer, legislator
9. John Henry (Cardinal) Newman	Banker	31. John Milton	Composer
10. Gottfried Wilhelm Leibnitz	Jurist, professor	32. Michelangelo Buonarotti	Nobility, no profession
11. Johann Wolfgang von Goethe	Jurist, professor	33. Thomas Jefferson	Surveyor, politician
12. Samuel Taylor Coleridge	Clergyman, schoolmaster	34. George Friedrich Handel	Barber, surgeon
13. Voltaire (François Arouet)	Notary	35. Galileo Galilei	Businessman, musicologist
14. John Quincy Adams	President of the United States	36. Benjamin Franklin	Tallow chandler, soap boiler
15. Alexander Pope	Linen merchant	37. Immanuel Kant	Strapmaker
16. Alfred Tennyson	Rector	38. Ralph Waldo Emerson	Minister
17. Georges Cuvier	Soldier and officer	39. Benjamin Disraeli	Writer
18. William Wordsworth	Attorney	40. Charles Dickens	Clerk, reporter
19. Walter Scott	Lawyer	41. Francis Bacon	Statesman
20. George Sand (Aurore Dupin)	Soldier	42. Friedrich Schiller	Army surgeon
21. Wolfgang Amadeus Mozart	Orchestra director	43. Joseph Priestley	Weaver
22. Felix Mendelssohn	Merchant	44. Michel Eyquem de Montaigne	Wineseller, poet
23. Henry Wadsworth Longfellow	Lawyer	45. Johann Kepler	Soldier, innkeeper
24. Victor Hugo	Lieutenant general	46. Robert Boyle	Lord High Treasurer of Ireland
25. Denis Diderot	Cutlery craftsman	47. Wilhelm Richard Wagner	Court clerk
26. René Descartes	Parliamentary councilor	48. Leonardo da Vinci	Notary, businessman
27. Humphrey Davy	Wood carver	49. Adam Smith	Jurist
28. Auguste Comte	Tax receiver	50. William Penn	Admiral
		51. Napoleon Bonaparte	Lawyer
		52. Alexander Hamilton	Businessman
		53. Charles Darwin	Physician

Table 15-1 (continued)

Genius	Father's Occupation	Genius	Father's Occupation
54. John Calvin	Attorney	65. Carolus Linnaeus	Clergyman
55. Ludwig van Beethoven	Musician	66. Abraham Lincoln	Carpenter
56. Richard Sheridan	Actor, theater manager	67. Joseph Louis Guy-Lussac	Jurist
57. Jean Jacques Rousseau	Watchmaker	68. Albrecht Dürer	Goldsmith
58. Isaac Newton	Farmer	69. Simón Bolívar	Military officer
59. Molière (Jean Baptiste Poquelin)	Tapestry maker	70. Franz Joseph Haydn	Wheelwright
60. Heinrich Heine	Businessman	71. Johann Sebastian Bach	Musician
61. Robert Burns	Peasant	72. William Harvey	Mayor
62. George Washington	Wealthy landowner	73. Martin Luther	Peasant, mineowner
63. Diego Rodríguez de Silva y Velázquez	Lawyer	74. Michael Faraday	Blacksmith
64. John Locke	Attorney	75. Gregor Mendel	Peasant

Source: C. Cox, *The Early Mental Traits of 300 Geniuses* (Palo Alto: Stanford University Press, 1926).

A genetic basis, however, predicts both direct transmission and sporadic origin of a trait, the frequency of the trait varying according to the number of genes involved. For quantitative traits, dozens or hundreds of genes may be involved which enable the cognitive and creative activities of the brain to function more efficiently. This demands an immense diversity of genotypes and a low probability of maintaining or transmitting such combinations of genes to the next generation. The genetic model also favors the appearance of talent among the progeny of the talented, but at frequencies comparable to those for other multifactorial traits, such as neural tube defects, cleft lip, and club foot. That is not much more than a five percent probability and thus most of the extraordinary talent and genius in humanity would, according to the genetic model, emerge sporadically from parents who lack these gifts.

Terman Studied Gifted Children for Several Decades and Identified Their Characteristics

During the 1920s Lewis Terman conducted a study of gifted children in the state of California whose IQs ranged from 140 to 200, with a mean of 146. Very high IQs were rare (out of 621 with IQs of 140 or above, 1 was 200, 2 were in the 190s, 12 in the 180s, and 28 in the 170s). These children shared many characteristics, especially middle-class home environments, well-adjusted personalities, and a love of school. They were followed for more than 30 years, and many of them became prominent, especially in science, law, medicine, engineering, government, and teaching. Fewer of them were in fine arts, music, dancing, and acting. They published thousands of articles and wrote hundreds of books. A substantial number received prizes and honors in recognition of their work. Terman's high-IQ group followed largely academic careers and filled the ranks of professional schools. Terman's "genetic studies of genius," as he called his projects, showed that high IQ is an excellent predictor of academic achievement and success in the traditional careers that are highly valued by the middle class.

There May Not Be a Strong Correlation Between Creativity and Intelligence

Talented individuals do not necessarily score high on IQ tests. Many do, but there are factors other than IQ which determine the creativity, drive, energy, and ambition which enable some individuals to achieve personal fame.

Victor and Muriel Goertzel studied eminent figures of the 20th century, using an approach analogous to Galton's. Instead of using biographical dictionaries, however, which reflect the bias of the editors, they chose a card catalogue in a small public library in Claremont, New Jersey, and selected persons who had two or more biographies written about them (Table 15-2). They eliminated sports figures on the assumption that their accomplishments were related more to motor skill than cognitive superiority.

Table 15-2   Occupations of Fathers of Eminent 20th Century People

Eminent individuals of the 20th century, like those in the previous four centuries, come from a variety of circumstances, mostly middle-class households. While some prominent individuals carry on a family tradition in music, art, or politics, others have talents and interests dramatically different from those of their fathers. As shown in Cox's study, this collection of eminent figures has a paucity of employed mothers. Women's liberation is a very contemporary movement. If the mothers' occupations prove to have as little influence on individuals' careers as do the fathers' occupations, the eminent individuals of the next generation will probably be no different in this respect from those in either of these studies. Of these 75 individuals, only 13 had working mothers: Anderson (washerwoman), Baden-Powell (artist), Bartók (teacher), Callas (actress), Copland (partner in a store), Curie (head mistress of a school), Edison (teacher), Lawrence (teacher), Shaw (music teacher), and Wright (teacher). For the accomplishment or field associated with each of these famous figures see Table 15-6.

Person	Father's Occupation	Person	Father's Occupation
1. Jane Addams	Businessman	3. Susan B. Anthony	Innkeeper
2. Marian Anderson	Iceman	4. Louis Armstrong	Laborer

Table 15-2 (continued)

Person	Father's Occupation	Person	Father's Occupation
5. R. S. Baden-Powell	Mathematics professor	30. Enrico Fermi	Railroad executive
6. Béla Bartók	Director of agricultural school	31. Henry Ford	Farmer
		32. Sigmund Freud	Textile mill owner
7. Clara Barton	Soldier, farmer	33. Robert Frost	Editor of San Francisco *Bulletin*
8. Alexander Graham Bell	Speech teacher		
9. St. Frances Xavier Cabrini	Farmer	34. Mohandas Gandhi	Prime Minister of Indian state
		35. Paul Gauguin	Journalist
10. Maria Callas	Chemist	36. George Gershwin	Small businessman
11. Enrico Caruso	Mechanic		
12. George Washington Carver	Slave	37. Thomas Hardy	Stonemason
		38. Ernest Hemingway	Physician
13. Pablo Casals	Organist	39. Adolf Hitler	Customs official
14. Paul Cézanne	Banker	40. Aldous Huxley	Editor, historian
15. Marc Chagall	Laborer in herring factory	41. Henrik Ibsen	Merchant
		42. Pope John XXIII (Angelo Roncalli)	Farmer
16. Charles Chaplin	Ballad singer		
17. Anton Chekov	Shopkeeper	43. James Joyce	Physician, actor, singer
18. Winston Churchill	Politician		
19. Mark Twain (Samuel Clemens)	Justice of the peace	44. Franz Kafka	Dry goods wholesale store owner
20. Aaron Copland	Department store owner	45. John Kennedy	Politician
		46. Martin Luther King	Minister
21. Marie Curie	High school teacher	47. D. H. Lawrence	Coal miner
22. Salvador Dali	Lawyer	48. Nikolai Lenin	School superintendent
23. Clarence Darrow	Minister, carpenter		
24. Claude Debussy	China shop owner	49. Charles Lindbergh	Congressman of the United States
25. Isadora Duncan	Businessman	50. Gustav Mahler	Vegetable store owner
26. Thomas Edison	Innkeeper	51. Thomas Mann	Grain merchant
		52. Mao Tse-tung	Rice merchant
27. Albert Einstein	Electrical engineer	53. Claude Monet	Grocer
28. Dwight Eisenhower	Mechanic, storekeeper	54. Friedrich Nietzsche	Minister
		55. Eugene O'Neill	Actor
29. William Faulkner	Livery stable owner; Treasurer, University of Mississippi	56. Pablo Picasso	Art teacher
		57. Giacomo Puccini	Musician

Person	Father's Occupation	Person	Father's Occupation
58. Franklin D. Roosevelt	Financier	68. Henri Toulouse-Lautrec	Independently wealthy
59. Bertrand Russell	Prime Minister of Great Britain	69. Harry S. Truman	Farmer
60. Margaret Sanger	Stonecutter	70. Booker T. Washington	Unknown; mother a domestic
61. Albert Schweitzer	Minister	71. Oscar Wilde	Physician
62. George Bernard Shaw	Manufacturer	72. Thomas Wolfe	Stonecutter
63. Jean Sibelius	Physician	73. Frank Lloyd Wright	Teacher
64. Joseph Stalin	Shoemaker		
65. Gertrude Stein	Vice president of streetcar company	74. William Butler Yeats	Artist
66. Dylan Thomas	Teacher	75. Emile Zola	Engineer
67. Leo Tolstoi	Estate manager		

Source: V. and M. Goertzel, *Cradles of Eminence* (London: Constable, 1965).

Some 400 figures were then traced, and school records, where available, were used. About 85 percent were in fields comparable to Terman's gifted children (Table 15-3). The other 15 percent were mostly in theater, films, music, or fine arts, or were known as popular personalities.

The mean IQ of those studied by the Goertzels was 127, considerably below the Terman standard for genius. Like Terman's children they grew up in middle-class homes, but unlike the children Terman studied a very large percentage (75 percent) of them came from troubled homes with such stresses as an alcoholic parent, divorce, an eccentric or neurotic parent, business failures, or a domineering, troublesome parent. (Of course, the vast majority of children with stressful home lives do not become eminent; indeed their outlook is often bleak for both professional and personal satisfaction in their adult lives.) Another major difference between the Terman and Goertzel studies was the attitude of the Goertzel eminent group towards school; about 60 percent of them disliked their primary and secondary school education.

Eminence Involves Several Factors and May Not Be Related to IQ Test Results

The Goertzels concluded that what is popularly called genius or extraordinary talent cannot be measured directly or predicted by IQ tests. High IQ test scores do predict academic success, but they do not predict eminence, creativity, or personality factors such as charisma, ambition, and energetic commitment to a career.

The biological and environmental factors which lead to eminence are not known. Eminence itself is a difficult characteristic to define, and it

Table 15-3   Comparisons of High-IQ and Highly Creative Individuals

The 1000 school children followed for some 30 years by Lewis Terman at Stanford University were selected for their high IQs and their exceptionally high classroom performance. Such academically motivated students continued to do well in high school and college, and successfully entered middle-class professions.

The Goertzels' study was based on a retrospective analysis, like Cox's, but was limited to 20th century eminent persons who had two or more biographies written about them. Such persons were more apt to be creative than to have high IQs. Eminence is related to neither precocity nor those academic motivations that lead one on to the conventional professions which depend upon high scholastic achievement. Poets, musicians, artists, politicians, and writers need not have been outstanding students while in school.

The major difference in the eminent group studied by the Goertzels was the home life, which involved controversy, financial reversals, and tension between the parents, with one parent being alcoholic, eccentric, or emotionally demanding. The individuals in both groups had parents who liked learning, and who had the money (at least during part of the child's formative years) to provide diverse stimulation.

Terman's High-IQ Study	*Goertzels' Eminence Study*
Stable home environment.	Troubled homes in 75 percent of the families.
Conventional middle-class values predominated at home.	Half the homes had a parent with opinionative, controversial views.
More than 80 percent had wealthy or middle-class economic status.	More than 80 percent had wealthy or middle-class economic status.
99 percent liked their schools and teachers.	60 percent disliked their schools or school teachers.
100 percent showed exceptional talent in school.	80 percent showed exceptional talent in school.
Mean IQ was 150.	Mean IQ was 127.
Doctor, lawyer, and engineer were preferred occupations.	Writer, artist, adventurer, and inventor were preferred occupations.

is puzzling why fame is so short-lived for all but the most outstanding figures of our civilizations. A popular actor of the 1920s may be all but totally unknown to the generation of the 1980s. Few of us could name more than a dozen of the 500 or more Nobel laureates. Only fellow professionals would be familiar with some of the names of the members of the National Academy of Sciences. Sometimes individuals become more famous after their death than they were during their own active careers, as was certainly true for Gregor Mendel and J.S.C. Bach.

The ephemeral qualities of genius and talent should not obscure the fact that some individuals do have extraordinary abilities which most of us will never have. Because a person's fame was short-lived in our memories does not mean the talent was any less outstanding. Unfortunately, compelling studies of either environmental or genetic determinants for these abilities have not yet been done, and few school boards in a democracy want to risk the public's tax money for programs that set apart and encourage the gifted children at the expense of the other children's education.

**It Is Difficult to Attach
a Label to the Mentally Retarded
Without Alienating a Person or Family**   If a child learns to speak later than others of the same age or fails to develop skills that are normally acquired at home or in school, the child is described by many terms, some merely descriptive and some laden with negative values. To avoid harmful associations with a label, the term *exceptional* is now used to distinguish these children in the public school systems. The term *mentally retarded,* however, is still used professionally by educators and psychologists when specific verbal and problem-solving (cognitive) skills are insufficiently developed. Mentally retarded has replaced the older label, *feeble-minded,* which was in use between 1900 and 1935. Certain professional terms such as *idiot, imbecile,* and *moron* have been replaced by a quantitative gradient of a single term: *severely retarded, retarded,* and *mildly retarded.* The terms *slow learner* and *borderline intelligence* have been retained for those whose performance is slightly below the average of their peers.

Clearly some terms are pejorative, wounding the parents and often the individuals, depending on the hopelessness that such terms convey. It is also abusive to attribute fixed biological determinants to all departures from the mean. The gifted on one side of the curve may then be left to fend for themselves on the assumption that their innate talent needs no special attention. Those with learning problems, on the other side of the curve, may be abandoned as unteachable because of an assumed innate defect that prevents them from learning. Such social abuse can and should be avoided. Nevertheless, it is clear that there is a biological basis for some learning disorders and that some children are limited in the skills they can acquire even with a lot of parental and professional attention.

**Only a Small Percentage
of All Individuals Are Retarded**   In racially homogenous, developed countries (e.g., Sweden, England, Australia) the number of school-age children with IQs under 50 is about 3.8 per 1000. The incidence of IQs below 50 at birth is about 5.4 per 1000. The fact that mental retardation is often accompanied by many birth defects affecting other organ systems results in the premature death of many retarded children before they attend school. If children with IQs between 50 and 70 are included, about 1–4 percent of school children are retarded, depending on the country studied; and for IQs between 71 and 85 another 6–9 percent of children need extra help to supplement their progress in school.

**Slow Learners and the Brain-Damaged
Have Different Histories**   Children in the 71–85 range almost never show evidence of brain damage. Their electroencephalograms are normal. In the 50–70 range there may be some with brain

damage or evidence of difficulties during birth, but the vast number of them are without detectable signs of brain damage.

Children in these two ranges are rarely encountered in middle-class or well-educated families; they are frequently encountered among the poor and the poorly educated. However, rich and poor alike share about the same frequency of severe retardation (3 to 6 per 1000).

Severely Retarded Individuals Are More Likely to Be Brain-Damaged
The severely retarded—those with IQs less than 50—are usually brain-damaged (Table 15-4). Their inability to learn is physiological rather than social. Some 23 percent of these children have Down syndrome (mostly trisomy-21) and thus are chromosomally abnormal with a nondisjunctional defect as the primary cause of their retardation. Another 21 percent have single-gene defects, familial polygenic defects, or consanguinity resulting in homozygosity for one or more genes affecting brain development. Disorders such as phenyl-

Table 15-4   Cause of Mental Retardation in Institutionalized Patients

The classification of institutionalized patients at the Fernald State School in Waltham, Massachusetts. The Fernald School is the oldest institution for the mentally retarded in the United States. It was founded in 1848. This survey, carried out in 1971 by Hugo Moser and Philip Wolf, shows that some 35 percent of mentally retarded individuals have chromosomal or known genetic disorders. About 26 percent became retarded due to a variety of environmental hazards (infections, trauma, loss of oxygen during birth). Retardation in those with abnormal neurological signs or seizures (23 percent) could be of genetic origin. Some 16 percent of the cases are of unknown origin.

Diagnosed Basis	IQ < 50	IQ > 50	Percentage of Total
Metabolic and endocrine*	38	5	3.1
Infections	75	17	6.7
Prematurity	105	39	10.4
Trauma	45	12	4.1
Toxicity and hypoxia	53	11	4.7
Chromosomal*	247	10	18.7
CNS malformations*	49	16	4.7
Neurocutaneous syndromes*	4	0	0.3
Multiple congenital anomalies*	64	16	5.8
Degenerative diseases*	5	7	0.9
Psychosis	7	6	0.9
Retarded, with neurological signs	230	48	20.2
Retarded, with seizures	31	10	3.0
Retarded, reasons unknown	124	98	16.1
Not retarded	0	6	0.4

Source: Adapted from Moser and Wolf, 1971, pp. 117–134.

*Primarily genetic (including chromosomal) in origin.

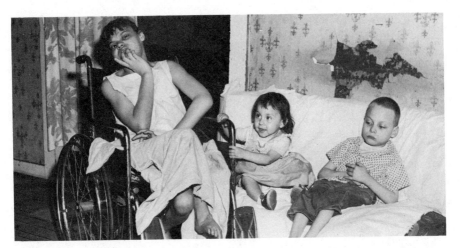

Figure 15-3 Phenylketonuria (PKU)
Children born with the autosomal recessive disorder phenylketonuria must be fed a special diet low in phenylalanine within 30 days after birth. Failure to do so may result in irreversible, serious brain damage and an IQ of 20–40. The three children shown here are homozygous for PKU. The girl in the wheelchair and her brother are retarded. Their sister has been fed a special diet since birth and she is alert and mentally normal. Some risks remain for this treated girl when she becomes an adult because regardless of the genotype of her husband, this treated person may produce microcephalic children with very low IQs (about 40–70) and congenital heart defects. The diet satisfactorily treated the child but it does not prevent a severe metabolic disturbance in the fetus whose development depends on the maternal nutrients crossing the placenta, which may contain excessively high levels of phenylalanine. Even when the pregnant PKU female faithfully uses a low phenylalanine diet throughout pregnancy, the fetus may receive excessive levels of phenylalanine and run the risk of microcephaly or heart defects. (March of Dimes Birth Defects Foundation)

ketonuria (PKU), galactosemia, and Hurler syndrome are examples of the many forms of hereditary defects associated with mental retardation (Figure 15-3).

Perinatal injuries caused by complications during delivery account for 6–9 percent of severely retarded children. Postnatal infections are responsible for 8–11 percent, and postnatal injuries, 1 percent. Prematurity (especially with low birth weight) accounts for about 3 percent, and kernicterus (which, we saw earlier, is a toxic poisoning of nerve cells from Rh blood group incompatibilities) causes another 1–2 percent. This leaves about 30–37 percent of the severely retarded without a known cause for their disorder. Some may have polygenic defects. Others may have acquired developmental defects of an unknown nature whose mark is not otherwise detectable by neurological or biochemical tests. Some of them, but probably not many, may reflect profound parental neglect or abuse. Since retardation is present in all races and social classes, including the royalty and the very rich, some underlying organic basis, rather than an environmental explanation, is likely.

The Causes of Two-Thirds of Severe Retardation
Can Often Be Determined   By   classifying
the various known causes of severe retardation, we can recognize many
that can be prevented or reduced in frequency.

Infections During Pregnancy   Rubella or German measles has about a
50 percent chance of causing birth defects if it strikes when the fetus is
in the first or second trimester. Deafness, mental retardation, blindness,
and congenital heart disease frequently result from such rubella infections. The damage may be caused by maternal antibodies entering the
fetal blood and attacking the fetal cells bearing viral antigens. The infection is prevented by immunizing children (or adults) against rubella. Unfortunately many parents neglect to bring their children in for rubella
immunization, and many schools neither enforce immunization nor provide it as a free service to students.

Cytomegalovirus can also cause birth defects and mental retardation,
but no effective immunization is available. Similarly some fungal infections (*Candida albicans*, mycoplasma) can cause fetal damage. The invasion by these fungal organisms occurs through a rupture in the amniotic membrane, which may arise from threatened miscarriage. Such
invasions are not readily preventable, and fungal organisms are not easily
treated with antibiotics.

Infection by the protozoan toxoplasma is preventable. This organism
is a sporozoan (similar in appearance to the malarial organism) whose
effect on the fetus can be serious. About 1 in 1000 pregnant females gets
infected with toxoplasma from eating undercooked meat, from accidental
ingestion of contaminants from dirty cat litter, or from eating food con-
taminated with mouse feces. It is not a good idea for a pregnant woman
to have a cat in the house, especially if the cat roams outdoors or catches
mice.

Syphilis is caused by a bacterial spirochete, *Treponema pallidum*,
and it can pass through the placenta and infect the fetus. Congenital
syphilis often results in deafness, defective tooth formation, and retardation. Since it is a venereal disease, it can be prevented by reducing
promiscuity, using prophylactics (e.g., condoms) in extramarital intercourse, and by prompt treatment with antibiotics when signs of venereal
disease are present.

Infections After Birth   Some infants infected with measles, herpes, or
certain mosquito-borne viruses may develop an inflammation of the brain.
This causes encephalitis which can leave a survivor with permanent
brain damage. Immunization against measles is readily available, but
neither herpes nor most of the rarer, mosquito-borne viruses that occasionally flare up during the summer are readily preventable. Mosquito
eradication, of course, can eliminate many of these viruses.

Acute diarrhea can cause brain damage by depleting electrolytes
(minerals) from nerve cells. The salmonella and coliform bacteria are

usually involved in the diarrhea itself but do not directly infect the nervous system. Clean hands and uncontaminated milk and foods will prevent such bacterial infections. Similarly, bacterial infections of the neural membranes (meningitis) by different bacteria can be prevented by sanitary habits and, in some cases, by treatment with antibiotics.

Twinning   There is more risk for single-egg (monozygotic) twins than for two-egg (dizygotic) twins. Some of this increased risk is due to the shared fetal blood supply found in some monozygotic twins. Retardation, which is a risk for both types of twins, comes from prematurity and low birth weight. Families with a history of dizygotic twins could reduce the risk or incidence of mental retardation by having fewer pregnancies.

Consanguinity   Intercourse of an incestuous nature, between parent and child, siblings, aunt and nephew, or uncle and niece, can result in homozygosity for autosomal recessive genes with monogenic or polygenic birth defects, including mental retardation. Although first cousins are usually exempt from state incest laws, such couples have been shown to have three times as many mentally retarded children as unrelated couples have. Cousin marriages vary in frequency according to religious and cultural traditions. Many prominent couples have been cousins, including Franklin and Eleanor Roosevelt, and Charles Darwin and Emma Wedgwood. Since first cousins have $\frac{1}{16}$ of their genes in common, the risks of homozygosity for a deleterious trait are raised slightly.

Blood Group Incompatibility   In Chapter 7 we learned that red blood cells have proteins on their membranes which reflect genetic differences. If transfusions are needed in surgery or for other disorders, the blood of the donor must be matched with that of the recipient. Failure to do so may cause severe illness or death from the destruction of the donated blood by the recipient's immune defenses against foreign proteins. If an Rh-negative mother has more than one Rh-positive pregnancy, we saw that this may cause the mother's antibodies to enter the fetal circulation and cause severe anemic jaundice with toxic impairment of the nerve cells (kernicterus). Such children, if they survive the anemia, may be brain-damaged. Prior to 1972, when C. A. Clarke devised the rhogam procedure to prevent such damage, an Rh-negative mother with a first Rh-positive child had a 17 percent risk of developing an immunization against Rh-positive blood. The risk now is less than 1 percent.

Nutritional Deficiencies   Low-protein diets during pregnancy may cause slower development of the fetus. Since infants with low birth weight have a higher risk of mental retardation (even those born to middle-class and affluent households) an adequate diet during pregnancy is essential. However, excellent prenatal eating habits have only modest effects on reducing the incidence of IQs below 50, because most of these cases can be assigned to causes not associated with nutrition. It is not known how many children in the 50–70 or 70–85 IQ ranges are the result of malnourishment during or after birth. Many other factors, such as infant stimulation, the quality of home life, and the values imparted by the parents are clearly involved.

Alcoholism   Consumption of alcoholic beverages by the mother in the first three months of pregnancy can cause mental retardation as well as birth defects (the fetal alcohol syndrome). Alcohol consumption in the last six months of pregnancy is not as likely to cause physical abnormalities or mental retardation, but it can lead to a lower birth weight, which is also hazardous to the baby's mental development.

The Social Aspects of
Retardation Are Complex   A child with a learning disorder brings about both economic and psychological problems for the family and society. State court rulings in the United States have made school districts responsible for the education of these children. Also social values have been changing, with more parents electing to raise retarded children rather than place them in institutions.

Whichever path parents choose, the economic cost is high. It is expensive to pay for institutional care and it is expensive to hire special tutors, teachers, psychologists to evaluate progress, and therapists to stimulate the minds and bodies of the retarded child. It is difficult for parents as well as society to devote substantial energy to the development of children whose prospects for normal or near-normal acquisition of essential skills (reading, writing, driving, holding a job with regular hours, relating to the general public, shopping, paying for services and goods) are limited.

Few children with IQs below 50 develop adequate skills to become normal, self-sufficient adults. In the 50–70 range there may be dramatic improvement, especially if the children are from homes marked by poverty or cultural isolation. And, since 3 percent of the population have IQs of 75 or less, and 90 percent of all the retarded fall into the 50–75 range, the efforts of society to educate these children are worthwhile. The difficulties of the children in the 50–75 range most likely are not biological, and even if their handicap is already limiting, the improved educational services may make them functional adults, capable of full-time employment and an unsupervised and fulfilling life.

Questions to Answer

1. What are the differences between precocity and genius?
2. What evidence would you use to support a major hereditary component of genius?

3. What evidence would you use to deny a major hereditary component of genius?

4. Are there differences between genius (defined by high IQ scores) and creativity? If so, what are the major differences?

5. What, in your value system, is eminence? What attributes does eminence have, if any, that fit neither genius nor creativity?

6. How are exceptional children (below-average performers) classified?

7. What are the frequencies of the following low IQs: (a) 71–85, (b) 51–70, (c) less than 50?

8. How are brain damage and IQ scores related?

9. What are the leading causes of brain damage?

10. What features would you include in a program for the education of children with lower IQs (below 75)? Discuss your reasons for selecting instructors, class size, mixed versus special classes, and other features you believe to be essential in your educational program.

Terms to Master

dizygotic twins
exceptional child
mildly retarded
monozygotic twins
retarded

rubella
severely retarded
slow learner
toxoplasma

Note: Table 15-5 on page 224 gives the major field or contribution of each of the people listed in Table 15-1 on page 210. Table 15-6 on page 225 gives the accomplishment or field associated with each of the people listed in Table 15-2 on page 213.

Table 15-5 Achievements of Geniuses Listed in Table 15-1

1. Philosopher, author, *On Liberty*
2. Inventor, steamboat
3. Author, fairy tales
4. Confederate general
5. Novelist, *House of Seven Gables*
6. Scientist, electromagnetic induction
7. Novelist, *Silas Marner*
8. Novelist, *Wuthering Heights*
9. Essayist, *Apologia pro Vita Sua*
10. Philosopher, mathematician, codiscoverer of calculus
11. Poet, dramatist, *Faust*
12. Poet, *Rime of the Ancient Mariner*
13. Novelist, *Candide*
14. President, U.S.A.
15. Poet, *Rape of the Lock*
16. Poet, *Lady of Shalot*
17. Founder of comparative anatomy
18. Poet, *Intimations of Immortality*
19. Novelist, *Ivanhoe*
20. Novelist, essayist
21. Composer, *Magic Flute*
22. Composer
23. Poet, *Courtship of Miles Standish*
24. Novelist, *Les Miserables*
25. Encyclopedist
26. Philosopher, mathematician, algebraic geometry
27. Chemist, miner's lamp
28. Sociologist, philosopher, positivism
29. Poet, *Don Juan*
30. Statesman, U.S. Senator
31. Poet, *Paradise Lost*
32. Painter, sculptor
33. President, U.S.A.
34. Composer, *The Messiah*
35. Scientist, law of falling bodies
36. Statesman, scientist, nature of electricity
37. Philosopher, *Critique of Pure Reason*
38. Essayist
39. Prime Minister of Great Britain
40. Novelist, *Oliver Twist*
41. Philosopher, *New Atlantis*
42. Poet, *Wilhelm Tell*
43. Chemist, discovery of oxygen
44. Essayist
45. Astronomer, elliptical orbit of planets
46. Chemist, physicist
47. Composer, *Parsifal*
48. Painter, inventor
49. Economist, *The Wealth of Nations*
50. Statesman
51. French emperor, general
52. Statesman, Secretary of Treasury, U.S.A.
53. Naturalist, *The Origin of Species*
54. Theologian, Protestant reformer
55. Composer, *Ninth Symphony*
56. Dramatist, *School for Scandal*
57. Philosopher, *Social Contract*
58. Scientist, *Principia*, laws of gravitation
59. Dramatist, *Tartuffe*
60. Poet
61. Poet
62. President, U.S.A.
63. Artist
64. Philosopher
65. Naturalist, taxonomy
66. President, U.S.A.
67. Chemist
68. Artist
69. Liberator of South America
70. Composer, *Water Music*
71. Composer, *Brandenburg Concertos*
72. Physician, circulation of blood
73. Theologian, Protestant reformer
74. Scientist
75. Scientist, law of genetics

Table 15-6   Activities of the Famous 20th Century Persons Listed in Table 15-2

1. Social reformer, settlement houses
2. Singer
3. Social reformer, suffragette, abolitionist
4. Jazz musician
5. Founder of Boy Scouts
6. Composer
7. Pioneer of nursing
8. Invented telephone
9. First saint from U.S.A.
10. Opera singer
11. Opera singer
12. Biochemist
13. Cellist
14. Painter
15. Painter
16. Actor
17. Novelist and dramatist, *The Cherry Orchard*
18. Writer, Prime Minister of Great Britain
19. Writer, *Tom Sawyer*
20. Composer
21. Physicist
22. Painter
23. Lawyer
24. Composer
25. Dancer
26. Inventor
27. Physicist
28. General, President, U.S.A.
29. Novelist, *Light in August*
30. Physicist
31. Inventor, auto industrialist
32. Founder, psychoanalysis
33. Poet
34. Liberator of India
35. Painter
36. Composer
37. Novelist
38. Novelist, *For Whom the Bell Tolls*
39. Politician, Chancellor of Germany
40. Novelist, *Brave New World*
41. Dramatist, *A Doll's House*
42. Pope
43. Novelist, *Ulysses*
44. Novelist, *The Trial*
45. President, U.S.A.
46. Civil rights activist
47. Novelist, *Sons and Lovers*
48. Premier, U.S.S.R.
49. Aviator
50. Composer
51. Novelist, *Magic Mountain*
52. Premier, China
53. Painter
54. Philosopher, *Thus Spake Zarathustra*
55. Dramatist, *The Iceman Cometh*
56. Painter
57. Composer, *La Bohème*
58. President, U.S.A.
59. Philosopher, *Principia Mathematica*
60. Birth control reformer
61. Humanitarian physician
62. Dramatist, *Man and Superman*
63. Composer
64. Premier, U.S.S.R.
65. Poet
66. Poet
67. Novelist, *War and Peace*
68. Painter
69. President, U.S.A.
70. Educator
71. Writer, *Importance of Being Ernest*
72. Novelist, *Look Homeward Angel*
73. Architect
74. Poet
75. Novelist, *Nana*

Part 5
Developmental
Genetics

The simple multiplication of cells would produce masses of cells but not an organism. That, in fact, is what happens when a tumor forms from a single cell. Normally, a fertilized egg proliferates a mass of cells which then forms tissues. What differences are present in the cells of different tissues? How did these differences arise?

A knowledge of eukaryotic developmental processes—growth, differentiation, and cell movement—is essential for understanding the events that lead to the formation of an embryo or fetus. Such knowledge also permits the study of abnormal development. Although some fairly good ideas of how genes control metabolism have been tested in bacteria, we still know very little about the developmental mechanisms of higher organisms.

By watching and observing we can get a reasonable idea of how one cell eventually produces an entire individual. If we go further and intervene in the development of a chick, frog, or mouse, our knowledge will be extended. Experimental embryology permits us to fuse cells, partition embryos, transplant parts of cells, and expose cells to unnatural environments. From these manipulations we can learn how birth defects arise, what causes twinning, why some changes are reversible, and how genes respond to chemical signals within the cell.

The chapters in Part 5 discuss a variety of developmental genetic problems. They will give you some insights into the human condition as you relate the gene and the chromosome to development. You will learn how readily the normal processes carried out by the embryo can be disturbed by radiation, chemicals, and mechanical obstructions. You will see why the embryo in the early stages of pregnancy is much more sensitive to damage than the embryo in the later stages.

Most important of all, you will have a much more substantial basis for evaluating the environmentalist movement. Why does pollution cause cancers and reduce life expectancy? Why is tobacco smoking at least as valid a concern to human geneticists as radiation exposure? Can one live a reasonably healthy life in a highly technical and industrial society?

Chapter 16
Embryo Manipulation and Cloning

Aldous Huxley's satirical novel, *Brave New World*, appeared in 1932. Although Hitler had not yet taken control of Germany, and Stalin's reign of terror had not yet begun, Huxley recognized the trend of some powerful states to disregard individual rights and democratic processes in favor of totalitarian dictatorships. He was alarmed by the way scientists and technology could be used by governments to achieve both control of the population and enthusiastic acceptance of the loss of freedom. In *Brave New World* the citizens are assigned status at birth (as test-tube babies) through eugenic selection, artificial twinning, developmental enrichment or impoverishment, and most of all by psychological conditioning.

Few who have read that novel can forget the clones of common laborers, mass produced by proliferating embryonic cells, each of which is subsequently cultured and developed into an identical twin or *clone*. This vision of cloning has been raised often in more recent times whenever scientists have developed new techniques for studying developmental biology. Human cloning involves two issues: Is it technically possible and is it morally desirable under any imaginable situation?

The Blastocyst Arises
from a Solid Ball of Cells
Embryo manipulation in small mammals such as the mouse is useful because it distinctly reveals what is happening in development. Observation alone may be inadequate to determine the events and processes taking place. After fertilization the mammalian zygote multiplies mitotically. The zygote does not have an external source for food, and after the mass of the cell is cleaved in two during the first mitotic division, each cell remains half the size of the original zygote.

Each new cell or *blastomere* divides again, several times, producing 4-, 8-, and 16-cell stages. These later blastomeres are considerably smaller than the original zygote, and while the cell cycle is typically mitotic, the G_1 phase does not lead to a *growth* of the cells. Eventually the cell mass will have to obtain nourishment from the mother's uterus to enable it to grow and continue its development. This process of embryonic mitosis without growth is called *cleavage*. About the time of the 16-cell stage, the embryo consists of a ball of cells resembling a raspberry. This *morula*, as it is called, then *differentiates* into components—an outer layer or *trophoblast* and an *inner cell mass*. The differentiated structure is called a *blastocyst* (Figure 16-1). How did these two tissues arise from the morula?

When an 8-cell mouse embryo is allowed to take up radioactive substances and is then surrounded by nonradioactive embryonic cells at the same stage of development, the blastocyst that subsequently forms contains a nonradioactive trophoblast. The reverse is true in the reciprocal experiment, where the inner cell mass is nonradioactive and the trophoblast is radioactive. This shows that the inner cell mass is determined environmentally by location. The inside cells of a morula form the inner cell mass and the outer cells form the trophoblast. Until that differentiation occurs, the inside and outside cells are freely interchangeable and

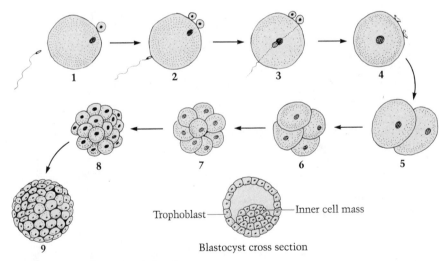

Figure 16-1   Cleavage of the Zygote

After ovulation the egg enters the oviduct where it meets a sperm (1) which causes the egg to complete meiosis (2) and produce an egg nucleus which moves centrally (3). This egg nucleus and the sperm nucleus fuse (4) and the zygote is formed. Mitosis without growth, or cleavage, begins (5), producing a 2-cell stage, followed by 4-celled (6) and 8-celled (7) embryos. At the 16-cell stage or morula (8) a cell movement occurs and the inside cells form an inner cell mass and an outer trophoblast. The blastocyst is composed of these two tissues (9). In the cross-sectional view it is cut open, revealing these two tissues.

will acquire the function of their new location. After differentiation has occurred, however, the functions of the trophoblast and inner cell mass are fixed. A different experiment is required to prove this.

Each of the Tissues of the Blastocyst
Has a Different Function
It has been known for more than 50 years that vertebrate embryos in early cleavage stages can be cut into equal portions, each of which will reorganize and form an identical twin. It is also known that two zygotes can be joined to form a single

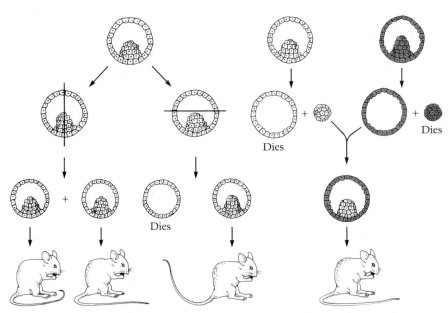

Figure 16-2   Embryonic Blastocyst Manipulation
On the left are simple vertical and horizontal cuts, dividing the blastocyst into two approximately equal pieces. In the case of a vertical cut each piece contains some of the inner cell mass and some of the outer trophoblast. Each of the two pieces forms an implantable embryo which develops into a normal mouse. The horizontal cut, however, shows that a trophoblast without inner cell mass, while capable of implantation, cannot develop into a normal embryo. The other part of the horizontal cut, bearing the inner cell mass and some of the trophoblast, readily develops into a normal mouse.

On the right is a more complex manipulation. The inner cell mass is removed from a trophoblast in both albino and pigmented blastocysts. Neither component alone can survive in the uterus to form a mouse. But if the albino inner cell mass is introduced into the pigmented trophoblast, the embryo develops and an albino mouse forms. If the reciprocal experiment is done, with a pigmented inner cell mass and an albino trophoblast, a pigmented mouse results.

These experiments demonstrate that the inner cell mass and not the trophoblast contributes to embryo formation. The function of the trophoblast is restricted to the implantation of the blastocyst.

individual. The early work with these ideas was done with frogs and salamanders, which have relatively large fertilized eggs. The technical difficulties of separating or uniting cells in these two species seemed less formidable than in mammals which have much smaller zygotes and embryonic cells.

In the mid-1960s several laboratories, especially that of R. L. Gardner and his colleagues in England, worked out techniques for isolating, handling, manipulating, and inserting embryos into the fallopian tubes or uteri of foster-mother mice and rabbits. The results of such experiments are exciting.

In mouse embryos the two tissues of the blastocyst can be separated by using a micromanipulator, which is a device attached to the microscope, and which has miniature tools scaled down to the cellular level. Isolated trophoblast tissue can implant in the uterus but isolated inner cell mass tissue lacks this ability (Figure 16-2). The function of the trophoblast cells of the blastocyst is recognition of uterine epithelium (the lining of the womb) and the subsequent implantation of the trophoblast cells into the uterine lining and their adhesion to it. When the inner cell mass is removed from the blastocyst and then implanted, neither implantation nor further development occurs. The failure of embryonic development implies that the embryo and its membranes cannot be provided by the inner cell mass without the support of an implanting trophoblast. When a blastocyst is cut in half symmetrically so that each portion contains trophoblast and inner cell mass tissue, a normal embryo can develop from either of the half blastocysts. This proves that at the blastocyst stage no irreversible differentiation of the cells in the inner cell mass has taken place and the cells have the potential for developing into an embryo. If the blastocyst is cut horizontally, and the upper hemisphere containing most of the trophoblast is discarded, the lower half, which contains the inner cell mass, will still be capable of successful implantation and full development. The isolated trophoblast, however, even when it is of a full size (but lacking in the inner cell mass), implants but fails to proliferate deeply into the uterine endometrium. This shows that the maintenance and further growth of the implanted trophoblast tissue depends on the inner cell mass.

Parts of Embryos
of Two Different Genotypes
Can Be Interchanged
Pieces from genetically different blastocysts can be surgically put together. Thus an inner cell mass from an albino mouse strain can be placed in a trophoblast taken from a pigmented mouse strain. The reconstituted blastocyst will invariably produce an albino mouse. If the reciprocal experiment is done, using an albino trophoblast and a pigmented inner cell mass, the mouse that develops will be fully pigmented.

If an inner cell mass of an albino is injected into a blastocyst of a pigmented strain, the two inner cell masses will merge (Figure 16-3). Twins are never produced by this procedure. Instead, the resulting mouse is a genetic *chimera*, consisting of two populations of cells—albino and pigmented in this case. The distribution of the pigmented and albino tissues in the chimera is not random (which would produce a salt-and-pepper pattern appearing grey at a distance) but irregularly patchy, somewhat like zebra stripes but with fewer and larger stripes apparently arising from the midline along the length of the spinal column.

When pigmented trophoblast cells are injected into an albino blastocyst, the embryo is invariably albino because trophoblast cells do not contribute to embryo formation. So, too, in the reciprocal cross, a pigmented inner cell mass does not accept albino trophoblast cells as embryo-forming cells, and thus a mouse is fully pigmented after completing its development.

These experiments prove that the two tissues, trophoblast and inner cell mass, are irreversibly differentiated. The trophoblast cells can only function in implantation, and the inner cell mass alone forms the embryo and contributes genetically to the newborn individual. Philosophically the experiments can be used to show the extension of reductionist methods to the early stages of development. Reconstituted blastocysts demonstrate that parts of the early embryo can be removed and replaced by comparable parts from another embryo. The chimeric embryos, especially those produced by the association of two genetically dissimilar inner cell masses, prove that a single individual, not two, forms from the resulting single inner cell mass.

Figure 16-3   Single-Cell Chimeras
If a pigmented cell is withdrawn from the inner cell mass of a blastocyst and transferred to an albino blastocyst, the mouse that forms may show one or more stripes of pigmented fur in its coat. If a trophoblast cell from the pigmented blastocyst is introduced into an albino blastocyst, no striped mouse occurs. This also illustrates the embryo-forming function of the inner cell mass and the absence of that function in trophoblast cells.

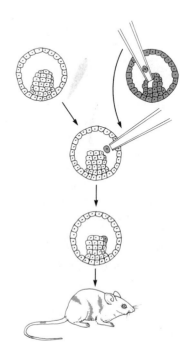

A Chimeric Mouse Can Be
Made by Embryo Aggregation

The experiments carried out in Gardner's laboratory show how experimental science can be used to manipulate the embryo to yield unambiguous answers to questions. A similar use of a different technique makes it much easier to obtain chimeric mice. This technique, *embryo aggregation,* was independently developed by Andrew Tarkowski in Warsaw, Poland, and by Beatrice Mintz in Philadelphia. In their experiments, genetically or chromosomally distinct strains of mouse embryos, at an 8-cell stage of cleavage, are aggregated together (Figure 16-4). The 16 cells continue mitotic division and form a single blastocyst. The resulting chimeric mouse reveals an abundant amount of information on early development.

Clearly half of all such fusions should produce either an XX embryo (potentially female) or an XY embryo (potentially male). The remaining half should be XX/XY chimeras. The chimeric mouse is rarely hermaphroditic (an individual with both male and female gonadal tissue or genitalia) in appearance. Only 8 hermaphrodites were seen among 560 chimeric mice. Some of the XX/XY chimeras are fertile females and some are fertile males. The *germ cells* or reproductive tissue are usually XY if the chimeria has testes and XX if the chimera has ovaries, but female XX/XY chimeras can sometimes pass on Y-bearing eggs. It is not uncommon for the germinal part of testes to be XY and the surrounding somatic part of the testes to be XX. In ovaries a similar situation can exist, with XX germinal tissue in an ovary whose somatic parts are XY. Once the testes or ovaries are formed, however, it is the hormones of these gonads

Figure 16-4   Embryo Aggregation Chimeras *The technique of embryo aggregation, developed by Mintz and by Tarkowski, is shown here with two 8-cell-stage cleavage masses, one albino and the other pigmented. When pressed together the 16 cells form a common morula which differentiates into a blastocyst. The mouse that emerges has striped fur because each stripe is descended from one or more initial cells of the same genotype along the length of the embryonic spinal column. Such chimeric mice are valuable for studying sex determination and the number of cells participating in the formation of each organ.*

Figure 16-5   A Chimeric
Mouse
*This unusual mouse had six
parents, two from a black
strain, two from an orange
strain, and two from an albino
strain. The resulting mouse,
formed by aggregation of three
embryos, is tricolor, showing
the fur color of all three
strains. (Courtesy of C.
Market)*

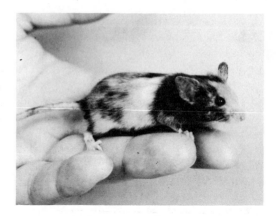

that determine whether other reproductive structures are differentiated
in a male or female direction.

The zebra striping of albino/pigmented chimeras produced by fusion
of the inner cell masses in Gardner's experiments is also found in many
of the chimeras produced by embryo aggregation (Figure 16-5). Only a
small number of cells is required to initiate a major organ structure. For
mouse coat color a few dozen pigmented cells are aligned along the midline and form the striped patterns. For the mouse retina, mosaic sectoring
produced by cell lineages of a defective retinal mutant indicate that about
ten cells are involved in the formation of an embryonic retina. The probable number can be inferred from the percentage of chimeric individuals
that do not show a mixture of both genetic (or chromosomal) types for a
given organ system. If a single cell determines a structure the chimeric
mice will be of one or the other type, never a mixture of both types.
When two cells participate, the mice will usually show mosaicism, with
both genetic (or chromosomal) types present in the tissue.

There Are Human Beings
Who Arose from Embryonic Aggregation   It is important to note
that chimeric mice produced experimentally arise from two sets of parents. They are *tetraparental*. It is very likely that some human individuals are also a result of embryo aggregation, either from two morulae or
from an abnormality of meiosis permitting a fertilized polar body to function mitotically and remain associated with the zygote. The human cases,
unlike the mice, are not tetraparental, but they do involve two fertilizations. Either mechanism could give rise to a chromosomally uniform
individual (if both sperms were X-bearing or if both were Y-bearing). Such
individuals have been found and they express two genotypes in some of
their tissues. For example, some of their red blood cells may be AB
Rh-negative, and some may be O Rh-positive. Those chimeras combining
XX and XY cells result in hermaphroditic births. Because XX/XY human
hermaphrodites also show mosaicism for blood groups it would be dif-

ficult to consider any mechanism other than some form of embryo aggregation responsible for their origin.

Cloning from a Single Zygote Can Be Done by the Transplantation of Its Cleavage Nuclei

Multiple identical twins can be produced artificially in vertebrates. This has been most effectively demonstrated in frogs and toads. The techniques, first developed by R. Briggs and T. J. King, are complex (Figure 16-6). The diploid nucleus of a fertilized egg is destroyed or removed by use of a microdissecting apparatus. Into the enucleated egg another nucleus is introduced, from an embryo in an early cleavage stage. Each of the nuclei of a morula can thus be inserted into a different enucleated egg, and these reconstituted zygotes will form normal, identical, adult frogs or toads.

The story of *nuclear transplantation* is useful in determining when the nucleus of a tissue loses its capacity for generating a complete development leading to a fertile adult. The more advanced the embryo becomes, however, the smaller are its cells and the less frequently are the cells dividing. This makes it technically more difficult to transfer a nucleus intact and at the proper stage of mitosis for it to initiate cell division in the reconstituted zygote. Toad cells are larger than frog cells and thus the success of nuclear transplantation is more assured for later stages of toad embryos than it is for frog embryos.

Briggs and King found that nuclei from the early cleavage stages (up to about the 64-cell stage) could readily be transplanted into enucleated eggs, where they would develop normally into tadpoles and frogs. When nuclei from later stages were used, as the blastomeres took on specific tissue functions the transplanted nuclei usually caused the embryos to abort. No fully differentiated adult frog tissue can be used for making clonal frogs.

However, using the toad *Xenopus*, which has larger cells, Gordon did succeed in obtaining some fully normal cloned toads from differentiated tissue, such as the gut cells of tadpoles. Adult tissues have a very low probability of forming a normal clone and for this reason it is doubtful that any adult human (whose cells are much smaller than a tadpole's) can make an identical twin from his or her tissue, at least with the techniques available today.

Mammalian cloning by nuclear transfer presents many difficulties. The mammalian egg is about $\frac{1}{100}$ the size of a frog's egg. Thus the micropipette must have a remarkably small diameter to introduce a nucleus. If the nucleus is undamaged when inserted into the enucleated egg it could have normal development. Despite these limitations, Karl Illmensee has claimed to have cloned a mouse from one of its morula blastomeres using techniques similar to those pioneered by Briggs and King (see Figure 16-7). While Illmensee's method has not been confirmed, the cloning of a mouse by slightly different techniques has been reported by Davor Salter and James McGrath of the Wistar Institute in Philadelphia.

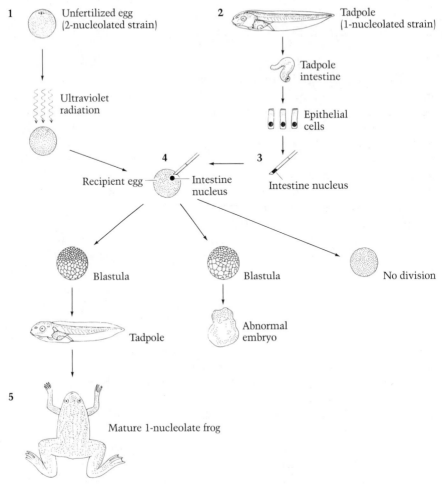

Figure 16-6 Nuclear Transplant Clones in Frogs

The techniques for nuclear transplantation in frogs (Rana pipiens) were worked out by Briggs and King in 1954. Later these techniques were applied by Gordon to a different species (Xenopus laevis) with larger cells and nuclei. In nuclear transplantation experiments the unfertilized egg (1) has its nucleus killed by ultraviolet radiation. The killed nucleus was of a genetic strain with two nucleoli per nucleus. A donor nucleus, from the gut of a tadpole (2) is isolated in a micropipette (3). This nucleus is from a genetic strain having only one nucleolus per nucleus. The isolated nucleus is injected into the egg (4) and development proceeds, forming a blastula. Sometimes development fails to occur or the blastula is abnormal. In those instances where a tadpole does form, a mature fertile frog results (5) whose nuclei all bear a single nucleolus. The mature frog is a clonal twin of the donor (2) tadpole. [Adapted from J. R. Whittaker, Cellular Differentiation (Dickinson, 1968).]

In plants nuclear transplantation is not necessary to achieve cloning. Isolated individual cells from finely macerated carrot roots can be cultured and, by laborious techniques, stimulated to form genetically iden-

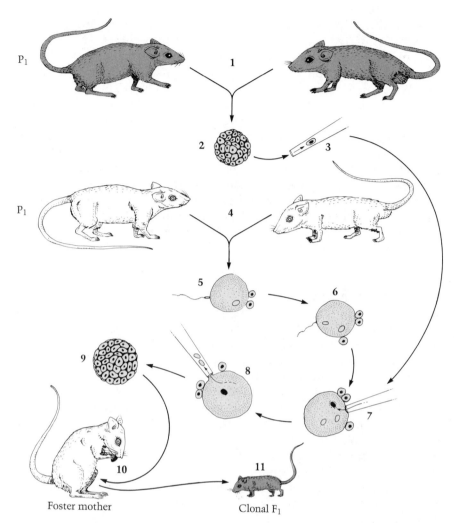

Figure 16-7 Cloning a Mammal

Mammalian cloning was reported in 1979 by Karl Illmensee. This work could not be repeated but a slightly different method was used to clone a mouse in 1983. In the diagram shown here the pigmented strain (1) of mice is used to generate an embryo in the morula stage (2). A cell is removed and the cell membrane destroyed in a narrow glass micropipette (3). Albino mice (4) are used to obtain a freshly fertilized egg (5) whose male and female haploid nuclei have not yet fused (6). The micropipette injects the diploid nucleus (7) from the pigmented strain and then sucks up the two haploid nuclei (8) in the egg. The resulting zygote produces a morula (9) which is implanted in a virgin albino foster mother (10) made pseudopregnant with hormones. She delivers a pigmented mouse (11) which is a clonal twin of the embryo (2) used as the donor nucleus source.

tical clones whose tissues and organs are as normal as the original plant. Many plants have a natural process for cloning—by buds, cuttings, or runners—but the realization that virtually any cell may generate a new plant if cultured properly implies that the genotype of the nucleus is not irreversibly altered when cells differentiate into tissues.

Human cloning does occur by natural (and still unknown) means in monozygotic twinning. Theoretically such twins could be artificially produced by removing the preimplantation early embryo, such as a morula, and partitioning it into several pieces. Each segment should produce an identical twin. If such cleavage blastomeres from a single morula could be cultured mitotically then larger numbers of twins could be produced. Since these cells do not contain an external source of nutrition until after implantation, the use of uncultured blastomeres would limit human clones to a few dozen individuals at the most from any fertilized egg. Nor is there a guarantee that an individual cell from a morula would have enough nutrients to produce both trophoblast and inner cell mass tissue sufficient for a successful implantation and development. Cloning in the *Brave New World* sense is not an immediate prospect.

There Are Both Industrial
and Human Applications of Embryo Manipulation   There are many situations where embryo manipulation is both desirable and useful. A valuable strain of cattle, for example, can be propagated by combined use of artificial insemination and superovulation. In the cow, a follicle-stimulating hormone can be injected to induce multiple egg release. The embryos can be removed and then cultured in test tubes. Other cows, of inferior quality, can be made pseudopregnant by injection of progesterone, and each of these cows can then be implanted with one of the cultured blastocysts from the superior cow. The resulting calves will inherit the selected genotypes of the prize cow and bull and show none of the traits of their foster mother.

A striking example of embryo manipulation was the temporary gestation of cattle embryos in the uteri of rabbits. These rabbits were flown from Australia to South Africa where the embryos were removed and reimplanted in pseudopregnant foster-mother cows. The South African cows delivered calves of the Australian breed. The embryo of a baboon can also be transferred to a pseudopregnant foster mother, and a normal infant baboon will be delivered.

As we learned in Chapter 5, embryo transfer was used in 1978 by Steptoe and Edwards in carrying out an oviduct bypass to achieve a successful pregnancy. Moral objections to this technique by both physicians and prospective patients have probably played a larger role in inhibiting its application to humans than have technical problems. Consider the hypothetical case of a woman who had tubal ligation and whose family was later killed in an accident. If she remarried, would it be ethical for her to have her own egg fertilized outside of her body by her husband's sperm and implanted into her oviduct or uterus? The only serious ethical objection to her use of this technique (as done by Steptoe and Edwards) would be that it is not an entirely natural process. Plastic surgery, prosthetic devices, preventative medicine, and organ transplants also involve some unnatural procedures, but because our attitudes on sex have a more extensive moral history than do our attitudes toward the rest of our body, sexual organ surgery is sometimes singled out for prohibitions.

Human cloning is far less likely to occur than oviduct bypass surgery because few individuals have a strong desire to have children with identical genotypes, given a choice. One situation, however, made this author realize that nothing is so strange that it meets with universal condemnation or indifference. A nurse once wrote a sad letter about the accidental death of her 13-year-old son. She hoped it would be possible to have some of her son's frozen tissue used for a nuclear transplant similar to those done by Briggs and King so that she could have an identical twin of her son. No doubt, if the techniques had been perfected and a physician had been willing, she would have tried this procedure, even with the foreknowledge that if it worked both she and her cloned son would have immense psychological problems coping with the past tragedy that led to his existence.

Even if the *Brave New World* model were technically feasible, identical twins would not have identical minds. Twins are unique individuals. A clone of Einstein would not necessarily become a great physicist. Diversity would be even more likely if the foster parents of such a clone lived in different cultures and had different occupations. Identical twins living in the same household frequently have different occupations and interests, and this fact makes clonal monotony improbable. From a genetic point of view human cloning on a large scale (where a few individuals produce numerous clones) is not desirable because it restricts the number of genotypes, and without abundant diversity the evolutionary opportunities for humankind would be diminished.

Questions to Answer

1. What is a blastocyst?
2. What experiment can you cite to show the functions of the trophoblast and the inner cell mass?
3. What is the fate of a pigmented inner cell mass cell injected into an albino blastocyst?
4. What is the fate of a pigmented trophoblast cell injected into an albino blastocyst?
5. If an albino and a pigmented 8-cell-stage embryo are aggregated, what is the fate of the resulting 16-cell unit?
6. What is the sex of chimeric mice? Discuss the possible reasons for your answer.
7. What is meant by the term tetraparental? How does such a mouse differ from a mosaic individual?

8. What is a clone?
9. How are frogs and toads cloned?
10. Are there natural clones? Discuss their origin.
11. What is the difference between the terms "test-tube baby" and "oviduct bypass"?
12. Discuss the difficulties you and your clone might experience if it were possible for you to produce a clone of yourself from one of your skin cells.

Terms to Master

blastocyst
blastomere
chimera
cleavage
clone
embryo fusion

hermaphrodite
inner cell mass
morula
nuclear transplantation
trophoblast

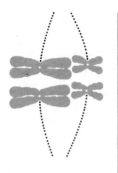

Chapter 17
The First Trimester:
Organogenesis

Development depends on several processes. *Growth* involves cell division and the formation of more cellular organelles as well as the components necessary for metabolism. *Differentiation* involves the "turning on and off" of genes so that the cells become specialized in their functions as tissues. *Morphogenetic movements* permit cells to migrate as sheets, roll up into tubules, or convolute in different ways, thereby forming the fields, shapes, and patterns characteristic of organ systems. Little is known about the mechanisms that control these three processes, although all of them can be studied by experimental techniques.

The remarkable events of development can be observed, from fertilization to birth, by following the embryos of mammals (including many human embryos obtained from hysterectomies or autopsies) and studying their preserved and stained tissues microscopically. This body of knowledge about embryonic development can also be applied to abnormal embryos whether they are naturally malformed or artificially made monstrous. Defects in development often provide knowledge of which stages go wrong and how birth defects arise.

The Blastocyst Forms Two Tissues
from Which the Embryonic Body Forms　　After the morula forms a blastocyst, implantation in the uterus occurs. The inner cell mass of the blastocyst differentiates into two layers, an *endoderm* and an *ectoderm* (Figure 17-1). The endodermal layer lines the cavity of the blastocyst, forming a *yolk sac*. The ectoderm lines a somewhat smaller cavity facing the implantation site of the trophoblast. This forms the *amniotic cavity*. The endodermal and ectodermal tissues in contact with each other form the *embryoblast*.

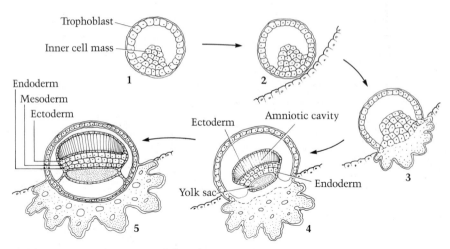

Trophoblast
Inner cell mass
Endoderm
Mesoderm
Ectoderm
Ectoderm
Amniotic cavity
Endoderm
Yolk sac
1
2
3
4
5

Figure 17-1   Implantation of the Blastocyst
When the blastocyst leaves the oviduct (A) it consists of an outer trophoblast and an inner cell mass. The trophoblast comes in contact with the lining of the uterus (B) and begins eroding the uterine cells. A special tissue, lacking cell membranes but filled with trophoblast nuclei, penetrates into the uterine layers (C). This tissue readily transports blood into vacuoles or spaces, and the food carried by the blood feeds the blastocyst. As this implantation proceeds, the inner cell mass differentiates into two tissues, ectoderm and endoderm (D), with a space above the ectoderm forming the amniotic cavity and a space below the endoderm forming the yolk sac. A third embryonic tissue, the mesoderm, forms between the ectoderm and the endoderm. It also lines the outsides of the amniotic cavity and yolk sac.

The Mesoderm Completes the Embryo
Which Then Forms a Series of Tubes   A third layer of cells, the
mesoderm, forms between the ectoderm and endoderm. These three embryonic tissues form all of the adult tissues and organs (Table 17-1).

As the three layers form, the ectodermal cells begin to form a thickened plate which rolls up into a tube. This *neural tube* runs along the
length of the embryo (Figure 17-2). It swells out in several bulges at the
anterior end, forming the future brain region, and remains relatively uniform in diameter for most of the rest of the length of the embryo. That
portion of the neural tube that is uniform in diameter will become the
spinal cord.

Some of the cells in the brain region form outpocketings on either
side. These new bulges are depressed by a movement of cells back toward
the center of the brain, forming a structure that looks like a caved-in
handball. These deformed bulges are the *optic cups,* and they form the
future eyes, the inner layer forming the future *retina* and the outer layer,
the future *choroid* layer of the eye. The *lens* and *cornea* of the eye,
however, are not formed by the tissues of the optic cup, but from the
adjacent ectodermal epithelium of the embryonic head.

Table 17-1   Major Derivatives of the Embryonic Tissues

The three embryonic tissue layers form all the organs of the body. Ectodermal
derivatives form primarily the nervous system, the sense organs, the skin, and the rest of
the integument. Endodermal components form the alimentary and respiratory tracts.
Mesoderm forms the bulk of the body, including the muscles, bones, cartilage, fatty
tissue, tendons, and circulatory system.

Ectoderm (nonneural)	Endoderm	Mesoderm
Skin, hair, nails, tear ducts, sweat glands	Oral cavity	Muscles
	Esophagus	Bones
	Tonsils	Heart
Teeth	Tongue	Blood vessels
Pigment cells	Parathyroid, thyroid, thymus glands	Blood
Anterior pituitary gland		Spleen
	Eustachian tubes	Gonads
Lips	Trachea	Kidneys
Anus (external portion)	Lungs	Fat
	Stomach	Connective tissue
Ectoderm (neural)	Pancreas	Tendons
Brain	Intestines	Cartilage
Spinal cord	Liver	
Nerves	Gall bladder	
Posterior pituitary gland	Colon	
	Primordial germ cells	
Ears	Anus (internal portion)	
Eyes		
Taste buds		
Olfactory organs		

The entire nerve system and the sensory organs have their origin
from the neural ectoderm that is derived from the embryonic neural tube
(Figure 17-3). Some additional cells, at the junction where the neural folds
join to form the neural tube, migrate to form the melanocytes or cellular
organs for pigment production and some of the tissues of the adrenal
gland. The rest of the ectoderm forms the skin, with its specialized organs
for hair production, sweating, lubrication (from the sebaceous glands),
and nail formation.

Incomplete Closure of the Neural Tube
Results in Birth Defects
Although the neural plate
folds and seals along the dorsal line, it occasionally fails to close completely. In such cases the embryo may develop abnormally and the newborn child may be impaired. In severe cases the cephalic or brain-forming
region fails to close and the child may be _anencephalic_ (Figure 17-4).
Such infants usually die within days of their birth. They may either have
a deformed _cranium_, where the portion of the skull containing the brain

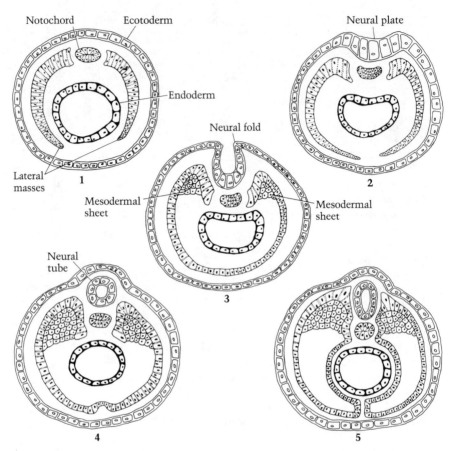

Figure 17-2 The Triploblastic Tissue Formation of the Vertebrate Embryo

The general body of a vertebrate can be simply described as a system of tubes. The outermost tube is ectoderm, forming the skin, but it produces a thickened area, the neural plate, *which rolls up and becomes the neural tube. The innermost tube is the endoderm from which the alimentary tract and its derivatives arise. Between the ectoderm and endoderm there is a tube of mesoderm which forms the lining of the pleural (lung) and abdominal cavities as well as the bulky musculature and connective tissues of the limbs, skeletal system, and circulatory system. In the diagram each stage of development is shown in cross section. The mesoderm forms between the ectoderm and endoderm (1) and then separates into a rod-like notochord and lateral masses. The thickening of the dorsal ectoderm forms the neural plate (2) which rolls into folds (3) that form the neural tube (4) at the same time as the mesodermal sheets on each side form a continuous layer between the ectoderm and endoderm (5).*

fails to form, or a head filled mostly with fluid rather than neural tissue. If closure fails to occur somewhere along the spinal cord, the child may have *spina bifida*. Such children may have nonrepairable nerve damage, even if the skin can be successfully repaired to prevent the entry of bacteria which cause infectious meningitis. The neurologically nonrepairable infants will be *paraplegic* for life and their life expectancy will

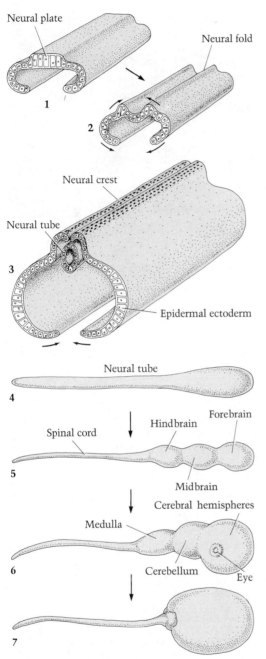

Figure 17-3 Development of the Nervous System

The human brain is an enlargement of the neural tube which is formed during the fourth week of pregnancy. This tube, at its anterior end, forms swellings and flexures which develop into the major parts of the brain—the forebrain, midbrain, and hindbrain. Human features begin to appear in the head about two months into the pregnancy. The forebrain has very few convolutions, or folds, before the seventh month of pregnancy.

In this sequence a neural plate (1) forms in the dorsal ectoderm and forms neural folds (2) which join together (3) forming the epidermal ectoderm, the neural tube, and a group of migratory cells, the neural crest (which form our melanocytes). The neural tube, seen longitudinally (4), enlarges at the cephalic region to form the three brain regions (5). The forebrain enlarges (6) into the cerebral hemispheres which eventually enclose most of the brain (7).

be reduced, especially if their paralysis involves the bladder and bowel. The high levels of α-fetoprotein (normally present inside the embryo), which escape into the amniotic fluid from the open neural tube, are detectable prenatally and parents can elect the medical abortion of a fetus with such a defect.

Figure 17-4 Anencephaly
When the neural tube of the embryo fails to develop at the anterior end, the brain (and sometimes the surrounding cranium) fails to form. Anencephalic infants rarely live more than a day. They also pose more of a risk to the mother because they have longer gestations and often are oriented in such a way that a breech delivery is necessary. Anencephaly, like most neural tube defects, can be successfully detected prenatally because of the elevated level of α-fetoprotein (a normal fetal protein inside the embryo) in the amniotic fluid. (Robert Desnick)

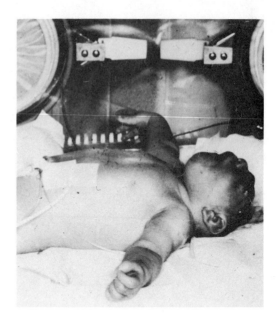

The Endodermal Gut Tube
Produces Many Internal Organs   As the embryo's ectodermal tissue becomes more complex, there are also accompanying changes in the endoderm which surrounds most of the blastocyst cavity and forms the yolk sac. Running parallel with the neural plate, the endoderm balloons out at the future head and future tail regions of the embryo, forming the gut tube. This gut tube forms a number of structures, including the mouth, esophagus, intestines, anus, lungs, and pancreas, which are involved with our digestion, respiration, and hormonal functions (Figure 17-5).

In the head region the gut tube forms outpocketings leading to the ears (our *Eustachian tubes*), the pharynx, the tonsils, and the thyroid and parathyroid glands. Further along, the gut tube sends out a slender elongation which bifurcates into two sacs, establishing the linings of the *trachea* and the *lungs*. The section of the gut tube parallel to the trachea forms the *esophagus*. The gut tube swells and twists to form the *stomach*. Several outpocketings of the tube form the *pancreas*, the *gall bladder*, and the *liver*. Most of the gut tube, however, forms the *intestines*.

The *mouth* and *anus* are formed by the perforation of the ectoderm in contact with either the oral or anal region of the gut tube. Our mouth and anus are thus of dual embryonic tissue origin. When we run our tongue along the base of our lower gums we are touching endoderm, but when we kiss someone on the lips we are touching ectodermally derived tissue.

Defects of the Gut Tube
May Lead to Malformations   Many embryonic defects can occur when the gut tube fails to develop normally. Children can be born with an

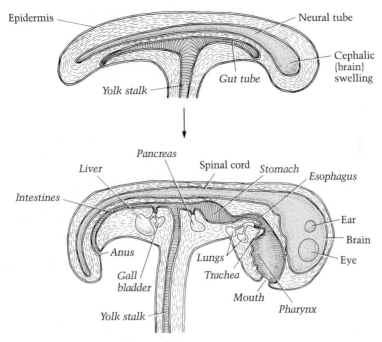

Figure 17-5   Derivatives of the Endoderm
The endoderm of the embryo is the innermost tissue. It rolls up into a tube, forming the future gut tract from the mouth to the anus. Along the way there are a number of outpocketings which form the trachea and lungs, the pancreas, the gall bladder, and the liver. There are also swellings which form the pharyngeal region, with smaller outpocketings forming portions of the Eustachian tubes, the tonsils, and the tongue, as well as the parathyroid, thyroid, and thymus glands. Additional swellings form the stomach, the various intestines, and the colon. The mouth and anus have a dual origin, each representing the region where the external ectoderm encounters the internal endoderm.

imperforate anus (Figure 17-6), a connecting channel between the trachea and esophagus, or cysts or clefts from the imperfect development of the pharyngeal endodermal pockets and arches. Most of these defects can be repaired, especially if the malformation is of moderate degree.

There is one serious genetic defect, however, which affects most of the endodermally derived organs. This is *cystic fibrosis*, the most common Mendelian recessive disorder in Caucasians, carried in heterozygous form by 1 in 35 Caucasians. The loss of regulatory control over secretion causes such individuals to develop mucus in the lungs, pancreas, and intestines (Figure 17-7). As a result pneumonia and emphysema develop in the damaged lungs, digestion becomes impaired, and absorption of nutrients is diminished. Secondary effects from the respiratory difficulties can lead to heart failure, but there is no primary defect in the heart. The malnutrition may also indirectly affect the function of other systems as the disease progresses. Portions of the reproductive tract derived from the urethra can also become clogged with mucus. Most cystic fibrosis

Figure 17-6 Imperforate Anus
When a baby is born the physician and nurse clean and examine the body. When the anus is absent, as shown here, the physician touches the spot where it should be. If there is a contraction, the underlying sphincter muscles are present and the surgical correction is easy and should lead to a normal anal function. If there is no contraction, as in the majority of cases, the prospects are not as good and a colostomy (the surgical construction of an artificial excretory opening from the colon) may be performed. (Courtesy of Cynthia Kaplan, M.D.)

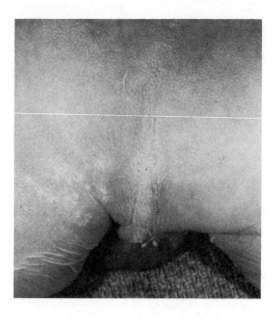

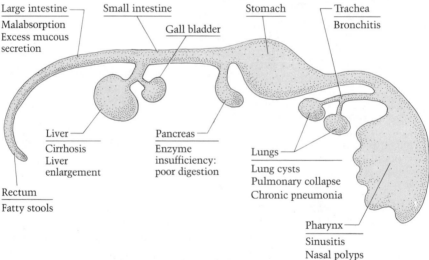

Large intestine — Small intestine — Stomach — Trachea
Malabsorption
Excess mucous secretion — Gall bladder — Bronchitis

Liver — Pancreas —
Cirrhosis — Enzyme insufficiency: poor digestion — Lungs —
Liver enlargement — Lung cysts
Pulmonary collapse
Chronic pneumonia

Rectum
Fatty stools

Pharynx —
Sinusitis
Nasal polyps

Figure 17-7 Cystic Fibrosis and the Endoderm
Cystic fibrosis is a disease which results from a defective secretion mechanism primarily in the tissues derived from embryonic endoderm. The accumulated mucus forms trapped plugs in the lungs, liver, and pancreas. Also the excess mucus coats the intestines and prevents absorption of nutrients. The mucus permits bacterial growth in the respiratory tract and destruction of the air sacs in the lungs, which causes emphysema-like symptoms and circulatory problems.

The collection of multiple effects of a single gene mutation is called pleiotropism. *One of the general features of gene function for a mutation is to lead to several tissue or organ disturbances. In addition to the endodermal defects noted for cystic fibrosis, there is a salt metabolism disturbance causing excess secretion of salt from sweat glands, which are of ectodermal origin.*

males, if they reach maturity, are sterile. Females, however, are fertile, because their reproductive organs have a mesodermal origin.

Most of Our Bulk Is
Produced by the Mesoderm
Most of our body is derived from the embryonic mesoderm. The mesoderm develops between the ectoderm and endoderm, organizing itself, at one site, into a rod-like structure which runs parallel to the developing neural plate. Neither the neural plate nor the rolled-up neural tube that it forms can develop in the absence of this mesodermal structure, called the *notochord*. The notochord is so basic to the vertebrate embryo that vertebrates are called *chordates* in taxonomic terminology.

From either side of the notochord a band of mesoderm moves laterally, splitting into two bands. The inner band wraps itself around the gut tube, forming thin membranes or *mesenteries* that hold our visceral organs in place, and the outer band forms the lining of the peritoneal cavity as well as the bulk of the body of the embryo beneath its outermost ectoderm.

Four buds burgeon, forming the *limbs*. The fingers and toes are partitioned by the alternating growth and death of the cells at the apex of these limb buds. The *muscles*, *bones*, *cartilage*, and *connective tissue* form from the mesoderm. So, too, do the *circulatory system* and the *blood*, as well as its organs of production, the *marrow*, *spleen*, and *lymph glands*.

The *kidneys* and their drainage ducts are mesodermal, arising initially as swellings just below the spinal column along either side of the sheet of mesoderm that forms the mesenteries.

Mesodermal Errors Frequently
Cause Malformations
Each of the organ systems produced by the mesoderm is controlled by many genes. Sometimes mutations prevent normal signals from performing the developmental events that are needed for the organ to form. As a result an infant may be born with malformations. If the limb buds do not develop, the child may lack arms or legs or they may be disproportionately reduced in size. Two examples of this are seen in *phocomelic* children with flipper-like arms and achondroplastic dwarfs whose limb bones fail to elongate.

Some of the mesoderm, in the vicinity of the embryonic kidney and in the posterior limb bud region, forms the male and female *internal* and *external reproductive* organs. Germ cells, which give rise to the sperm and eggs, are derived from the endodermal cells in the embryonic yolk sac.

Defects of the mesoderm can be systemic, as in the *Marfan syndrome*, an autosomal dominant disorder causing bone elongation; loose, elastic skin and joints; dislocated lenses; and imperfectly joined layers of the blood vessels leading to dissecting aneurysms. In this condition blood forces itself between muscle layers in the aorta, the major artery leading out of the heart to the rest of the body. The outer wall of the aorta then balloons out and may burst if surgical repair is not attempted.

Other mesodermal defects may be organ-specific. Infants may be born with only one lung because of a failure of one of the lung buds to continue its development. A baby may be born without kidneys or with defective drainage ducts to or from the bladder. There may be extra fingers and toes or fusion of two or more digits.

Environmental Agents Sometimes
Cause Birth Defects
Most birth defects are genetic, resulting from either single gene mutations or polygenic mutations. Any organ system may be defective (Figure 17-8). In large municipal hospitals the pediatric ward will have several babies in intensive care at any given time for heart defects (holes between the ventricles or auricles, malformations of the major vessels, malformations of the valves), neural tube defects, umbilical cord ruptures, organ system failures (lung, liver, kidney, bladder, pancreas, intestine, esophagus), or unusual formation of the head or limbs.

When a single-organ defect is present, such as a cleft lip or palate, it is probably a polygenic or simple Mendelian defect, although there are exceptions to this. If many organs are malformed, the problem is often environmental or chromosomal in origin. Environmental defects may arise from maternal exposure to x rays during early pregnancy; from ingestion of anticonvulsants such as dilantin for control of epilepsy; from over-the-counter drugs such as thalidomide, an unproven antinausea drug which caused babies to be born limbless or to have flipper-like arms; or from chemotherapeutic agents such as aminopterin for arresting tumor growth.

It is a good idea for pregnant women, especially in the first three months of pregnancy, to avoid smoking, drinking alcohol, exposure to x rays, or any prescription or over-the-counter drugs not absolutely essential. Weight-loss diets not medically supervised should be discontinued because any protein or vitamin deficiency could seriously affect the embryo's organ system development.

Chromosomes Cause Birth Defects
by Altering Normal Development
Whenever an extra chromosome is present, as in trisomy-13, -18, or -21, there may be an excess of gene products and regulatory signals from the genes that are present in triple dose. As a result there will be many defective organ systems because each chromosome probably has several genes participating in each major organ system. If key regulatory genes are involved in triple dose, a major abnormality can arise. It is not surprising then that infants with

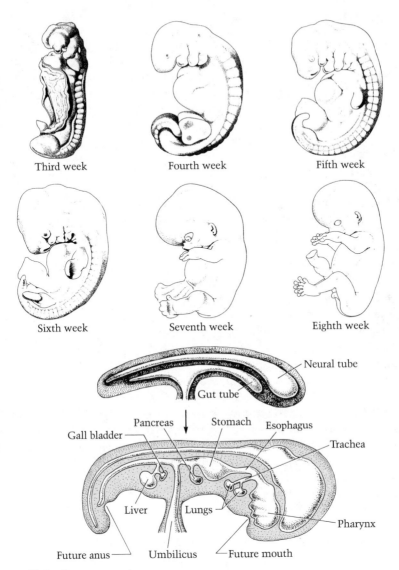

Third week

Fourth week

Fifth week

Sixth week

Seventh week

Eighth week

Neural tube

Gut tube

Pancreas   Stomach   Esophagus

Gall bladder

Trachea

Liver   Lungs

Pharynx

Future anus   Umbilicus   Future mouth

Figure 17-8 Organogenesis
*Human embryos form their organ systems at different stages of development.
The three-week-old embryo has a clearly defined spinal column but a poorly
shaped head. By the fourth week the head is prominent and eyes are
developing. Limb buds appear in the fifth week but do not form digits until
the seventh week. The external portions of the ears also appear by the seventh
week. During these early weeks of development the embryo is extremely
sensitive to chromosomal defects, viral infections, and chemical teratogens
(agents like thalidomide, which produce malformations). Below the six
external views of the embryo is a longitudinal section through a three-week-
old embryo showing the tubular system established from the neural ectoderm
and from the endoderm. A five-week-old embryo shows more organ
development from these initial structures. (Drawings of embryos from "The
Thalidomide Syndrome" by Helen B. Taussig. Copyright © 1962 by Scientific
American, Inc. All rights reserved.)*

these autosomal nondisjuctions have numerous birth defects. Each of these syndromes has its own cluster of major defects.

In trisomy-13, the absence of a forebrain results in a secondary defect—a cleft lip and palate. This is because the forebrain is essential for inducing the facial ectoderm to differentiate properly. The smallness or absence of eyes, however, is a primary defect, caused by failure of the optic cups to develop normally. The heart condition and defects in other vital organs may either be primary or secondary effects. Similarly in trisomy-18 or -21 there will be numerous mismatchings of tissue movements and threshold regulatory signals which the triple dose of genes brings about.

Occasionally partial deletions or duplications of other chromosomes may occur, producing syndromes rarely encountered before. It is worthwhile, if parents have a child with multiple congenital abnormalities, to request that the child be checked by means of a chromosome count and a banding study. This would identify whether the child is a mosaic or whether all of its cells are affected. If the child is a mosaic, having cells with normal chromosomes as well as cells with abnormal chromosomes, then it is not usually necessary for the parents to be tested. Nonmosaic infants, however, may indicate a germ cell or cryptic mosaic origin in one of the parents. If the parent is mosaic, some of the cells may have an abnormal chromosome number. In such cases, cells from more than one tissue type should be sampled (blood is mesodermal and skin is ectodermal).

Birth defects are detectable *in utero* for only a small number of the 2800 known syndromes and abnormalities that are recognizable at or shortly after birth. Improvements in metabolic analysis of birth defects, in fiber optics for *fetoscopy*, and in ultrasonic resolution may greatly increase the percentage of abnormal fetuses detectable prenatally.

Questions to Answer

1. Discuss the origin of the neural tube.
2. What are the major derivatives of the endoderm?
3. Discuss the origin of the embryonic tissues prior to organogenesis.
4. To what extent can you consider your body to be a "sandwich" of mesoderm between layers of ectoderm and endoderm?
5. Discuss the nonneural derivatives of the ectoderm.
6. What failure in early development do you think would cause an imperforate anus?
7. What is the embryological basis of spina bifida and anencephaly?
8. Which of the following are mesodermal derivatives: heart, bones, spleen, liver, kidneys, hair, lungs?

9. Define the following terms: neural plate, notochord, anencephaly.
10. Why would it be reasonable to consider cystic fibrosis as a disease of the endoderm?

Terms to Master

anencephaly
differentiation
ectoderm
endoderm
fetoscopy
imperforate anus

mesoderm
morphogenetic movements
neural tube
notochord
optic cups
spina bifida

Chapter 18
Singletons and Twins

Twins have been recognized since antiquity. Many of you may be familiar with the Old Testament account of the trickery used by Jacob to deprive his twin brother Esau of his birthright. The Greeks celebrated Castor and Pollux as the twin offspring of Jupiter and Leda (Leda was seduced by Jupiter in the guise of a swan). The Romans attributed the founding of Rome to Romulus and Remus, twin sons of Mars, who were suckled by a wolf. Shakespeare, whose wife bore him twins of unlike sex, used such a pair, Sebastian and Viola, in *Twelfth Night* and another pair of identical twins in *A Comedy of Errors*.

In some parts of the world twins are welcomed with joy as omens of good luck. This is particularly true in Mali and in Nigeria. In many other parts of Africa, as well as in the Australian aborigine tribes, Japan, and parts of India, twins have always been regarded as a misfortune, and sometimes the hostility toward the twins was also directed toward the parents.

There May Be a Social Selection
For or Against Dizygotic Twinning It is not surprising that this social selection has altered the birth rates for twins. The rate is highest among the Yorubans in Nigeria, who bear 45 sets of twins per 1000 pregnancies. It is lowest in Japan, where 7 sets of twins occur among 1000 pregnancies. In Europe and the United States the average is about 12 sets of twins per 1000 pregnancies (Table 18-1).

Twins of the same sex may look as dissimilar as singleton siblings, or they may be so alike that only a family member or friend can tell them apart readily (Figure 18-1). An identical twin, as we have already seen, is actually a clone of the other twin. Such twins have an identical

Table 18-1   Frequency of Twin Pairs per Thousand Births

The frequency of DZ (dizygotic) twins is more variable than that of MZ (monozygotic) twins. Thus in Nigeria there are eight pairs of DZ twins for every pair of MZ twins. In Japan, however, there are more MZ twins than DZ twins. There is good reason to believe that multiple ovulation can be selected genetically. The cultural attitudes towards twins and multiple births in the recent past in Nigeria and Japan support this view.

Birthplace	Total	DZ	MZ
England	12.3	8.8	3.5
United States (white)	11.3	7.1	4.2
United States (black)	15.8	11.1	4.7
Gambia	16.6	9.9	6.7
Nigeria	44.9	39.9	5.0
India	16.8	8.5	8.3
Japan	6.8	2.9	3.9

Source: Adapted from S. J. Strong and G. Corney (New York: Pergamon, 1967).

Figure 18-1   Identical Twins
Identical twins (monozygotic or MZ twins) retain a lifelong similarity of appearance. Two-egg twins (dizygotic or DZ twins) do not. In this photograph note the striking similarity in facial features of these male MZ twins. (Courtesy of Harvey Stein)

genotype because they arise by a partition of two or more cells derived by mitotic divisions of the original zygote. Such identical twins are also called one-egg or *monozygotic twins*. In technical publications they are called MZ twins.

Opposite-sex twins and dissimilar same-sex twins are usually derived from separately fertilized eggs. Thus they are variously called fraternal, nonidentical, two-egg, or *dizygotic twins*. They are usually designated as DZ twins.

The Hensen-Weinberg Rule Can Be Used to Determine Twin Frequencies

The frequency of MZ and DZ twins can be determined directly by clinical and biochemical tests, or indirectly by the Hensen-Weinberg rule (Table 18-2). In this calcula-

Table 18-2   Rules Governing the Frequency of Twins

Hellin's rule applies to spontaneously occurring multiple births. Whatever the factors governing multiple ovulation and multiple partitioning may be, they act independently because the same probability applies to each additional multiple birth. The rule does not apply to artificially induced twins where the use of clomiphene or other hormones resulted in multiple ovulation.

The Hensen-Weinberg rule assumes that half of all two-egg twins are boy-girl twins. If the total of all twins is known and the frequency of boy-girl twins is known, then the formula $T - 2D = M$ can be used to calculate the frequency of one-egg twins. The data for the United States at the time when all twins were of spontaneous origin, shows that a little over 25 percent were MZ twins.

Rule

Hellin's rule	*Expected*	*Obtained*
1. n = frequency of twins	1 in 86	1 in 86
2. n^2 = frequency of triplets	1 in 7396	1 in 7604
3. n^3 = frequency of quadruplets	1 in 636,056	1 in 669,922

Hensen-Weinberg rule

1. Unlike-sex twins must be DZ twins = D.
2. For each two eggs fertilized, $\frac{1}{4}$ are both male, $\frac{1}{2}$ are male and female, and $\frac{1}{4}$ are both female.
3. Thus 2 × (pairs of unlike-sex twins) = $2D$ = total DZ twins.
4. The total of all twins = T.
5. $T - 2D = M$, where M is the frequency of MZ twins.

For 1890–1910 (before attempts to induce ovulation) in the United States

1. 717,904 twins = T
2. 264,098 = D = male and female twins
3. 717,904 − 2(264,098) = M
4. M = 189,708 = 26.4%
5. Since sex ratio is normally 105 male to 100 female, M corrected = 25.6%.

tion it is assumed that half of all DZ twins are of unlike sex. Thus the total of all twins minus twice the frequency of twins of unlike sex should equal the MZ frequency. Both methods yield the same results. In the United States and Europe there are three to four MZ twin pairs and seven to ten DZ twin pairs per 1000 pregnancies. When the world's twin frequencies are studied, a surprising fact emerges. Only DZ frequencies seem to vary substantially; the MZ rate remains relatively unaffected in high or low twin frequency cultures. Among the Yorubans in Nigeria, 95.5 percent of all twins are DZ, but among the Japanese 57.3 percent of all twins are MZ twins. In both cases the MZ frequency per 1000 pregnancies is roughly the same.

The genetic mechanism for DZ twinning is not known but it is probably multifactorial and it may involve the response of ovarian follicles to hormonal controls. In families who have DZ twins there is a greater chance that multiple ovulation will occur in later generations. There are also some environmental factors conducive to DZ twinning, including maternal age, older mothers being more likely than younger ones to have twins. Even less is known about the environmental conditions that produce MZ twins. Such twins, however, are independent of parental age.

The Embryonic Origin of Twins
Can Sometimes Be Told
from Their Membranes   Clearly, DZ twins arise from the
maturation and rupture of two follicles (or less often, the release of two eggs from one follicle). These two eggs, after fertilization in the oviducts, may end up in separate locations in the uterus, resulting in two embryos, each with its own set of surrounding membranes (Figure 18-2). If the implanting eggs end up near each other in the same part of the uterus, they may form a fused set of outer membranes.

The relation of the membranes to the twins is complex but the components participating in this relationship are few in number (Figure 18-3). The singleton embryo is cushioned by a fluid-filled membrane, the *amnion*. The embryo may have a small, temporary *yolk sac* between its umbilical stalk and the amnion. Surrounding the amnion, the embryo, and the yolk sac is the *chorion*. It is the chorion that provides the exchange of liquids and gases between the mother and the fetus.

The region of the chorion that presses against the uterine wall contains a network of the embryo's blood vessels branching off from the umbilicus. The mother's blood vessels and the embryo's blood vessels remain separate, with all metabolic exchanges occurring across the chorionic membrane of the fetus and the uterine lining of the mother. The resulting structure is called the *placenta*, and is the region for nutritional and respiratory exchanges. A third membrane, the *allantois*, collecting metabolic wastes (mostly urea) from the embryo, presses against the chorion at its placental junction. Since both MZ and DZ twins have separated unfused allantoides, we shall omit this membrane in discussing the differences among twin placentas.

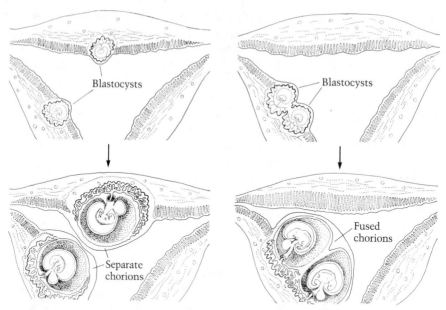

Figure 18-2   Origin of Two-Egg Twins

Dizygotic or two-egg twins are derived from separate fertilizations of two eggs. The two eggs may be both from one ovary or from separate ovaries. If the two blastocysts implant in the uterus some distance apart from each other, each develops a separate chorion and amnion. If the two implant close together, the chorions will fuse. Most DZ twins have separate dichorionic placentae.

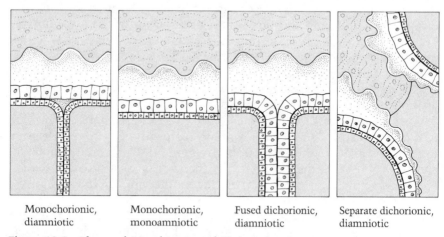

Figure 18-3   Placental Membranes and Twins

At birth, the obstetrician will frequently check the membranes of the afterbirth, the placental membraneous sac which protected the twins during pregnancy. A single monochorionic structure, containing one or two amnions, is always associated with MZ twins. Twins found with two chorions, fused together along one side or completely separated, may be either MZ or DZ twins. Thus other factors must usually be taken into account to prove whether the twins were MZ or DZ in origin.

If the two blastomeres of the first mitotic cleavage separate, they may later implant, like DZ blastocysts, in different parts of the uterus (Figure 18-4). Each identical twin will then have a separate set of membranes. This condition establishes *separate dichorionic placentae*. If the two blastocysts from the separated blastomeres implant near one another, their chorions will fuse along one side and a common placenta will supply both twins. Such twins have a *fused dichorionic placenta*. MZ and DZ twins implanting this way would have indistinguishable membrane compositions.

Most MZ twinning, however, occurs within the blastocyst, either in the inner cell mass or in the embryo derived from it (Table 18-3). If the embryonic tissue mass is partitioned into two pieces, they will develop a single chorion but each embryo will establish its own amnion. This leads to a *monochorionic diamniotic placenta*. Only MZ twins have such a structural arrangement, but it accounts for only about 65 percent of all MZ twinning. Dichorionic twins (fused or separate) account for almost all the rest of the MZ twins. A very small number of MZ twins (about

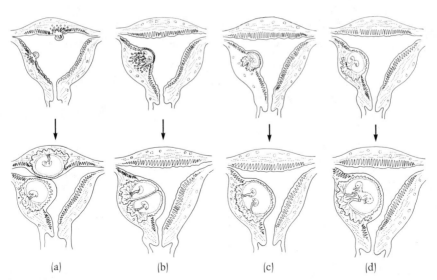

(a)   (b)   (c)   (d)

Figure 18-4   Origin of One-Egg Twins
Monozygotic (MZ) twins arise from a single fertilized egg. Three fates are possible for such an egg if it should form twins. (a) The blastomeres may separate into two masses, each forming its own blastocyst. These blastocysts can implant nearby or far apart, thereby producing placentae similar to those found for DZ twins. Most MZ twins, however, form after the single blastocyst implants. (b) The inner cell mass, an embryonic tissue mass, separates into two pieces, each forming a separate embryo usually enclosed in separate amnions but encased in a single chorion. (c) Later separation of the early embryonic tissue mass results in twins within a single amnion. (d) Very late separations of the embryonic mass can lead to conjoined (Siamese) twins.

Table 18-3   Embryonic Membranes and Twinning

When twins are born the membranes are useful in determining the zygosity of the twins. DZ twins never have a single chorion, whether that chorion contains one or two amnions. Unfortunately only about 60 percent of all MZ twins are monochorionic. The double chorion, separate or fused, can be found in either MZ or DZ twins. The Hensen-Weinberg rule showed that in the United States and Europe about 75 percent of twins with a double chorion are DZ twins. If the twins are not monochorionic, other means must be used to determine zygosity, such as blood groups, HLA antigens (present in the skin of individuals), and the similarities of facial and body features.

Membrane Structure	MZ	Percentage	DZ	Percentage	Zygosity Unknown	Total	Percentage
Separate dichorion	12	15	69	85	4	85	42.5
Fused dichorion	13	23	44	77	9	66	33
Monochorion diamnion	38	100	0	—	6	44	22
Monochorion monoamnion	1	—	0	—	0	1	0.5
Unknown	1	—	1	—	2	4	2
Total	65	36.4	114	63.6	21	200	—

Note: The percentage of MZ and DZ for membrane structure excludes the cases of unknown zygosity.

Source: Adapted from S. J. Strong and G. Corney (New York: Pergamon, 1967).

Table 18-4   Selection and Twinning in Nigeria

Zygosity of twin pairs

Area	%DZ	%MZ
Yoruba	95.5	4.5
Hausa	82.3	17.7
Other N. Nigerians	73.7	26.3

Membrane structure of twins

Area	Dichorionic MZ	DZ	Monochorionic MZ	DZ
Yoruba	0	63	3	0
Hausa	4	51	7	0
Other N. Nigerians	8	56	12	0

Note: The Yoruba and Hausa cultures in Northern Nigeria have regarded twins as good luck omens. Note that DZ twinning is more frequent among these two cultures than in other cultures of Northern Nigeria which do not worship twins, such as the Fulani, Idoma, and Igalla cultures.

Source: Adapted from I. MacGillivrany, P.P.S. Nylander, and G. Corney (Philadelphia: Saunders, 1975).

2 percent) have embryonic separation leading to *monochorionic mon-oamniotic placentae*. This is a hazardous arrangement. The two fetuses are in contact with each other, and the chances that their umbilical cords will twist and cut off circulation is high. The frequency of formation of this MZ monoamniotic placenta is much higher than 2 percent, but most such pregnancies abort spontaneously. The risk of congenital malformations of one of the twins is very high for this type of twin birth.

The relation of embryonic membranes to DZ or MZ twinning may be seen in the studies of Nigerian twins (Table 18-4). Most of the Nigerian twins must arise from multiple ovulation because the fertilization of two eggs leads to dichorionic membranes. Only MZ twins, as we have seen, can form monochorionic membranes.

Conjoined or Siamese Twins Arise
Late in the Embryo's Development

The rarest type of twinning, and the type that occurs latest in the development of the embryo, involves the imperfect separation of the embryo about 15 days or more after zygote formation. This produces conjoined twins, popularly called Siamese twins. Most conjoined twins are inseparable (technically called *teratologies*) and do not survive. The separation of conjoined twins is a medical rarity, since the twins may be sharing a vital organ (e.g., there may be a common heart in breast-to-breast twins).

An MZ Twin Diffusion Syndrome
May Arise and Can Cause
Dissimilarities in Birth Weight

When twins share a common placenta, especially within a single chorion, there is a high probability that the smaller vessels branching from one twin's umbilical arteries and veins will encounter those of the second twin and anastomosis will take place (Figure 18-5). Such anastomoses or connections enable blood from one twin to circulate in the other. When a tissue mass, such as the inner cell mass or the early undifferentiated embryo, cleaves into two pieces, one segment may be larger. The bigger piece may develop somewhat faster and thus establish its organ systems earlier. If anastomosis occurs between a larger and a smaller fetus *in utero*, the larger twin may utilize more resources at the expense of the smaller twin. In extreme cases the smaller twin can die *in utero* or abort, or it may develop without a heart or head, existing until birth as a parasite of its twin. More often in the case of dissimilar MZ twins, the larger one is born first and appears ruddy, and the smaller one looks pale and anemic. If the quantity of hemoglobin in the fetal blood is reduced by 35 percent or more from a normal level, the child has an *MZ twin diffusion syndrome*.

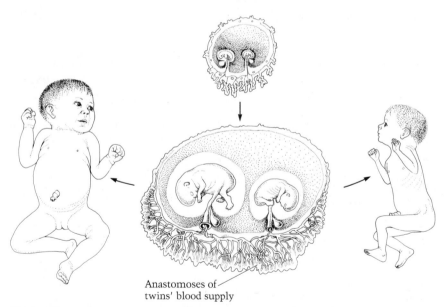

Anastomoses of
twins' blood supply

Figure 18-5   The MZ Twin Diffusion Syndrome
Monozygotic twins with a single chorion may have connections, anastomoses, between their blood supplies in the placenta. If the embryonic mass separates into two unequal pieces, the larger developing fetus may deplete some of the blood of the smaller fetus. At birth one identical twin may be robust and ruddy, while the other may be emaciated and pale. Such physical differences usually disappear within six months after birth. There is an ongoing controversy about the amount of mental damage (i.e., lower IQ), if any, that is done to the smaller twin from the diffusion of the fetal bloods.

It Is Debatable Whether
the MZ Twin Diffusion Syndrome
Affects IQ

Although the weight difference between twins disappears a few months after birth, there may be long-lasting effects on the smaller twin. According to Munsinger, there may be a mean IQ decrease of about 11 points, apparently the result of intra-uterine anoxia and malnutrition. Since two-thirds of MZ twins are monochorionic and one-third of those have anastomoses of their vessels, some 22 percent of all MZ twins have a prenatal bias in the formation of their nervous system. The larger twin's IQ is not raised. The mean IQ of the larger twin is 99.2, with the smaller twin of such pairs averaging 87.8. Where no major weight difference exists, the IQs of the heavier and the lighter twins are, respectively, 101.9 and 98.0, showing no significant difference between the twins.

IQ differences in twins of differing birth weight, however, have been questioned by Kamin who finds not more than a five-point decrease in IQ for the smaller birth weight twin when verifiable weight differences are used. Kamin disputes minor IQ decreases resulting from malnutrition or anoxia, and believes most differences in IQ reflect postnatal social, familial, and educational practices rather than prenatal physiological differences between the one-egg twins.

From a biological perspective, we can now see why there are more hazards in the development of MZ twins than DZ twins. DZ twins, having dichorionic structures, do not experience a diffusion syndrome and their neurological development is the same as for singletons.

More Than Two Embryos May Be
Present in a Human Uterus

Rare cases of triplets, quadruplets, and quintuplets have been recorded since ancient times, but larger numbers of embryos or babies reported in historical times (litters of 12 or 365!) are hoaxes. The frequency of multiple births is easily determined by using *Hellin's rule* (see Table 18-2). If n is the frequency of twins (i.e., 1 in 86 births) then n^2 is the frequency of triplets, n^3 of quadruplets, and n^4 of quintuplets. Triplets in the United States and Europe are thus found about once in 7500 births. Quadruplets occur once in about 700,000 births, and quintuplets about once in 60 million births. In recent times multiple births, especially three or more, have appeared more frequently than Hellin's rule predicts. This rise in frequency comes from the treatment of infertile females with a compound called chlomiphene, which induces multiple ovulation.

Any uterus with more than two embryos in it is crowded. It is no surprise that such children are often born prematurely and weigh less than five pounds (2150 g), which is the weight considered essential for life outside a uterus or special incubator. The mortality rate of such infants is very high, and only in the 20th century have triplets, quadruplets, and quintuplets had good chances for survival.

Triplets can be of MZ origin, tri-ovular, or a combination of an MZ pair and a second fertilized egg. Similar variations exist for quadruplets and quintuplets, although hormonally-induced multiple births are always multi-ovular.

Establishing Zygosity of Twins
Is Not Always Easy

Newborn babies often look alike, even when born to different parents. How then can twins be classified as MZ or DZ? We have seen that the examination of the placenta is helpful but it is not sufficient except for monochorionic *afterbirths* (the expelled fetal membranes), which are always an indication of MZ twins (or only some 23 percent of all twin pairs born). To distinguish dichorionic MZ and DZ twins, genetic, biochemical, and clinical studies are needed. Clearly almost all unlike-sex twin pairs are DZ, one being from an XX zygote, the other from an XY zygote. Same-sex dichorionic twins can be classified by blood groups, physical features, including birth marks, enzyme studies, and in some cases by skin graft tests (because only MZ twins can exchange skin grafts without rejection). Since MZ twins have a common genotype, they should match for blood groups and the properties of their proteins. They should also accept each other's skin graft. Occasionally one twin later in life may need a kidney transplant, and the zygosity of the twin has to be determined accurately if the kidney is to function normally the rest of that person's life. Usually, skin grafts are not done unless there is some medical problem requiring an organ transplant.

Although 20 percent of MZ twins may look strikingly different at birth because of the diffusion syndrome, they become more alike in appearance as they mature (Figure 18-6). DZ twins, however, become progressively more unlike each other as they age. MZ twins also have similar fingerprint patterns (loops, arches, or whorls), but they never have identical fingerprints.

MZ twins arising from embryonic separation may show a mirror-image symmetry, with hair whorls and other differences seen on the left side of one twin and on the right side of the other. In extreme cases a major organ such as the stomach, heart, or aorta, may be shifted to the right side of the body in one twin. Handedness, however, is not strongly related to this mirror-image symmetry effect.

There Are Fallacies
About What MZ Twins Can Do

MZ twins do not have thought transference, and each develops a separate personality, even when some personality traits are similar. Popular novels like *The Corsican Brothers* by Dumas create the illusion that one twin feels the pain or joy of the other twin, even when they are separated from one another, but this has not been found for the 60 or more separated MZ twins and the hundreds of nonseparated MZ twins who have been studied for their various char-

Figure 18-6   Symmetry in Identical Twin Features
If a photograph of identical twins is cut vertically from crown to chin, and the left side of one twin is joined to the right side of the other twin, the reconstituted picture still resembles the original twins (see Figure 18-1). This is not true for same-sex twins who are of two-egg origin. (Courtesy of Harvey Stein)

acteristics. While MZ twin pairs have sometimes achieved fame in similar fields (e.g., identical twins Abigail Van Buren and Ann Landers, and marine explorer twins Jean and Auguste Picard), other MZ twins choose dissimilar occupations, and occasionally one twin will elect an advanced degree or profession while the other will consider his or her education complete upon graduation from high school. These diverse occupations and lifestyles seen in some MZ twins do not support genetic determinism for their social traits.

**Twins Have Been Studied Frequently
in the Nature-Nurture Controversy**   Galton was the first to suggest the use of identical twins for analysis of hereditary and environmental components in human traits. Hundreds of studies have since been done which have attempted to assign precise percentages to the hereditary components in complex traits such as intelligence, talent, longevity, general health, and personality. All of these traits are currently the sub-

ject of ongoing research and considerable controversy from other scientists who dispute these claims. Physical features such as height, weight, fingerprint patterns, facial features, and susceptibility to specific diseases (e.g., heart disease, arthritis, and cancers) have shown more convincing, or less disputed, evidence that heredity plays a major role in producing similarities and high risks for these traits.

Although scientists will continue to use twins for behavioral research, we should recognize certain defects in the use of twin studies:

1. About 20 percent of MZ twins have nonidentical prenatal environments which may alter their development.
2. Parental response to twins may be more uniformly applied to each MZ twin than to same-sex DZ twins.
3. MZ twins may be more aware of their clonal identity and for this reason may, at an early age, emulate each other more than same-sex DZ twins do.
4. Separated MZ twins are rarely encountered today because the social services which try to keep families together are more prevalent than they were in the first half of the 20th century. Thus separated twin studies usually involve very small sample sizes.

While the first point in this list tends to raise the reliability of heritability estimates, the other three points tend to cast doubt on the accuracy of such measurements. These, and other complexities of upbringing, make twin studies far from ideal for resolving the nature-nurture controversy.

Questions to Answer

1. Why would you expect that DZ twins would be more likely to be subject to genetic control than MZ twins?
2. If the frequency of twins in Nigeria is 1 in 30, what is the expected frequency of triplets? Of quadruplets?
3. What type of amniotic membrane structure would you expect in MZ twins?
4. What type of amniotic membrane structure would you expect in DZ twins?
5. Can identical twins look less alike at birth than two-egg twins of the same sex? Support your position.
6. How would it be possible for MZ twins to be of unlike sex?
7. What are conjoined twins and what is their probable origin?
8. Why are twin births more risky than singleton births?
9. What is the MZ twin diffusion syndrome?
10. Discuss the differences between the attributes described in science fiction clones and those found in real-life identical twins.

Terms to Master

afterbirth
amnion
chorion
conjoined twin
dizygotic

Hellin's rule
Hensen-Weinberg rule
monozygotic
MZ twin diffusion syndrome
placenta

Chapter 19
Sex Determination: The Seven Levels of Human Sexuality

Although society has changed many of its stereotypes about masculine and feminine behavior, there is almost universal agreement that humanity exists as two sexes, male and female. Sexual identification is usually made at birth by examination of the external genitalia, and it is recorded on a birth certificate, thus establishing a *legal sex* for the child. Fortunately, most male and female infants develop normally—biologically and psychologically—and accept their sexual identification. In a small number of cases this classification fails and the infant cannot be properly assigned a legal sex at birth. In a larger number of cases the biological and psychological or cultural identifications of the sexes are at odds, resulting in individuals who are homosexuals, transvestites, or transsexuals.

If we study the components of human sexuality, we can establish seven levels of sexual development. Any one of these, if modified, can lead to ambiguous sexual identification. Sexuality is complex because there are six major organ systems involved in its development: the germ cells, the gonads, the internal genital structures, the external genitalia, the brain, and the glands whose hormones determine reproductive function. These six biological components of human sexuality and the psychosocial concept of gender constitute what can be called the seven sexes of humans. We can refer to seven sexes in humans because the term sex can be applied either as the sum of all the functions we identify with a male or a female or to the particular structures or functions involved in the composite we recognize as a male or a female. When referring to these components, we use the terms chromosomal sex, genetic sex, gonadal sex, and so on.

The Reproductive Cells and
Gonads Have Separate Origins

The sperm or eggs of a mature adult are derived from the primordial germ cells of the embryo. These cells are formed about the sixth to tenth week after fertilization and they multiply in the upper portion of the embryonic yolk sac (Figure 19-1). The *gonads*, which eventually become *testes* or *ovaries*, have a different origin. They develop as a bulge near the embryonic kidney and consist of an inner core, called the *gonadal medulla*, and an outer rind, called the *gonadal cortex*. They are like an empty house, whose tenants (the primordial germ cells) have not yet moved in. About the sixth to eighth week of embryonic development the primordial germ cells leave the yolk sac and crawl like amoebas to the embryonic gonads, which they invade (Figure 19-2). If the primordial germ cells are chromosomally XX, they proliferate in the cortex, the medulla degenerates, and the gonads become ovaries. If the primordial germ cells are XY, they multiply in the medulla, and the cortex atrophies, producing testes. Whether a child is born with testes or ovaries, the gonad itself is of dual origin, the sperm or eggs having

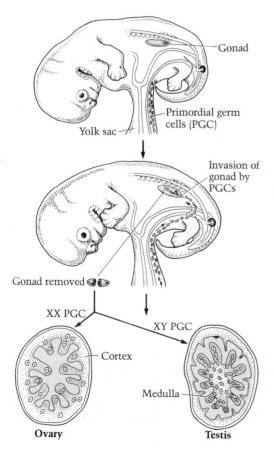

Yolk sac

Gonad

Primordial germ cells (PGC)

Invasion of gonad by PGCs

Gonad removed

XX PGC

XY PGC

Cortex

Medulla

Ovary

Testis

Figure 19-1 The Invasion of the Gonad by Primordial Germ Cells

The primordial germ cells multiply in the upper reaches of the yolk sac and migrate to the gonad, a mesodermal structure which forms near the embryonic kidneys. The XX primordial germ cells proliferate in the cortex or outer rind of the gonad. The XY primordial germ cells proliferate in the medulla or central core of the gonad. The adult gonad thus has a dual origin: its reproductive component is derived from the endodermal primordial germ cells and its glandular component is derived from mesoderm.

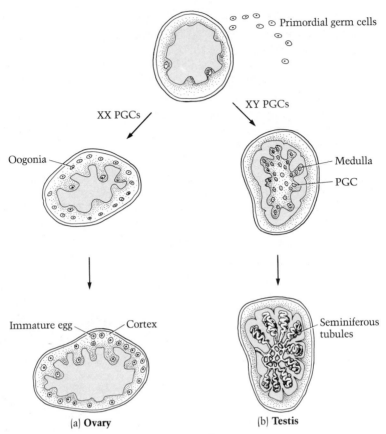

Figure 19-2   The Dual Origin of the Gonads
The sexually neutral embryonic gonad (with inner medulla and outer cortex) is invaded by primordial germ cells (PGCs). In the female (a) XX PGCs proliferate in the cortex, the medulla atrophies, and the gonad becomes an ovary. In the male (b) the XY PGCs proliferate in the medulla, the cortex atrophies, and the gonad becomes a testis.

their embryonic origin not from the gonad but from the yolk sac. The yolk sac is part of the embryonic endoderm and the gonad itself is mesodermal.

Gonads Cannot Form Without
an Invasion of Primordial Germ Cells   If there are no functional primordial germ cells because of some birth defect, the gonads may not develop into adult-sized ovaries or testes. When a zygote is formed bearing a single X chromosome and no Y, the condition (which we learned is called Turner syndrome) results in an absence of functional germ cells and ovaries that degenerate in the fetus. Poorly developed gonads ("streak gonads") may form instead of ovaries. Testes cannot form in the absence of a Y chromosome.

Not all of the Y chromosome is male-determining. There is a region which leads to a detectable protein, called the HY antigen, which is usually found in male tissue but occasionally is present in female tissue. XY individuals whose Y chromosome fails to produce this HY antigen are phenotypically females at birth, similar to Turner females but lacking the dwarf stature, webbed neck, and other nonsexual characteristics of the disorder.

Chromosomal Sex Disorders May Include
Abnormalities Unrelated to Sex

At this point we can identify two of the seven sexes of humans: *chromosomal sex* and *gonadal sex*. Normally a male has an XY and a female has an XX chromosomal sex. But the XO Turner syndrome, with streak gonads, and the XXY Klinefelter syndrome, with sterile, pea-sized testes, suggest what is probably going on in the embryo. The XO does not produce functional germ cells and the XXY produces inadequate signals in the medulla for the gonad. In Klinefelter syndrome a modest growth of the medulla and a degeneration of the cortex occur but the XXY primordial germ cells do not multiply effectively and they do not develop meiotically. Both Turner and Klinefelter adults are permanently sterile.

The gonadal sex arises from the reciprocal relation of the cortex and the medulla of the gonad. The primordial germ cells, if functional, will initiate cortical (ovarian) or medullary (testicular) dominance. Presumably these tissues respond to products released by the proliferating XX or XY primordial germ cells in the embryonic gonad. Although XYY and XXX chromosomal sexes also involve abnormal sex chromosome numbers, the XYY primordial germ cells usually produce fertile functional testes, and XXX primordial germ cells usually produce fertile functional ovaries. At least at this level of sexual development the XXX or XYY chromosome sex is within a normal range.

The Internal Genitalia Arise
from the Müllerian and Wolffian Ducts

Two duct systems accompany the development of the embryonic gonads—the *Wolffian ducts* and the *Müllerian ducts* (Figure 19-3). Since the potential male and female both have a common set of undifferentiated gonads and ducts, the embryo at this stage is often considered sexually neutral. After the invasion of primordial germ cells, however, the gonadal sex is determined, with testes or ovaries as the outcome of medullary or cortical growth. In the testes, but not the ovaries, a hormone is released called the *Müllerian Duct Inhibitor* (MDI). It is specific for preventing the differentiation of the Müllerian duct into oviducts (fallopian tubes), uterus, and the upper third of the vagina. Since the female lacks MDI, her Müllerian ducts form these internal genital structures.

In the male, MDI prevents the development of internal female genitalia, but has no other effect on sexual development. The Wolffian ducts

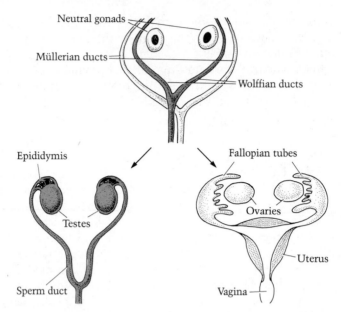

Figure 19-3   The Complementary Relation of the Internal Genitalia
At about the eighth to tenth week of development, the human embryo has neutral gonads with two adjoining duct systems, the Wolffian ducts (dark) and the Müllerian ducts (light). About the 13th week the gonads differentiate in response to the invasion of primordial germ cells. If these cells are XY, the gonads become testes (left) and release Müllerian Duct Inhibitor (MDI) which causes the disappearance of the Müllerian duct. A second hormone from the testes, testosterone, converts the Wolffian duct into the semen-collecting apparatus (epididymis and sperm duct).
*   In the developing female embryo neither of these hormones are released. The absence of testosterone causes the Wolffian duct to degenerate, and the absence of MDI permits the Müllerian duct to develop into the oviducts (Fallopian tubes), the uterus, and the upper third of the vagina. Some biologists believe the female reproductive system is not induced by any hormones, but is the constitutive state of the embryo.*

respond to a second male hormone, *testosterone*, also released by the testes. In response to testosterone the Wolffian duct forms a network of collecting tubules and storage vesicles for transporting sperm from the testes to the penis. The epididymis and sperm duct are the most apparent structural changes brought about by testosterone. MDI has no effect at all on the Wolffian duct.

This difference in function of testosterone and MDI has been demonstrated in mammals by surgical removal of the neutral gonads. If an agar cube, impregnated with testosterone, is implanted near the ducts, the Wolffian duct will form male internal genitalia (the collecting tubules, sperm duct, and epididymis). The Müllerian duct, in the absence of gonads, will form oviducts, a uterus, and the upper third of the vagina, regardless of the presence or absence of testosterone. In both these experiments it does not matter if the embryo is XX or XY. Chromosomal sex normally leads to gonadal sex, but it is the gonadal sex which produces the *internal genital sex*. One striking feature of mammalian sex determination is the control or regulation of embryonic structures. Regardless of XX or XY constitution, the embryo will become a female unless the male hormones MDI and testosterone prevent this. MDI *turns off* Müllerian duct differentiation, and testosterone *turns on* Wolffian duct differentiation.

The Male and Female External Genitalia Arise from Common Embryonic Genital Rudiments

There is a common meeting ground for the Wolffian and Müllerian ducts when these join their mates in the posterior region of the embryo. They enter the *cloaca* which eventually provides the exit for eggs, sperm, and urine. Where the cloaca comes in contact with the posterior ventral surface of the embryo, a bulging occurs consisting of a *genital tubercle*, surrounded by *genital folds*, and a larger pair of *genital swellings* (Figure 19-4). As in the case of the gonads and the internal genitalia, the rudiments for the external genitalia are identical in the early XX or XY embryo.

The same male hormone, testosterone, that causes Wolffian duct differentiation causes the external rudiments to form the male genitalia. In the absence of testosterone these rudiments form the female's external sexual apparatus. The female's genital tubercle enlarges slightly to form a *clitoris*. The genital folds form the *labia minora*, and the genital swellings become the *labia majora*. These are the external structures leading to the *vagina*, whose lower two-thirds is derived from the differentiation of a cloacal structure called the *urethra*. That urethral portion of the vagina joins the Müllerian duct portion of the vagina, which connects it to the uterus and oviducts.

In the male, testosterone causes the genital folds to elongate and join together forming part of the *penile shaft*. The genital tubercle enlarges as it is pushed forward by the developing folds and becomes the *penile glans* or head of the penis. The rest of the penile shaft, whose internal lumen is connected to the sperm duct, is derived from the urethra. The genital swellings also enlarge, forming a sac which joins together in a common *scrotum*. Much later, the testes, which are internal, descend, entering the scrotum through the inguinal canals of the embryo. The differentiation of those visible male and female structures constitutes the *external genital sex*.

Hormones Govern Puberty and the Sexual Activity of Mature Adults

Although boys and girls are clearly differentiated by their external genitalia, there are major changes that occur as they mature during puberty (between the ages of 10 and 20). In males increased production of testosterone results in larger size, deeper voice,

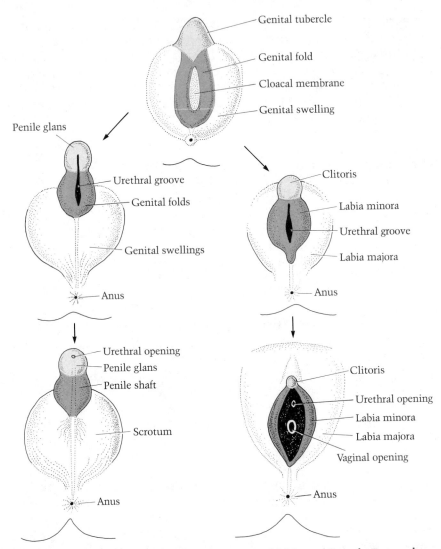

Figure 19-4 The Common Embryonic Origin of Male and Female External Genitalia

The genital tubercle of the early embryo enlarges to form the glans or head of the penis in the male. It remains small and is recessed in the upper apex of the vagina in the female. The embryonic genital folds move outward and curl around to form a penile shaft in the male (along with some of the urethral tissue), but in the female they form the labia minora. The genital swellings form the scrotum in the male and the labia majora in the female.

and hairiness (especially facial hair). The male hormones also cause hair to grow in the pubic region and in the armpits. They also stimulate sperm production in the testes. Puberty normally begins between the ages of 12 and 14 for boys.

In girls this pubertal change starts earlier, usually between the ages of 10 and 12, and it is also hormone-dependent. The ovaries, in response to pituitary hormones, begin producing eggs, the menstrual cycle commences, and the ovarian and pituitary hormones stimulate breast development and pelvic bone enlargement. There is also hair growth in the pubic region and in the armpits. The pubertal changes in males and females produce the *secondary* or *mature sex.*

Mutations Affecting Sexuality May
Cause Pseudohermaphroditic Disorders   Occasionally there are errors other than nondisjunction which interfere with the normal development of the embryo. Fortunately most of these are rare. Their analysis provides us with some valuable insights into the way genes control human development. In the more startling cases there is an inconsistency between gonadal sex (testes or ovaries) and the external genital sex (penis with scrotum or vagina with labia). Individuals with such ambiguities are called *pseudohermaphrodites.*

An XY Zygote Can Occasionally
Result in a Female Baby   There is a gene on the X chromosome that controls the response of some of the body's tissues to testosterone. If this gene is mutated, the testes still develop normally, MDI is released, and there is no formation of uterus, oviducts, or the upper third of the vagina. Testosterone is also released, but it has no effect on the external genital development because the cells are insensitive to testosterone. The internal differentiation of the Wolffian duct, however, does occur and the testes have an accompanying set of collecting tubules with the epididymis and sperm ducts. At birth such individuals appear as females, their external genital sex being used to determine their legal sex. They are raised as girls and think of themselves as females (Figure 19-5). At puberty they develop breasts, but they have little pubic or axillary hair.

This X-linked recessive disorder is called *congenital insensitivity to androgen syndrome* (CIAS), formerly called testicular feminization syndrome. The chromosomal sex of such females is XY. Because their gonads form testes, these are sterile females who never menstruate. They lack the internal female sexual organs and they have a short vagina, consisting of the lower two-thirds derived from the urethra. CIAS females have normal male levels of testosterone in their blood. They will not masculinize even when given additional testosterone in large amounts. CIAS females have a high incidence of undescended or partially descended testes, which are frequently removed. This is an important surgical procedure because undescended testes have a 25 percent chance of becoming cancerous.

It is important to note that CIAS females think of themselves as females. Since their hormones are from their testes and they do not have ovarian hormones, neither sex hormones (gonadal sex) nor the XY condition (chromosomal sex) determines this *psychological sex.* The mas-

Figure 19-5 Congenital Insensitivity to Andro-
gen Syndrome
Ordinarily XY zygotes develop
into males. In the congenital
insensitivity to androgen syn-
drome (CIAS) a gene mutation
on the X chromosome makes
most of the body tissues un-
responsive to testosterone.
Thus the fetal external geni-
talia develop as female struc-
tures. The tissues do respond
normally to the Müllerian
Duct Inhibitor released by the
testes. The internal genitalia
consist of a shortened vagina
with no uterus or oviducts.
The testes usually remain in
the body cavity until puberty.
A CIAS female, as shown
here, has a feminine body. The
mothers of XY CIAS females
are fully normal. CIAS fe-
males are male pseudoher-
maphrodites—that is, they
have male gonads and female
genitalia. (Courtesy of R. J.
Gorlin, D.D.S.)

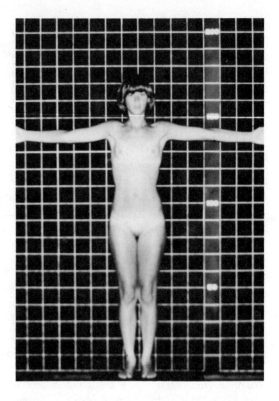

culine or feminine psychological identification one has is primarily social in origin.

CIAS represents another level of sexuality, *genetic sex*, involving those genes that produce hormones and regulatory responses as well as genes that control the structure and formation of the embryonic components in the sexual apparatus.

An XX Zygote Can Sometimes Develop into a Baby Boy or an Infant with Ambiguous Sexuality

There are other forms of genetic sex defects leading to pseudohermaphroditic conditions. In some infants an autosomal recessive mutation causes the adrenal glands to enlarge and convert much of their steroids into male hormones. The condition, *congenital adrenogenital syndrome* (CAS), causes a modification of the external genitalia in the female (XX) embryo somewhat later than normally occurs in the male (XY) embryo. Since the XX embryo has ovaries, developmental progress (until the adrenals begin releasing testosterone) has been typically female. Internally the ovaries, oviducts, uterus, and vagina, especially the upper portion, are normal. But as the testosterone level builds up, the clitoris and labia become modified, producing a phallus-like structure with a partial or complete scrotum (Figure 19-6). The degree of masculinization is variable and depends on the quan-

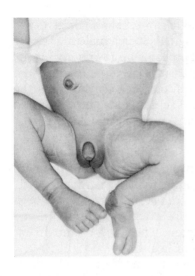

Figure 19-6   Congenital Adrenogenital Syndrome

An autosomal recessive mutation causes cells in the adrenal cortex to synthesize male hormone from the steroid compounds present in them. In an XX fetus this causes development of an enlarged, phallus-like clitoris and scrotal sac instead of labia majora. These XX female pseudohermaphrodites have ovaries with masculinized external genitalia. Their internal genitalia are female; they have a uterus and oviducts as well as a vaginal tract. In some of the mutations, the enzyme defect also disturbs salt balance, in which case adrenal surgery might be required with artificially supplied hormones used to maintain the infant. (Courtesy of Dr. David D. Weaver)

tity of testosterone produced as well as the time and duration of the hormone release in the embryo. Since at least 12 weeks will have elapsed before CAS XX females are exposed to the male hormone, it is less likely that the external genital sex will resemble that of a normal male; more likely the external genitalia will be ambiguous. Such a child may have a vaginal opening between the halves of an incompletely formed scrotum, and the penis may have an incompletely closed genital fold or urethra resulting in a *hypospadias* (any opening of the urethral passage other than its normal location on the glans).

The adrenally produced male hormone may inhibit cortical development of the ovary in such females, and as a result breast formation may not occur at puberty. Others may have functional ovaries which do ovulate and produce monthly cycles and stimulate breast development.

Many Sexual Disorders Are
Surgically Correctible in Infants   Although the sexual alterations in CAS children may create psychological problems and require surgery to eventually shift the XX child to a functional female status, the adrenal condition itself may be life-threatening, especially if the adrenal gland enlarges and pours out other adrenal hormones, or if the metabolic defect affects salt balance in the tissues. In such cases surgical removal of the adrenals or hormone therapy could save the life of the child.

If an XY embryo receives excess adrenally derived male hormone, the child may be born with some signs of secondary sexual development, including pubic hair and a large penis. This form of CAS is sometimes called *macrogenitosomia praecox* or the infant Hercules syndrome.

Some Other Forms of Abnormal
Sexual Development May Occur   If the embryonic testis does not make MDI, or if the Müllerian duct does not respond to MDI, then an

XY individual will develop into a male who has internal female genitalia (oviducts, uterus, and an upper third of a vagina). Such a male may never know that this situation exists and he will probably be a perfectly fertile, functional male. Occasionally such a uterus may develop an abnormal growth of the uterine lining (endometriosis) or a cancer, and thus the male's disorder is revealed. Also the internal genitalia may be discovered accidentally during an autopsy for some unrelated cause of death.

Sometimes, mitotic nondisjunction may occur during the early cleavages of an XY zygote, resulting in an XO/XY mosaic. Two cell lines may also arise from the fusion of two zygotes in the oviduct, and the implantation of the chimeric XX/XY blastocyst could produce a hermaphrodite. In contrast to the pseudohermaphrodite, a *true hermaphrodite* has both ovarian and testicular tissue. This can cause a wide range of external genital development, as in the CAS children. The XX/XY chimerism, however, does not affect longevity or overall health.

In some hermaphrodites the legal sex may be male but at puberty an ovary or ovatestis may begin functioning. If there is a functional uterus a menstrual cycle may occur with a bloody urine corresponding to that cycle, discharged through the penis. There are a few instances of adult hermaphrodites with a functional penis and a functional vagina surrounded by a bifid scrotum. Unlike many animal species with hermaphroditic life cycles, no true human hermaphrodite is capable of self-fertilization.

Pseudohermaphrodites have a normal chromosomal sex, but either a genetic or environmental defect causes a noncorrespondence of gonads and external genitalia. During early pregnancy, a mother with an XX fetus may have received steroid hormone therapy or she may have a tumor of an adrenal gland which may cause it to release male hormones. The external genitalia of the embryo could then masculinize, as in CAS infants, but after birth this male hormone source would be removed and all subsequent development would be female. Such XX pseudohermaphrodites could have surgery to remove an excessively large, phallus-like clitoris or scrotal tissue.

How Parents Should Handle
a Baby with a Sexual Disorder   It is important for parents and medical personnel to request a delay in entering the legal sex of a pseudohermaphroditic child on the birth certificate and to request a detailed examination of the child to best determine which sex the child should have for the rest of its life.

Failure to do so could create problems later in life. There are some reported instances of XX pseudohermaphrodites, who were surgically corrected and raised as females, but who were denied marriage licenses or church weddings because the birth certificate revealed a male legal sex which was at odds with the adult sexual status. Rather than sanction an alleged homosexual union, some officials will refuse such individuals a permit to marry. Such individuals may be able to change their legal sex on a birth certificate but only after expensive and lengthy litigation, and there are no guarantees that what seems sensible to the person requesting the change will strike the court in the same way.

Psychological Sex Is More Subject to Culture Than to Our Genes

Although the biological levels of sex are in agreement for most human beings, agreement between the psychological and biological levels of sex is not always present. The number of individuals with such a disparity varies with the culture and the era.

The most well-known form of such ambiguity is *homosexuality*. The definition of homosexual varies and three types should be distinguished. Those who have had an occasional homosexual activity at some time in their life during or after puberty, but who consider themselves heterosexual, form the largest percentage of males or females who may be classified (perhaps erroneously) as homosexual. A second category involves individuals who occasionally (or frequently) enjoy homosexual relations as adults, but who participate in heterosexual relations as well. These individuals are called *bisexual*. The third category involves exclusive homosexuals, individuals who have no interest in heterosexual experiences. These individuals may sometimes, but not always, adopt mannerisms characteristic of the opposite sex, thus appearing to others as effeminate men or masculinized women. Very often they do not adopt such mannerisms and there is no way that their social behavior in public is recognizably different from that of heterosexuals.

The sexual preference of male homosexuals cannot be changed through hormone therapy. Additional testosterone may increase their sexual activity, but if it does it will still be activity with other men because that is their psychological preference.

There is no evidence that homosexual behavior is genetic and none that indicates a major disturbance in hormone activity, although hormonal disturbances *in utero* have not been ruled out. It is an apparently acquired trait produced by the socializing experiences of the family, and possibly by the peers encountered by the individual who encourage or share their homosexual tendencies. In recent years psychologists have rejected the classification of homosexuality as a pathology. There may be homosexuals who need psychotherapy for conflicts and guilts associated with a homosexual lifestyle (such problems also exist in heterosexual lifestyles), but they are not a necessary consequence of homosexuality.

Gender Roles Are Not Likely to Involve Innate Tendencies

Psychological sex, while primarily an acquired identification, is not likely to be randomly chosen, even in cultures tolerant of homosexuality. Role models of sexual behavior are present throughout our childhood and patterns of heterosexual preference are likely to remain strong. What are more likely to change are the spurious gender roles assigned to occupations, legal rights, and social functions. There is nothing intrinsically masculine or feminine about cooking, taking out the garbage, playing sports, studying medicine, law, engineering, or business, or becoming a ballet dancer. The disproportionate sexual identification associated with these activities reflects our values and biases, not our biology.

Transvestites and Transsexuals Have Intense Gender Conflicts

There are some individuals, *transvestites*, who identify so strongly with the opposite sex that they attempt to hide their biological sex by their choice of clothing and other external cultural symbols of sexuality. They may even believe they are of the opposite sex but are trapped in the wrong body. Such individuals, if male, may grow long hair that is given feminine styling, remove facial hair with electrolytic or chemical treatments, wear stylish dresses and jewelry, and use perfume and other cosmetics. Transvestites restrict their sexual discrepancy to the outward appearances of the opposite sex. They do not normally undergo surgery or hormone therapy to alter their biological sex.

A more radical attempt at sex reversal involves biological sexual conversion. It is more commonly done by males (committed to the idea that they are females) who undergo surgical removal of their testes, scrotum, and penis and then have a surgically-constructed vagina introduced instead. Such surgery involves several operations and it is both expensive and time-consuming. The castrated male then takes estrogen, a female hormone, to stimulate breast formation (Figure 19-7). Such *transsexuals* do not have a uterus or ovaries and thus they remain sterile and incapable of achieving pregnancy or of having a menstrual cycle. Furthermore, some long-range follow-up studies have shown that no greater adjustment to their major psychological problems was made after surgery than would have been made using conventional psychological counseling. For this reason some hospitals have discontinued sex-change operations as a valid therapy for transsexual individuals.

Minor Sexual Abnormalities Are Common and Usually Correctible

The seven sexes of humans illustrate the flexibility of sexual development. There are both minor and major alterations of normal development. We have explored most of the major abnormalities, but we should bear in mind that these are rare, involving less than one percent of all births. Minor abnormalities in development such as hypospadias, undescended testes (*cryptorchidism*), delayed onset of menstruation, development of very small or very large breasts, and other variations in biologically determined levels of sexuality, are much more common than the major abnormalities previously

(a)   (b)

Figure 19-7   The Transsexual Conversion
Richard Raskind (a) after surgery became Renee Richards (b). (Wide World Photos)

discussed. Most of these conditions are readily repaired if they are considered problematic. If they are not of any medical concern, they still may provoke psychological problems by affecting the individual's self-image.

From a biological perspective, the development of human sexuality shows how genetic, chromosomal, hormonal, and morphological features are related. More obvious for sexual development than for that of our other organ systems is the role of culture in modifying the functions of the sexual organs to fill psychological needs.

From a personal perspective it is important to question how many of the sex roles assigned to us are unessential for a healthy acceptance of our sexual identity. There may even be some sex roles that are actually harmful to the overall mental health of the individual barred from full participation in an activity mistakenly assigned a specific gender.

Questions to Answer

1. Discuss the dual embryonic origin of the gonad.
2. What are streak gonads? What conditions produce them? What is the sexual phenotype of individuals with them?

3. What is chromosomal sex? How does it differ from gonadal sex? Can the chromosomal and gonadal sex disagree? If so, how?
4. Discuss the fate of the Müllerian duct in males and females.
5. Discuss the common origin of the scrotum and the labia majora.
6. What is a pseudohermaphrodite? How does such an individual differ from a hermaphrodite?
7. What is the congenital insensitivity to androgen syndrome? How does it arise?
8. What are sex roles? Show how they have changed for: cooking, playing Little League baseball, sewing, hair length, use of cosmetics, ballet dancing, and construction work.
9. Distinguish among the terms homosexual, bisexual, transvestite, and transsexual. What are the social advantages or disadvantages of removing homosexuality from the categories of psychological pathology?

Terms to Master

chromosomal sex
external genital sex
genetic sex
genital folds
genital swellings
genital tubercle
gonadal cortex
gonadal medulla
gonadal sex
HY antigen

internal genital sex
Müllerian duct
Müllerian Duct Inhibitor
primordial germ cells
pseudohermaphrodite
psychological sex
secondary sex
testosterone
Wolffian duct

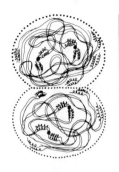

Chapter 20
Induced Birth Defects:
Teratogenesis

In 1960 a disturbing observation was made in Germany and Great Britain. Obstetricians delivered babies whose ears, arms, or legs were missing or curiously deformed. These striking features were often accompanied by other defects as well. A limbless baby (*phocomelia*) is normally very rare, and the only previously encountered case in Germany was written up in a mid-19th century medical journal. Suddenly there were hundreds of cases throughout Germany and the epidemic seemed to be spreading to other countries—to Scandinavia, for example, and even to faraway Australia. Yet no such cases were observed in nearby Poland or France.

Thalidomide Induced Severe Malformations
in Developing Embryos
Eventually one German pediatrician and an Australian physician recognized a feature common to all of these births. The mothers had taken a sedative, *thalidomide,* to control morning sickness. When the doctors' claims were made known, first by correspondence with the German manufacturer of the drug and its British distributor, the companies vigorously denied any relation of their product to this tragic outcome. The outcry from the newspaper publicity, when physicians reported their findings publicly, was so overwhelming that the manufacturer pulled the product off the market, still protesting its harmlessness.

Thalidomide Was Barred
from Distribution
in the United States
Thalidomide was not a prescription drug. It was sold as an over-the-counter drug because it was considered a safe, effective sedative, nonhabituating and without side effects. In fact, how-

ever, it was not very effective as a sedative. Most of the original animal tests and later attempts to verify its sedative effects failed. In sugar solutions it was toxic, and when elderly patients used it they complained of prolonged constipation and a long-lasting numbness or tingling in their hands or feet.

Fortunately, in the United States, Frances Kelsey of the Food and Drug Administration refused to approve the drug for public use because she wanted further tests done. She was not impressed by the quality of earlier tests of thalidomide, either for its effectiveness or for its safety. For her caution she was berated as an "officious bureaucrat" by the American licensee, who was anxious to market thalidomide as fast as possible.

Many Errors of Judgment Were Made in the Sale of Thalidomide

Thalidomide was used for pregnant women because there were no effects on babies of *nursing* mothers who used it. No trials were actually carried out with *pregnant* women, but testimonials from pregnant women who did use it to suppress morning sickness were then used by the manufacturer to promote its use for this problem.

Tragically, few of the thalidomide families received adequate compensation for their suffering or expenses. No case in Europe came to a court's verdict because the drug company managed to settle out of court by forcing long delays, by using statutes of limitations, and by using its very ample legal staff to press for legislative help on the grounds that the tragedy was "an act of God" rather than an instance of human error, corporate poor judgment, or faulty testing procedures.

Organogenesis Is the Most Vulnerable Stage for Malformations

To understand what happened when thalidomide was taken by pregnant women we need to know the timetable for organogenesis in humans. The first week is spent in cleavage. During the second week implantation occurs with the first embryonic tissue formation. The body plan of the embryo emerges in the third week with the beginnings of the neural tube and gut tube. It is at this point that organogenesis begins. Close to the end of the first month the arm and leg buds begin to appear (Figure 20-1). As Table 20-1 demonstrates, the build-up of the major organ systems takes place throughout the second month. After the second month, about day 57, organogenesis is complete and the embryo is considered a fetus. The already-formed rudimentary organs become enlarged and more complex during fetal development.

Severe Damage Before Organogenesis Would Be Fatal to the Embryo

Imagine any damaging substance that prevents cell growth early in development. If the cell growth is disrupted shortly after fertilization, cleavage might not occur. Thus no

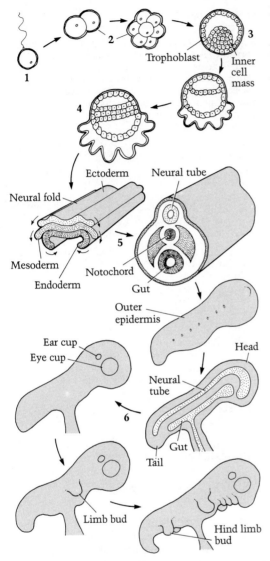

Figure 20-1
The first sixty days of pregnancy are shown in this schematic form. Note the major subdivisions: (1) fertilization, (2) cleavage, (3) blastocyst formation, (4) tissue formation, (5) neural tube and gut formation, and (6) organogenesis.

blastocyst would form. If the damage occurred during implantation no embryonic ectoderm, endoderm, or mesoderm might form, and this would surely be fatal to the embryo. If the cell divisions stopped or became irregular during the formation of the neural tube or gut tube, the embryo would abort. Very likely it would not be until these systems and the circulatory system were established that damage to cells might lead to a birth defect.

Of particular sensitivity in the second month would be the development of the ears, arms, legs, external genitalia, and facial features. These could become malformed if elongation, filling out, or complexity were prevented during this critical stage (four to seven weeks after fertilization).

Table 20-1   The First Two Months of Pregnancy

The sequence of events in the first two months of pregnancy is shown here. Note how the first three weeks are devoted to embryonic tissue formation and embryonic body plan. Specific organs rapidly develop in the next two weeks, with internal organs developing as rapidly as external ones. Finer details, such as the formation of fingers and toes, occur about the sixth or seventh week of pregnancy. Once the basic organs have formed in the embryo, they undergo more complex growth and development throughout the fetal stage. Early damage to organ-forming tissue is thus more likely to produce birth defects than damage occurring after the fetus is formed.

Day(s)	Event
1–5	Fertilization through blastocyst formation
6–10	Implantation
11–15	Endoderm and ectoderm tissues formed and organized as double-layered embryonic disc
16–19	Triple-layered embryo forms neural plate and tube
20–21	Brain swellings and neural tube formation; gut tube begins to form outpocketings; circulatory system develops
22	Heart begins to beat
23	Eye and ear primordia form
26	Arm buds form
27	Leg buds form
30	Lens, optic cups, nasal pits form
32	Paddle-shaped hands form
34	Paddle-shaped feet form
38	Upper lip forms
40	Fingers separating from webs
44	Eyelids form
45	Toes forming
47	Gender still neutral, with sexual primordia external
50	Limbs elongate
53–55	External genitalia, either male or female, develop
57	Fetal period begins

When women took thalidomide as a medication for nausea, especially in the second month of their pregnancy, the limb and ear development of their embryos was severely affected (Figure 20-2). Table 20-2 shows the relationship between specific defects in the babies and the times when their mothers began to use thalidomide. It is not possible to demonstrate whether women using thalidomide before but not after the second month aborted their embryos or gave birth to babies with heart, brain, or other internal defects. Such women would probably have used thalidomide for different reasons (e.g., as a sedative) because they might not have known that they were pregnant. Because morning sickness occurs in the first trimester of pregnancy, it is not likely that thalidomide was used widely in the later stages of pregnancy. Even if it were, it is not likely that the fetus would have been damaged because organogenesis would have been completed.

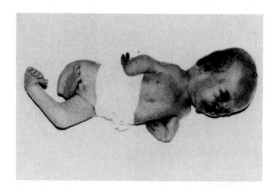

Figure 20-2
The photograph illustrates a typical response to thalidomide. The severity and organs involved reflect the time of first exposure, the duration of exposure, the quantity of thalidomide that entered the embryo, and the response of the embryonic tissue to thalidomide. (Courtesy of R. J. Gorlin, D.D.S.)

Table 20-2   Teratogenic Damage

From interviews and medical records of women who were given thalidomide, a rough map of teratogenic damage was constructed. Note that those embryos less than 34 days old either received no thalidomide (because morning sickness had not begun), had defects that were not unique to thalidomide (e.g., congenital heart defects), or aborted because of the severity of damage. Thalidomide taken after organogenesis would not reverse the development of organ systems already formed. The ears, arms, and legs form in that order, and thus exposure in early organogenesis is more likely to produce armless and earless babies than is exposure in late organogenesis.

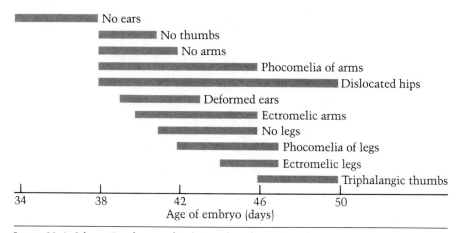

Source: M. A. Salmon, *Developmental Defects and Syndromes* (Aylesbury, England: H. M. & M. Publishers, Ltd., 1978).

Other Teratogens Can Cause Birth Defects

Thalidomide is not the only teratogenic agent that can affect humans. Many more have been identified. In Chapter 3 we saw that x rays, in large doses, can seriously disturb the developing embryo or fetus. Many deformed babies were born in the nine months following the Hiroshima and Nagasaki atomic bombings. The malformations, as you would predict from a knowledge of human organogenesis, were most severe for those exposed in the first trimester of preg-

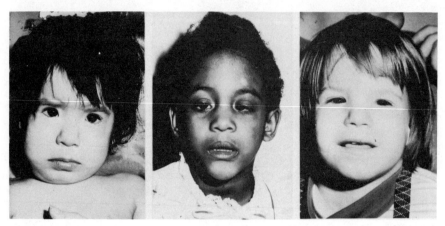

Figure 20-3
Shown here are children whose mothers had been exposed to alcohol during pregnancy. They have facial abnormalities, mental retardation, and cardiac defects. (From P. Streissguth et al., Teratogenic effects of alcohol in humans and laboratory animals, Science, *209, 1980, 353–61. Reprinted by permission.)*

nancy and least severe for those exposed during the last trimester of pregnancy. Please remember that a woman in her second or later month of pregnancy is not at risk for teratogenic effects for her child from a chest x ray; however, there would be a *very slight* risk of gene mutation to a subsequent generation or even of cancer for that child.

Women who are on necessary medication, such as barbiturates and other anti-convulsants for epilepsy, have a higher risk of having a child with a birth defect. For women taking the anti-convulsant phenylhydantoin the risk of birth defects is very high—about 40 percent—with cardiac defects being the most commonly encountered. When a woman has cancer or some other systemic disease such as lupus, requiring chemotherapy, the embryo or fetus can be severely malformed from exposure to these potent chemicals. Even well-known chemicals such as ethyl alcohol cause birth defects (Figure 20-3). In general, women who are in the early stages of pregnancy (or think they might be) should avoid alcohol, coffee, prescription drugs (unless discussed at length with a physician who is knowledgeable about teratogens), and occupational or household chemicals (such as paint solvents, laboratory reagents, pesticides, and herbicides).

Some Teratogens Are Viral or Cellular Parasites

Unfortunately not all teratogens are avoidable. Some women become exposed to viruses such as rubella (German measles) which can cause defects of organogenesis in the first trimester. Other viral infections, such as cytomegalovirus, can also cause birth defects. Toxoplasma, a cellular parasite, can be inhaled from cat feces and cause damage to the developing embryo.

**Any Woman of Reproductive Age Should
Always Be Asked about Pregnancy
Before Receiving x Rays or Prescription Drugs**   It is most unfortunate that the woman who does not look pregnant, and who may not yet know that she is pregnant, is the most vulnerable to teratogenic exposure. Dentists, physicians, and x-ray technicians may not think to ask a female about the possibility of her being pregnant, and most of the public not involved in the health professions are rarely well enough informed to know what to say to protect themselves.

The responsibility for safety, however, should ideally be threefold. Health professionals should routinely provide lead aprons for everyone and should always ask a female of reproductive age if she might be pregnant and not know it. Also no medication should be given, except for the most compelling reasons, during the first trimester of pregnancy. The second line of defense should be the federal agencies who evaluate the tests of new drugs. We have to balance the need for new and more effective medications with the long-term, careful testing of their safety. This is not an easy task because economic as well as health issues are involved. The third line of defense is you. You have to learn to ask questions for your own safety. You must recognize that knowledge of and concern about teratogenic agents may be low for those who manufacture, distribute, sell, prescribe, and use products that have the potential to damage embryos.

Questions to Answer

1. If a sudden appearance of malformed babies were reported in Canada, Australia, New Zealand, and Great Britain, but not in the United States, Europe, Asia, or Africa, what hypotheses would you consider least likely and most likely to account for this event?
2. Discuss the relation of organogenesis to the time sequence of the first trimester of pregnancy.
3. Why would a dental or chest x ray, even if not carried out with the protection of a lead apron, not be teratogenic?
4. If a woman is exposed to rubella in her sixth month of pregnancy, is she at risk for teratogenic effects? Discuss your reasons.

Terms to Master

phocomelia
rubella
teratogen

thalidomide
toxoplasma

Chapter 21
Cancer: The Errant Cell

A normal adult cell usually does not divide unless it is in a tissue that must be replaced regularly, such as the epidermis, the lining of the digestive tract (from mouth to anus), the blood cells, the reproductive cells, or the inner lining of the circulatory system. In the rest of the body most of the cells divide rarely or not at all. Also, cells in a well-defined tissue, such as muscle or kidney tissue, stick to their neighboring cells. If they are accidentally dislodged they will not be able to adhere or multiply at any other site in the body. Nor will such cells be capable of movement within their tissue. An intestinal lining cell, for example, will not leave its site and migrate, even to nearby cells of the same tissue.

The Properties of Normal Cells
Differ from Those of Cancer Cells
In a cancer cell all of the properties of normal cells are lost. The *contact inhibition* which normally prevents adjacent cells from multiplication and migratory movement is abolished and the cancer cells divide, some rapidly (every few days) and some slowly (every few months). The cancer cell also loses the adhesiveness that permits cells to stick together. Cell recognition, which permits cells of one tissue (e.g., liver) to distinguish themselves from cells of all other tissues (e.g., kidney) may also be abolished. In normal tissues a mixture of separated kidney and liver cells will reaggregate into recognizable masses of kidney tissue and liver tissue. A tumor cell, however, once it dislodges from its primary cancer (e.g., in the cervix), may grow within a tissue that is quite different from its origin (e.g., in the lung). Most cancer cells undergo both *metastasis*, which is detachment and growth at a distant site, and loss of tissue specificity.

**Cancer Causes Sickness by Starving,
Eroding, and Crowding Normal Tissue**   A cancer is a malignant tumor. It is malignant because it causes disease and death. The sickness comes from destruction of normal tissue, competition for a common nutrition, mechanical blocking or compressing of a vital organ, or interference with normal tissue function. Not all malignant tumors are metastatic. A brain tumor, for example, may kill by occupying so much cranial space that the brain cannot function, yet no metastasis to other organs may have occurred.

Benign tumors still retain some normal properties. They do not lose their adhesiveness and do not reappear as metastases when the original tumor is surgically removed. Ordinarily, they do not cause debilitating sickness or death.

**Cancers Are Named
for the Tissue in Which They Arise**   Cancers arising in connective tissues (e.g., tendons, bones, cartilage, fat cells) are called *sarcomas*. Cancers of epithelial tissues (e.g., the inner or outer lining of organs; membranes; epidermal skin) are called *carcinomas*. Cancers of embryonic or undifferentiated cells have a terminal suffix, *-blastoma*. For example, retinoblastoma is cancer of the embryonic sensory retinal cells; neuroblastoma is cancer of the embryonic nerve cells, especially the neural crest; and nephroblastoma is the involved embryonic kidney cells of Wilm's tumor. *Melanomas* are cancers of the pigment-producing melanocytes.

**Cancer Cells Can Arise at Any Time
in the Life Cycle But Usually
They Are More Frequent in Older Persons**   In general the cancers of embryonic or undifferentiated cells are childhood cancers. They often appear six months to two years after birth and rarely appear after puberty begins. Adult cancers vary in their relation to age and sex (Figure 21-1). Stomach cancer, for example, first appears at about 40 years of age and its incidence increases rapidly and continuously with age. Lung cancer among smokers also begins to rise starting at the 40th year, accelerating very rapidly to about age 65, after which the frequency begins to decline. In uterine cancer, there is a slow rise in post-pubertal females which continues until it plateaus at age 45 and remains constant. A similar rate is seen for breast cancer, which rises rapidly between the ages of 30 and 45 and then continues to rise at a reduced rate.

**The Cancer Incidence May Vary in Populations
and Careful Interpretation Must Be Made
to Avoid Erroneous Conclusions**   As a major cause of death cancer has moved from seventh place in 1900 (with 64 deaths per 100,000 persons in the United States) to second place in 1970 (with

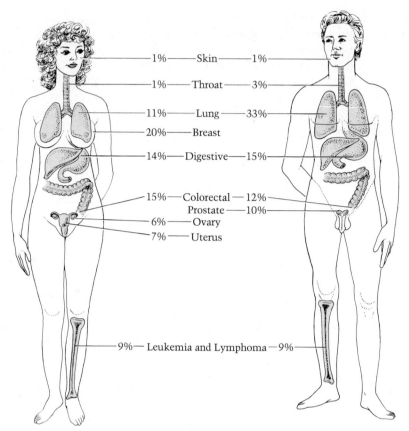

Figure 21-1   Sex Differences and the Site of Cancer Origin

The most common forms of cancer in the female and male, as well as their major sites, are indicated. Cancers of the breast, uterus, and ovaries account for 43 percent of all female cancers. By contrast, only 10 percent of male cancers are sex-related, and those are primarily in the prostate gland. Males, who have smoked cigarettes more frequently and for more years than females, have a high incidence of lung cancer (33 percent of male cancers versus 11 percent of female cancers). Other cancers, such as skin, digestive tract, and blood cancers, occur equally in both sexes.

163 deaths per 100,000 persons in the United States). Partly this reflects the increase in longevity produced by the elimination of infectious diseases and better medical care. It also reflects an increase in induced cancers from cigarette smoking, automotive air pollution, and the immense amount of chemicals introduced into agriculture and industry. Cancers of the skin, breast, buccal cavity, and uterus have shown little change in incidence since the 20th century began, but lung cancers have risen dramatically. Certain cancers are also distributed unevenly when geographical comparisons are made. Cancers in uncommon sites such as the bladder, are high in industrial and mining areas where the working population is exposed to minerals such as arsenic and chromium. Similar concentrations of unusual cancers are found in those exposed to asbestos fibers, nickel, aromatic amines, coal tars, and petroleum products.

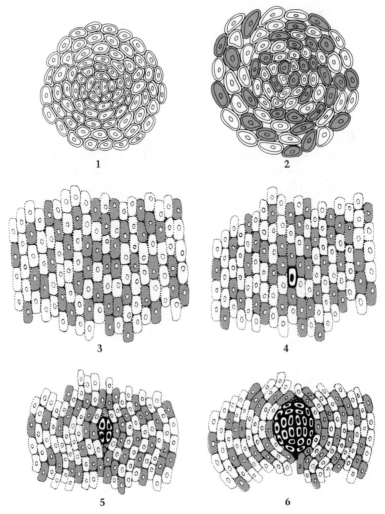

Figure 21-2 The Single Cell Origin of Cancer

*In female mammals only one of the two X chromosomes in a cell is
functional. The other is inactivated by remaining tightly coiled and bound to
the cell membrane as a chromatin spot. This inactivation does not occur
during cleavage (1). It takes place about the time of implantation when the
inner cell mass contains several dozen cells (2). The functional X remains
functional in each descendant cell. A female heterozygous for a deficiency of
the enzyme glucose-6-phosphate dehydrogenase (G6PD) will have a mixture of
cells, some with the normal enzyme present and some lacking the enzyme (3).
If one of those cells, in this example (4) containing the normal allele G6PD +,
becomes altered to form a tumor in the breast, then all of its descendant cells
will have the same type of X found in the original tumor cell (5). As the years
progress, the tumor becomes large enough to be detected as a lump. The
biopsy of such cancer tissues reveals (6) that all of the cells produce G6PD +.
Among women who are heterozygotes for G6PD deficiency, half have tumors
with the enzyme present and half lack the enzyme in the tumor. No mixture
of the cell types is found in such tumors, yet all of these women have both
types of cells present in the surrounding noncancerous breast tissue. This
demonstrates the single cell origin of the tumor.*

The Origin of Cancer Is
a Single Altered Cell

Most cancers begin with a change in a single cell (Figure 21-2). This can be shown in several ways. Sometimes all the tumor cells have an abnormal chromosome number which is not present in the unaffected cells of that individual. For females who are heterozygous for the enzyme defect glucose-6-phosphate dehydrogenase (G6PD) deficiency, the Lyon hypothesis (Chapter 6) predicts that postcleavage embryonic cells begin to inactivate one X or the other, resulting in some cells that are (G6PD +) and others that are (G6PD −). Once such a cell has an inactivated X, all of the mitotic descendants retain a copy of that inactivated X which they, in turn, maintain as the inactive X. In heterozygous females with a single primary tumor (e.g., cervical cancer, breast cancer) all of the cancer cells have only one of the two X chromosome types inactivated, while surrounding normal tissue has both types of cells. However, if the heterozygous female has multiple primary tumors, as in the dominant disorder familial polyposis, some tumors have one X inactivated and other tumors have the other X inactivated.

Similarly, in one particular kind of leukemia, the chronic myelogenous form, all of the tumor cells, but none of the normal cells, contain a translocated chromosome 22. This was named the *Philadelphia chromosome* after the city in which it was first discovered.

It Takes a Long Time
for an Altered Cell to Produce Symptoms
or a Noticeable Tumor

The single cell origin of a cancer has many implications. It means that a cancer may take many years to develop into a noticeable tumor (see Figure 21-3). The rate of division of such cells is rarely as fast as once a week, and more often it is once every month or two, and some cancer cells divide only about

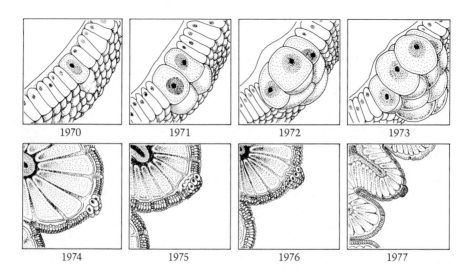

1970	1971	1972	1973
1974	1975	1976	1977

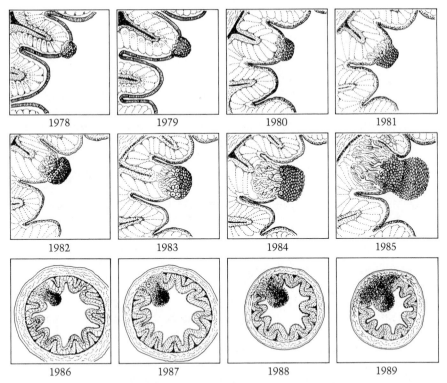

Figure 21-3   The Long Delay Between a Cancer's Origin and Its Symptoms
*In this year-by-year diagram of the growth of a cancer we begin with the
single cell origin of a cancer cell in 1970. The cancer cell arises in a gut
epithelial (lining) cell and is probably a glandular type of modification
(adenocarcinoma cell) of the original gut tissue. The tumor cell divides slowly,
about once every six months. Thus in the early years (1970–1974) the total
number of cells produced is only about 500, and is smaller than the size of
the period at the end of this sentence. The tumor becomes detectable when it
has a diameter of 1 mm or more (about 1980). As the tumor mass grows to
several thousand cells (1981–1983) it becomes vascularized and may invade
the deeper layers of the gut muscle and connective tissue. There might be
sufficient vascularization around the tumor for several millimeters to call
attention to its presence during an examination, but at this stage the person
might feel vigorous and symptom-free and so might not be alarmed. The
tumor in 1980 would have over ½ million cells. By 1986 it would have
enlarged to the size of a pea, and bleeding, pain, or vague discomfort might be
the first symptoms of the tumor's presence. Tumor cells rarely divide faster
than once every three months, and at least 30 rounds of division would be
needed to produce a visible mass.*

twice a year. If a tumor cell divides once every two months, at the end
of two years the tumor will have a little over 2000 cells, which would
not be much larger in diameter than the period at the end of this sentence. At the end of four years it will have more than 8 million cells, a
segment of tissue still too small to cause symptoms. Since the quantity
of tumor cells will increase 64-fold every year at this rate, the tumor's

growth will seem unusually rapid once a symptom, such as a small visible lump, appears.

Viral and Mutational Theories of Cancer Have Been Proposed

The single cell origin of tumor cells also suggests that a unique combination of events has taken place to transform the normal cell to the cancerous state. This combination of events could be either an unusual, successful entry of viral DNA into the cell nucleus or a series of mutations which altered the basic properties of adhesion, cell-specificity, cell recognition, and contact inhibition. Both theories, viral and mutational, have their advocates and there is no question that certain cancers can be assigned unambiguously to each mechanism.

A Viral Basis for Some Cancers Has Been Demonstrated in Birds, Amphibians, and Small Mammals

It has been known since the early studies of Peyton Rous, beginning in 1911, that viruses can cause cancers. Rous identified a virus in certain sarcomas of chickens. The filtrate of tumors from these chickens, injected into a normal chicken, produced the *Rous sarcoma.* Later in the 1930s similar viral-induced tumors for breast cancer were found in mice. Surgical removal of the fetuses from their tumorous or tumor-prone mothers and their suckling on tumor-free mothers prevented mammary cancer among them. However, if they were nursed by their own mothers they eventually developed breast cancer.

When viruses were analyzed biochemically in the 1940s and 1950s some of the tumor viruses had an RNA chromosome and other types of tumor viruses used DNA for constructing genes (Table 21-1). RNA viruses were found to cause leukemia in birds (avian leukemia virus), mice (murine leukemia virus), and cats (feline leukemia virus). They were also the cause of the Rous sarcoma in fowl (Rous sarcoma virus, or RSV).

DNA viruses were found to cause papillomas (wart-like tumors) in rabbits, polyomas (multiple cancers) in hamsters, kidney cancer (Lucké's) in frogs, and various solid tumors in small mammals. Many of the viruses responsible for these tumors were studied intensively, chromosome maps were worked out to identify the location of their genes, and many functional studies were made for the proteins they manufacture (Figure 21-4).

It Is Suspected, But Has Not Been Proven, That Human Cancers Arise from Specific Viruses

So far no virus has been isolated which is unambiguously the cause of human cancer. The best candidates for such cancer viruses, however, are the herpes DNA viruses. Many college students are infected with a form of herpes

Table 21-1   Tumor-Inducing Viruses

Tumor-inducing viruses have been demonstrated in smaller mammals, poultry, and some of the larger mammals such as cattle and horses. The RNA viruses, in general, are associated with blood cancers (leukemias and lymphomas). DNA viruses are more likely to be found in solid tumors. A virus may be benign in one animal and cancerous in another. Rabbit papilloma, for example, has a mild course in wild rabbits, but when introduced to laboratory white rabbits (by rubbing the shaved skin with virus extracted from the harmless warts of wild rabbits), the papillomas form malignant growths which kill the rabbit. SV 40 has no harmful effect on the monkeys that harbor it but produces cancer in hamsters and other small mammals. Human wart virus produces tumors (warts) in the skin but these are not malignant. Human adenovirus, like SV 40, can cause cancer in small mammals but it is not malignant in humans. Peyton Rous, in 1911, was the first to demonstrate that a virus can cause cancer. Many attempts have been made to identify viruses that cause human cancers but the results have been indirect. The long delay between infection and appearance of a cancer is one of many difficulties scientists face in proving a viral basis for a human cancer.

Viral Organism	Nucleic Acid	Viral Organism	Nucleic Acid
Fowl leukemia	RNA	Bovine papilloma	DNA
Rous sarcoma (fowl)	RNA	Rabbit papilloma	DNA
Fowl lymphoma	RNA	Mouse mammary carcinoma	DNA
Mouse leukemia (Gross)	RNA	Equine skin papilloma	DNA
Mouse leukemia (Friend)	RNA	Polyoma	DNA
		Simian virus (SV 40)	DNA
Human wart	DNA	Human adenovirus	DNA

virus, the Epstein-Barr strain, that causes mononucleosis. This same virus, or a mutant form of it, is probably transmitted by mosquitoes in equatorial regions of Africa, producing a cancer, Burkitt's lymphoma, which affects the facial lymph nodes in some of the infected children. The Epstein-Barr viral proteins are present in large quantities in such children, but the intact virus itself is not. Burkitt's lymphoma rarely occurs in Europe or the western hemisphere and seems to be indigenous to equatorial African countries. It is not known whether it is the viral strain or the particular genotype of the susceptible Africans that is responsible for the expression of this cancer.

Viral Cancers Can
Involve Complex Mechanisms   The Rous sarcoma virus is one of the most thoroughly studied tumor viruses. The virus buds off from infected cells with a protective membrane that is derived, in part, from the cell membrane. In electron microscope pictures these stalked and budding viral elements are called C-bodies. The RNA of RSV has a molecular weight of three million which would provide enough nucleotides to code about five to ten genes. The viral RNA introduced into the cell cytoplasm

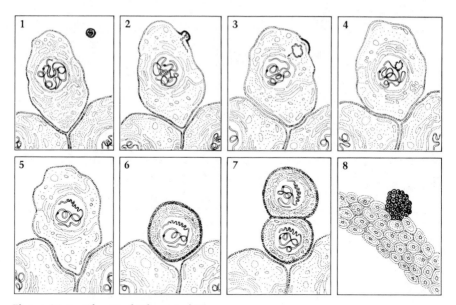

Figure 21-4   The Viral Theory of Cancer

A DNA virus, such as polyoma, infects a cell by inserting its chromosome into the host cell. In this sequence, the approaching virus (1), adheres to the host cell (2) and introduces its chromosome into the cytoplasm (3). The viral chromosome passes through a nuclear pore and comes in contact (4) with a chromosome of the animal cell. The ring DNA opens up and is joined to the chromosome (5) by crossing over. Genes on the viral DNA are turned on and alter the function and morphology (6) of the cell, converting it to a tumor cell. The tumor cell divides (7) and multiplies over the ensuing years to form a tumerous mass (8) in the tissue from which it originated.

replicates in a unique way by producing an enzyme, *reverse transcriptase,* which copies DNA from RNA (the reverse of the usual process in eukaryotic cells). The ends of this RSV DNA are then attached to each other by an enzyme to form a circular DNA (Figure 21-5). The circular DNA may enter the cell nucleus through a pore in the nuclear membrane and attach itself to a specific site on one of the chicken chromosomes. A form of crossing-over between the virus and the chicken chromosome splices the viral DNA into the chromosome. Some of the genes in the viral chromosome may be "turned on," either by regulatory products of a viral gene or by the host cell.

A complicating factor in the relation of RSV to cancer is the accidental detachment of viral chromosomes that are missing one or more of the RSV genes. Such defective viruses exist in infected flocks of fowl. The RSV virus has two cancer genes, called *oncogenes* by geneticists: *sarc* (for the sarcoma) and *leuk* (for a blood disorder, leukosis). If the *sarc* gene is missing, the flock will suffer from the infectious blood disease. If both the *leuk* and the *sarc* genes are missing there will be small, associated viruses that do not cause cancer although they may multiply in

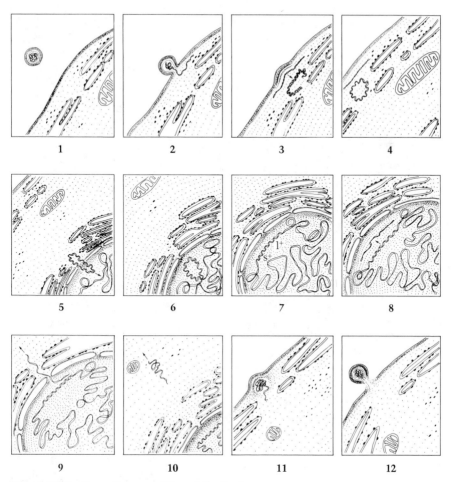

Figure 21-5   Rous Sarcoma Virus Life Cycle

The Rous sarcoma virus (RSV) is an RNA virus (1) which infects poultry. The viral protein adheres to the cell membrane (2) and the RNA is introduced into the cytoplasm. There the RNA produces an enzyme, reverse transcriptase, which makes a DNA (3) whose ends are tied together by another enzyme, a ligase (4). This circular DNA enters through the nuclear pore (5) and inserts into a chicken chromosome (6). The integrated DNA transforms the cell by releasing proteins from its decoded oncogenes (cancer genes), sarc and leuk, represented as s^+ and l^+ (7). The viral DNA also makes RNA copies of its DNA (8) but sometimes leaves the nucleus (9), enters the cytoplasm where it condenses (10), and becomes wrapped in a portion of the cell membrane (11). The stalked RSV (12) then detaches itself and can infect other cells.

infected cells. The defective viruses are called Rous associated viruses (RAV). The leukosis strain is RAV-1, and the nonpathogenic strain is RAV-0.

Note that when the RSV DNA is incorporated into a eukaryotic cell, the cell may "turn off" the gene controlling viral multiplication, but the genes for the leuk and sarc functions may be "turned on." Such a cell may then lose its specificity by altering its surface proteins and by resuming mitosis. The cell is then a tumor cell and can grow and metastasize.

If an integrated viral DNA has both its replicating mechanism and its tumor genes "turned off," then the cell may function normally for many years, but if one or more gene mutations in its eukaryotic or viral genes caused the oncogenes to function, then the cell would shift to a tumor state. Agents such as x rays, chemical mutagens, and ultraviolet light could directly alter the DNA during acute or chronic exposure and cause the quiescent virus to function and transform the cell to the malignant state. Other agents, such as minute fibers and dust particles, may irritate the cells, causing excesses or deficits of metabolic products which trigger the regulatory controls "turning on" the oncogenes.

Clearly a strain that lacks the virus cannot develop Rous sarcoma, even if the chickens are exposed to heavy, nonlethal doses of x rays. But some strains of chickens normally contain the proteins characteristic of RSV or RAV viruses. In such flocks the potential for cancer is transmitted genetically because all the cells, including the sperm and eggs, have this sequence of genes in a specific chromosome. Here x rays and other carcinogens can induce cancer. Such strains may have acquired that sequence in a reproductive cell infected in a distant generation, or the RSV or RAV viruses may have arisen from a normal sequence of genes in such a strain.

The Mutational Theory of Cancer
Involves Many Possible Mechanisms
The mutational theory of cancer is based on several lines of evidence (Figure 21-6). There are several categories of cancers, such as single gene cancers, familial polygenic cancers, probable polygenic cancers, and predisposing genetic factor cancers. Also, agents that are known to induce mutations usually induce cancers. Some induced cancers show structurally rearranged chromosomes, such as translocations or deletions. All of these findings are consistent with a mutational theory of cancer cell origin.

Directly inherited cancers are rare. They are usually autosomal dominant disorders and probably involve regulatory genes. In the dominant disorders, retinoblastoma and Wilm's tumor (also called nephroblastoma) the embryonic cells are involved. In familial polyposis, also a dominant single gene cancer, a later onset (15 to 40 years) occurs in the large bowel (colon). Nothing is known about the causes of the early onset in retinoblastoma or Wilm's tumor and the late onset in individuals with familial polyposis.

Retinoblastoma Is a Single Gene Cancer
of the Eye in Children
Retinoblastoma is a cancer of the embryonic sensory retinal cells. The child shows tumor growth in one or both eyes between the ages of six months and four years. It is

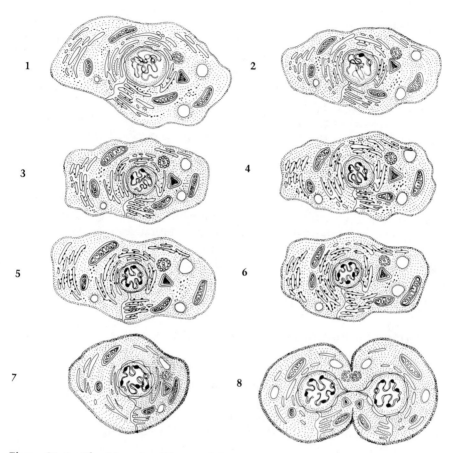

Figure 21-6   The Mutation Theory of Cancer
Although a few cancers are known to be inherited as Mendelian monogenic traits, the vast majority of cancers do not fit so simple a model. Common cancers, such as those of the breast, cervix, colon, prostate, and lungs, are believed to involve several genes (four or five) before the cell takes on all the properties of a tumor. The normal cell (1) experiences a mutation (2), represented by a black dot. Additional mutations over the years (3, 4, 5) increase the number of metabolic changes in the cell. A fifth mutation (6) changes the surface proteins (7) and the combined alterations change the morphology of the cell and restore its capacity to divide (8).

rare for the tumor to originate after early childhood. The tumor may cause the retina to detach from the choroid layer, or it may be large enough to reflect tangential light so that a white or yellow glowing color in the iris is seen, as is often noted at night when a startled cat or dog is crossing a street as a car approaches. The tumor kills by moving along the optic nerve to the brain. It can be cured (there is about an 80–95 percent success rate) by removal of the eye, radiation therapy, chemotherapy, or coagulation of the tumor by concentrating a light focus with a lens or laser.

Retinoblastoma can arise sporadically, and there is a new mutation rate of about 1 in 25,000 births. If the mutation arises *after* fertilization, the child will be mosaic and may have the tumor in only one eye. If the mutation arises *before* meiosis begins in a reproductive cell, the child will be bilaterally affected. In the bilateral form of retinoblastoma there is an additional ten percent risk that bone tissue or some other tissue unrelated to the eye will become cancerous later in life.

One possible mechanism for retinoblastoma (and other directly inherited cancers) is the *two-hit hypothesis*. If a person inherits the retinoblastoma mutation, all of the cells contain this gene. This is the first hit or mutation. In embryonic retinal cells the chances that another mutation (the second hit) will affect this particular regulatory system are low, about one in two million. Since there are many millions of retinal cells there will be multiple primary tumors in the eyes. The mean number per eye in a bilateral case is three, but the range is highly variable, following a random but predictable pattern. In most unilateral cases, however, a single cell must have both rare events occur in it, and thus the onset of the tumor would be later and it would rarely be transmitted. These predictions are supported by analysis of affected individuals.

Deletion of a Region in Chromosome 13
Can Also Cause Retinoblastoma
When a chromosome containing the normal allele of the retinoblastoma gene is broken so that a strip of genes including it is removed, the resulting chromosome has a deletion (Figure 21-7). From studies of some children with multiple congenital anomalies who later developed retinoblastoma, the retinoblastoma gene was mapped in the long arm of chromosome 13 (the altered chromosome is called 13q−). Most cases of retinoblastoma, however, do not have this deletion, and thus gene mutation is more likely to be involved in the cause of this disease. The deletion of the region bearing the gene supports the two-hit model because any embryonic retinal cell lacking that gene has a much higher probability of becoming tumorous (1 in 2 million rather than 1 in 4×10^{12}, the product of the independent probabilities for two hits in a single cell).

There Are Other
Single Gene Cancers in Humans
Wilm's tumor of the kidney is very similar to retinoblastoma because it usually arises sporadically and has an early onset. Its cure rate is not as high, though, because the tumor is usually much larger (and metastatic) when its symptoms are first expressed. Unilateral involvement (one kidney affected) is more common than bilateral (two kidney) involvement. Chemotherapy and x rays are used, and there is a 50 percent survival rate for those whose tumor is recognized before the child is two. Bilaterally affected individuals rarely survive. The hereditary nature of this cancer is based on its occasional transmission from unilateral survivors to their offspring, a high concord-

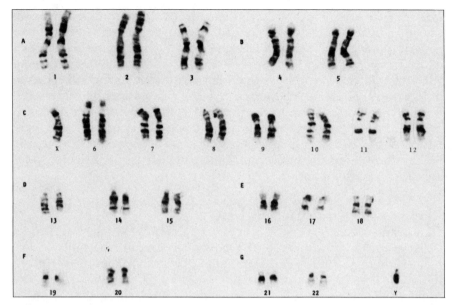

Figure 21-7 Deletions and Retinoblastoma
A small deletion may arise as a result of unequal crossing-over, yielding a small duplication in the homologous chromosome. Several such deletions are expressed in the childhood eye cancer, retinoblastoma, but are also accompanied by other abnormalities, such as mental retardation, absence of thumbs, or blocked intestine, depending on the size of the deletion. The karyotype shows a missing band in the long arm of the right member of the 13th pair. Although this case of retinoblastoma arose from a deletion, the chromosome 13s in most individuals with retinoblastoma show no abnormality in structure. (Courtesy of Dr. Carolyn Trunca)

ance in monozygotic twins, and occasional instances of a gonadally mutant but somatically normal parent who has two or more affected children.

Familial polyposis has a later onset, enabling many affected individuals to reach maturity and raise families before the colon outpocketings or polyps become malignant. There may be thousands of polyps, the number increasing with age, and their malignancy rate is high. The disorder, like Wilm's tumor and retinoblastoma, is an autosomal dominant and is present in 1 out of 8300 births. The malignancy is usually an adenocarcinoma, a slowly multiplying but highly metastatic tumor which is very unresponsive to radiation and present chemotherapeutic treatment.

Another autosomal dominant disorder produces tumors, usually benign, but which can turn cancerous or be life-threatening without being metastatic. This disorder, called *neurofibromatosis* or von Recklinghausen's disease, occurs about once in 3300 births. Affected individuals have numerous coffee-colored spots on their skin and may develop multiple tumors of the skin or membrane layers of bones and nerves. The expression of the disease varies widely, from grotesque deformities such as we saw in Figure 6-9 to apparently normal phenotypes. There is about a two percent chance that a person with this disease will develop a malignancy in one of the tumors.

Although there is only a slight, but significant, chance that neurofibromatosis will form malignancies from its tumors, there are some recessive disorders which have a very high risk of doing so. Unlike the dominant traits leading to tumors, these recessive disorders are not regulatory defects. Most of them appear to be defects of enzymes that repair broken chromosomes or that contribute components essential for keeping the chromosome intact.

Chromosome-Repair Enzyme Defects
Can Lead to Cancers
Xeroderma pigmentosa is a lethal condition in which affected individuals cannot repair ultraviolet-induced thymine dimers (adjacent thymines which are abnormally linked to one another). As a result, the skin becomes mottled with numerous tumors and thickenings. Skin cancers are frequent and inevitable, with most victims dying before they reach their 30th year.

In normal cells there are at least four enzymes which participate in the excision of a damaged segment of a DNA strand, the replication of complementary DNA, and the ligation or joining of the copied section to its contiguous DNA strand. The UV-specific dimer excision enzyme is missing in individuals with xeroderma pigmentosa and thus any outdoor exposure to light is likely to induce such dimers. Their presence in dividing cells leads to gene mutations, chromosome loss, and chromosome rearrangement.

Children with the *Bloom syndrome* have stunted growth, narrow faces, and elongated skulls. They are sensitive to sunlight, developing a butterfly rash along the sides of their nose and cheeks. Throughout their bodies their cells contain many chromosome rearrangements, especially translocations. Cancer frequently occurs between their 10th and 40th years.

In *Fanconi's anemia* there is a high incidence of chromosome breakage in the cells of affected individuals. Homozygotes usually die of anemia as their bone marrow cells cease to multiply, but if they survive their anemia, they may develop leukemia.

The *Louis-Bar syndrome*, also called ataxia telangiectasia, causes a paralysis expressed by a wobbling gait. It is a progressive disease accompanied by numerous vascular nodules of the skin, especially the hands, face, and neck. Cancer of the lymph is common and death usually occurs before maturity. As in the Bloom and Fanconi defects, there is a very high incidence of abnormal chromosomes in the Louis-Bar syndrome.

Most Familial Cancers Are
Polygenic or Multifactorial
Familial polygenic cancers do not involve fragile chromosomes. Colorectal, breast, prostatic, and liver can-

cers may run in families. The probability of transmission is variable as it is with neural tube defects, cleft palate, and other polygenic defects. Environmental factors also play a role in the expression of these cancers and the age of onset.

It has long been known that mutagens and carcinogens are related (Table 21-2). Recent bacterial and viral assay systems have demonstrated that 90 percent of known carcinogens are also mutagens. The fact that mutagens and carcinogens are usually the same agents suggests a common route for their effects. The induction of a cancer cell differs from

Table 21-2   Common Industrial Carcinogens or Suspected Carcinogens

The relationship between occupation and cancer has been known since the 18th century when chimney sweeps in England were found to have a high incidence of scrotal cancer. This was a preventable disorder, requiring no more than a daily scrubbing with soap and water to remove the coal tars that accumulated in the groin. This table presents a few of the more familiar agents that can be hazardous to those working with them or exposed to them in unusually large quantities.

Agent	Most Common Cancer Sites	Population at Risk
Ionizing radiation	Skin, thyroid, tonsils, thymus, blood (leukemia)	Patients receiving medical x rays (especially for therapy); persons receiving accidental overexposure (x-ray technicians, industrial metal workers); uranium miners
Ultraviolet radiation	Skin	Persons exposed to excessive sunlight (farmers, gardeners, athletes, lifeguards)
Asbestos	Lungs, mesothelium	Asbestos miners; construction workers
Nickel	Nasal cavity, sinuses, lungs	Nickel miners; refiners; plating factory employees
Chromium	Lungs	Metal workers
Gold, silver	Lungs	Engravers (inhaling metal dust)
Tar, wax, or paraffin (petroleum)	Skin	Oil refinery workers; mechanics
Pitch, anthracene, coal tar	Skin, pharynx, lungs	Road repair and construction workers; lumber industry workers; workers in chemical factories
Aromatic amines, benzol	Bladder, urinary tract, blood (leukemia, lymphoma)	Workers in chemical factories; dye workers; manufacturers of solvents such as paints and plastics; printers; mechanics

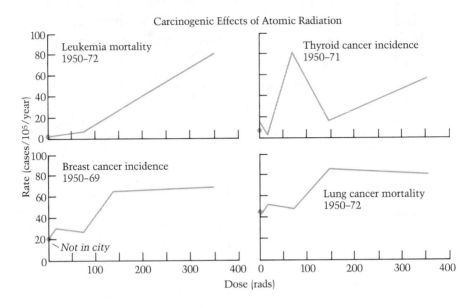

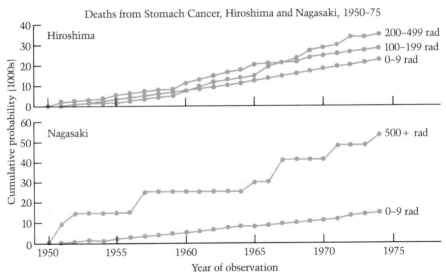

Figure 21-8   Radiation Exposure and Cancer

The 1945 atomic bombs which were exploded in Hiroshima and Nagasaki have had long-term effects on the survivors. Not only has leukemia appeared among the exposed children and adults, but also solid tumors such as tumors of the thyroid, breast, lung, and stomach. The leukemias and thyroid cancers involve all ages in the population, but the breast cancers involve women who were 10 years old or older at the time of exposure. The lung cancer involves adults who were 35 years old or older at the time of exposure. Note also that stomach cancer was increased by exposure to the radiation. At very high doses (where less than half those exposed survived) a five-fold increase in stomach cancers began to appear about six years after exposure. The sharp climb in stomach cancers has continued without leveling off some 30 years after exposure.

the induction of a single gene mutation because most cancers do not increase linearly with the dose of radiation (Figure 21-8). Rather, they require several hits to restore mitotic activity, to alter surface antigens (proteins that permit cell recognition), and to deplete cell membrane proteins that allow normal cells to adhere to contiguous cells of the same tissue. The multiple hits from prolonged exposure to environmental mutagens support the polygenic basis of familial cancers, especially if the number of additional hits in a high-risk family is less than those required in low-risk families.

There Are Three Major Therapies for Cancer

It is likely that there is no single common biological or molecular property for all human cancers. There are probably numerous gene sequences that have the capacity to transform a normal cell to a tumorous state. The somatic genotypes for polygenic cancers may also be highly heterogeneous. These factors make it unlikely that a single vaccine or a single specific drug will ever cure all cancers.

At present there are three major types of cancer therapy (Table 21-3). The most obvious treatment is surgical removal of the tumor. If the tumor is encapsulated and has not metastasized it may be success-

Table 21-3   Survival from Cancer: Five Years Without Recurrence

Success in the treatment of cancer depends on many factors. Early detection is most important because the cancer is more likely to be localized in its early stages. If the cancer has become widespread through metastasis, survival beyond five years is unlikely, though not impossible. Tissue-specific treatments, such as endocrine therapy and radioactive iodine treatment, may favor destruction of some cancers (such as thyroid cancer). Other factors important in the treatment of cancer include tissue type (rapidly dividing cancers are more likely to respond to radiation or chemotherapy), the organ involved, and the ease with which surgery can be done. The five-year survivals in this table represent all stages of cancer of a given type. Cancers that are detected early have higher percentages of survivors than those widespread cancers that are detected late. For the childhood cancers (leukemia, Hodgkin's disease, Wilm's tumor, neuroblastoma) more aggressive chemotherapy initiated soon after diagnosis has resulted in dramatic increases in the numbers of five-year survivals. The long-range prospects for these individuals—that is, the prospects of living 30 or more years after treatment—are not yet known. Solid tumors have had the fewest improvements in five-year survival rates because these tumors tend to be slow-growing and unresponsive to radiation and chemotherapy.

Primary Site	Percentage Who Survive Five Years	Primary Site	Percentage Who Survive Five Years
Esophagus	4	Eye	77
Lung (bronchus)	14	Brain	22
Stomach	14	Cervix	64
Bone	36	Skin (melanoma)	75
Rectum	50	Breast	68
Colon	48	Thyroid	87

Source: Adapted from Schollenfeld and Fromeni, 1982.

fully removed. If it is embedded or intermingled in a tissue, both the tumor and some of the surrounding normal tissue may be excised. If no remnants of tumor cells are left behind the cancer is cured. Early tumors are more likely to be successfully removed by surgery than are the later, more massive tumors.

Unfortunately, if a cancer cell is visible to a pathologist it is highly likely that the cancer has metastasized, even if the primary tumor removed by the surgeon is small (e.g., a melanoma or an osteosarcoma). If metastasis at remote sites is likely to be an early feature of such tumors, then the second type of therapy, chemotherapy, is essential. Chemotherapeutic agents will be chosen depending on the type of cancer that is present (fatty or nonfatty, rapidly dividing or slowly dividing). The agents used will be as specific as possible, to cause maximum damage to the cancer cells while doing as little damage as possible to the normal cells. Some agents (e.g., alkylating agents such as nitrogen mustard) break chromosomes and prevent rapidly dividing tumor cells from completing mitosis. Some agents are fat-soluble (e.g., CCNU, a mutagen with a cyclic organic residue attached to it so that it can embed preferentially in nerve cells which are very fatty) and they can thus be used for treating brain tumors whose rich fat content will make them concentrate the mutagen. Ideally, pharmacologists would like to discover unique chemical properties of tumor cells, so that cell-toxic agents could be synthesized specifically to concentrate in such cells. This can be done in only a few cases presently. Thyroid tumors, for example, will take in large quantities of radioactive iodine. Since iodine is a rare component in other cells and does not accumulate in most cells, the intense radioactivity in the tumor cells selectively kills them and causes minimal damage to surrounding tissues.

Chemotherapy works best on blood cancers because the individual can be given lethal doses of the substance but survive under medically protected conditions (such as blood transfusions and antibiotics) until the marrow and circulatory system begins producing its own cells again from nontumorous tissue. If the cancer cells are rapidly dividing and the normal cells are not, then the treatment may kill the dividing or cancerous cells, while the normal, slowly dividing cells will not enter breakage-fusion-bridge cycles and they will have an opportunity to mobilize their repair enzymes and once again participate in mitosis when they do divide.

X rays or other ionizing radiations may be used to treat localized tumors. They are not useful for metastatic cancers because the doses needed to kill all tumor cells exceed the doses needed to kill a human with whole-body radiation. In the tumor cells radiation induces chromosome breakage, leading to fatal breakage-fusion-bridge cycles.

Relapses May Occur If All
the Cancer Cells Are Not Destroyed

There is one feature of blood cancers which makes cures difficult and future relapses likely: some of the tumor cells may be in the cerebrospinal fluid. Very little of the mutagen can pass through the membranous barriers (sometimes called *blood-brain barriers*) of the central nervous system and back into the circulatory system. But, a surviving cancer cell, months or years later, may have multiplied sufficiently to once again enter the circulatory system and cause a rapid return of the fatigue, fever, and other symptoms of an active cancer.

Some chemotherapists use a very aggressive treatment to prevent this survival of isolated cancer cells in Hodgkin's disease (lymph cancer) or leukemia. They use chemotherapy and whole-body radiation, so that even the tumor cells in the cerebrospinal fluid will be impaired. The risk involved in this procedure is that such massive destruction of normal lymph or bone marrow tissue may occur, along with the destruction of tumor cells, that aplastic anemia (the inability of the marrow to make new red blood cells) or a failure of the body's immune system may result. A balance must be struck between the individual's survival and the tumor's.

For local tumors, such as brain tumors, radiation can be directly applied either as x rays (usually from cobalt-60) or as beams of atomic particles (e.g., neutrons). With this type of irradiation there is no radiation sickness syndrome to the rest of the body as there is with chemotherapy and whole-body radiation treatments, both of which cause chromosome breaks in normal cells as well as in tumor cells. The dividing cells of normal tissues with these induced breaks enter the breakage-fusion-bridge cycle and die as a consequence of this.

Although chemotherapy and radiation can induce new tumors, these are not likely to manifest themselves for ten years or more, and the long-term risk is preferable to the reduced life expectancy that would exist without the treatment.

Many tumors are well-advanced when symptoms first appear and for such individuals no therapy is effective. By this time metastasis may be widespread and surgery would only remove a minor portion of the sites where the cancer is present. Treatment in such cases is palliative; it may relieve some of the symptoms and the acute illness or pain of the cancer, but it does not extend life expectancy for more than a few months, if at all.

Current Research on the Nature of Cancer Cells May Lead to New Treatment Strategies
Several new approaches to cancer research have been developed in recent years. Solid tumors, for example, release a substance which promotes blood vessel formation. If the genes controlling this activity could be repressed, or if the release of this product could be blocked, then solid tumors would starve and die from a cut-off blood supply. At the very least they would be unable to grow.

Another new approach is immunological. Tumor cells have their own specific proteins, and the normal cells from which they are derived lose some of their surface antigens when they become tumorous. If the tumor-specific antigens could be identified, antibodies might be manufactured to react with them or bring in highly lethal toxins, thereby destroying them or making them incapable of further growth. There is a strong suspicion that tumor cells normally form in our tissues but that our immune system usually recognizes and destroys such cells. If the body's own immune system could be stimulated to react against developing tumors then suitable agents might be designed to promote an individual's immune response. Spontaneous regression of tumors may involve such alterations in the immune system.

It would be foolish to predict the imminent arrest of cancer, given the many biological features we just considered. Cancers can involve mechanisms of viral-host interactions, multiple mutations, environmental carcinogenesis, familial predilections, failures of the immune system, and accessory developmental changes in surrounding tissues. The relatively slow multiplication rate of tumor cells makes the time interval between tumor cell formation and noticeable symptoms a matter of years or decades. Various biological factors like metastasis, blood-brain barriers, and the site of origin of the cancer influence the potential success or failure of treatment. Nevertheless, almost all of this knowledge about cancer is the result of research over the past 25 years, and it would be unduly pessimistic to take a fatalistic attitude towards the future treatment and prevention of cancer.

Questions to Answer

1. What is contact inhibition? Why would this property not be present in cancerous tissue?
2. Discuss the differences among the following terms: carcinoma, sarcoma, retinoblastoma, and melanoma.
3. Tumors may be benign or malignant. What is malignancy?
4. Cancer has moved from seventh place (1900) to second place (1980) among the major causes of mortality. Discuss the reasons for this increase.
5. Demonstrate the single cell origin of cancers.
6. Why would cancer not be seen in a population exposed to a nuclear accident if medical teams examined the population in the early weeks or months following the accident?
7. Should a chemical or nuclear industry today be liable for medical damages if its employees show an excess of cancers 20 to 30 years from now? Defend your reasoning.
8. What evidence supports a viral basis for certain cancers?

9. What evidence supports a mutational or genetic basis for certain cancers?
10. What are the major medical treatments for cancer? What is the reasoning employed in each of these treatments?
11. What is a metastasis?
12. What changes in your lifestyle would you consider if a close relative (parent, sibling, grandparent) had: (a) colorectal cancer, (b) breast cancer, (c) cervical cancer, (d) lung cancer?

Terms to Master

benign tumor

metastasis

Burkitt's lymphoma

oncogene

contact inhibition

reverse transcriptase

malignant tumor

Chapter 22
Smoking and Lung Cancer: A Self-Inflicted Biohazard

About 40 percent of those of you who are reading this book are cigarette smokers. Because my father smoked, I used to believe that this was just a matter of personal choice. My tolerant attitude towards smoking was not even shaken when I read the Surgeon General's Report (1968) which established a link between smoking and such diseases as lung cancer, emphysema, and heart disease. After all, if people wanted to take that risk, wasn't that their business? And who was I to elevate myself to being my "brother's keeper"?

Unfortunately, two deaths from lung cancer occurred in my family—my oldest brother and my first wife. Ben was an engraver and I often watched him carve intricate designs into gold rings and silver platters. His hand-rolled cigarette was perpetually by his tools. Ben was never told the cause of his lung disease, although he suspected it. He was hopelessly incurable from the day of his diagnosis and he rapidly declined, with considerable pain, for seven months. Each visit I made testified that so much more life had melted from his flesh and the cancer had withered his once robust body. He kept smoking even during his last days of life.

My first wife was also a cigarette smoker. Helen was a poet and like many writers facing the anxiety of blank paper and bottled-up creativity, she relied heavily on cigarettes to help remove the barriers to her unborn poems. In her case the cancer in her lungs had already metastasized to her brain and her first symptoms were neurological. She would miss the ashtray and grind her cigarette in the table cloth, or she would fail to see a teaspoon immediately next to her coffee cup. Helen survived 17 months after diagnosis, and despite her knowledge that smoking and lung cancer are related, she would send me out to hunt for a carton of cigarettes to bring back to her hospital room.

I seriously began to read the literature on smoking and health when I first learned that my brother Ben had lung cancer. This will not be a pleasant chapter, especially if you smoke, but I must write it because that tiny little boxed-in statement you see in cigarette advertisements and on your cigarette packs does not adequately inform you of what is happening to your lungs when you smoke.

Although there are many health hazards associated with tobacco smoking, this chapter will explore in detail the relationship between smoking and lung cancer. In effect, your lungs may serve as an experiment in genetics. After you have smoked for many years and exposed your bronchial cells constantly to known and probable mutagens produced by cigarette smoke, some of your bronchial cells will experience the multi-hit combination leading to a tumor cell. I believe lung cancer is a valid subject for the genetic and biochemical study of cancer cell formation.

The demonstration of a causal relationship between smoking and cancer is indirect. Also, it tweaks our consciences and offends our lifestyles if we smoke and are forced to read an unpleasant account of what may happen to us. Whether you smoke or not, at least you will have seen some of the more rigorous arguments for its carcinogenic effects. I do not use arguments involving lifestyle or morality (e.g., smell, dirt, tobacco breath, or infringement on nonsmokers' rights) because they are not essential for a chapter on smoking and lung cancer.

The Passage of Smoke to the Lungs
Destroys Many Lining Cells
Normally when a nonsmoker breathes, the air passes down the trachea, gets shunted through the bronchi into the lungs, and ends up in one of the billions of microscopic air sacs (alveoli) that exchange the air's oxygen for the carbon dioxide present in the capillaries surrounding these air sacs. Air, as we well know, is often polluted with dust particles, and as the air passes down the trachea and bronchi, the dust particles are caught in the fringe-like projections present in the outer cell layer lining these passageways. The particles are eventually shoved back, trapped in mucus, and coughed up to be spat out or swallowed, depending upon our personal and cultural habits.

If a person has been smoking for a year, this first line of defense against dust is battered down (Figure 22-1). The lining layer of cells dies and the cells flatten out to form a scaly layer of squamous cells, serving a function similar to that of a callous. Combustion products of smoke are deposited on the cell surfaces and seep into the tissue, stimulating cell divisions in response to the toxic effects on the outermost cells. After several years the lining of the major passages becomes thick with many dozens of layers of squamous cells.

Tumor Cells in the Bronchial Tissue of Smokers
Are Induced After Several Years
and Are Not Immediately Apparent
During these many years of smoking some of the combustion products will cause chromosome breaks and gene mutations in some of the cells. As long as

Figure 22-1 Cigarette Smoking and Lung Cancer

The normal bronchus (a) is lined with a single layer of columnar cells. These have a brush border whose cilia sweep particles back to the throat. After a year of smoking the brush border is obliterated and the columnar cells are flattened into scale-like cells called squamous cells. The squamous cells accumulate over the years and form thick, irregular layers (b). Some of these cells may become cancerous about 10 to 20 years after regular smoking but the microtumors they form are free of symptoms and they usually remain trapped by the surrounding tissue. If there is further damage to the bronchial tissue the tumor mass may spread (c), invading the bronchial tissue and lungs. Students whose bronchi and lungs are in stage (b), should they quit smoking, will have their brush border columnar cells restored to a normal function about a year after quitting, with considerably reduced risks of getting lung cancer. (Courtesy of John J. Godleski, M.D.)

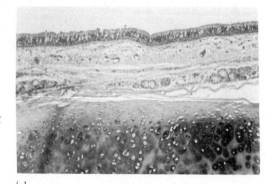

(a)

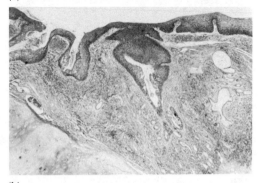

(b)

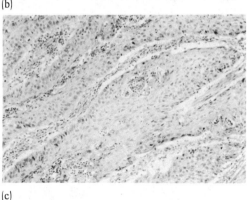

(c)

the cells are nondividing these individual assaults will have little effect on the smoker's lungs. But cancer may be a multi-hit event, and eventually a cell may have the changes necessary to convert it to tumor status. As we saw, these changes include a loss of tissue specificity, a loss of adhesiveness, a resumption of cell division, and a resumption of cell migration. In the case of a smoker the tumor cell may be so hemmed in by surrounding normal cells that it cannot multiply readily, and only a tiny colony of cancer cells may be present for many years after the initial conversion of a normal cell into a tumor cell.

However, if there is a shifting of cells brought about by further damage to the surrounding cells from smoking, then some of the cancer cells may break through the membranes or dead squamous layers and rapidly form a tumor. These cells can also detach from the tumor and enter the bloodstream, ending up in the brain, liver, or bone marrow where they often divide more rapidly than at their primary site in the lung.

Unfortunately, throughout the early stages of tumor cell formation and early tumor growth there are no symptoms at all. The smoker is oblivious to the tumor, and when symptoms first appear, years later, that tumor has either metastasized or grown too large to be effectively treated by surgery, radiation, or chemotherapy.

There Are Several Types of Lung Cancer and There Is Little Chance of Surviving Any of Them

There are several different forms of lung cancer which can be distinguished from each other by the size and shape of the cancer cells and by the location and spread of the tumor in the body. They all share one feature in common—survival is a long shot. About 95 percent of all individuals diagnosed with a primary lung cancer will not live an additional five years. Medical attention may add several months to the lives of those people, but no cure presently exists. Why, then, are there any lung cancer survivors? The lucky five percent may have a tumor cell that still retains its adhesiveness and thus no metastasis occurs. Also, when part or all of the lung bearing the tumor is surgically removed, the patient survives, especially if smoking is abandoned for the rest of that person's life.

Lung cancer often kills by blocking off the air passages when the tumors become large and protrude into the bronchi or trachea. They can also kill by eroding large amounts of lung tissue, resulting in severe pneumonia or hemorrhaging from damaged blood vessels. The x rays of advanced lung cancer patients often show numerous regions of destroyed tissue which prevent the lungs from exchanging oxygen and carbon dioxide.

Those Who Stop Smoking Dramatically Reduce Their Risks for Lung Cancer

Only 20 percent of all lung cancer patients have tumors that can be scheduled for surgery and of these people, less than one-third will survive five years or more. There is a lag of ten or more years between the start of smoking and the conversion of a normal cell to a tumor cell. Then it takes another ten years or more for that tumor cell to establish a colony of cells that breaks free, either to metastasize or to form a detectable tumor. If you are a typical college student who smokes, you have been smoking about five years, and it is unlikely that a tumor cell has been induced yet by your cigarettes. If you are an older student (such as a graduate teaching assistant) you may already have one or more tumor cells trapped within normal tissue. By quitting now you could save your life for two very

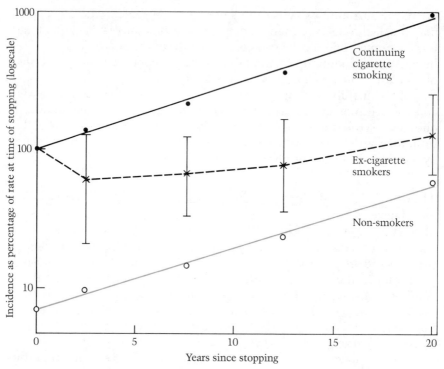

Figure 22-2 Quitting Smoking and Lung Cancer Risk
Lung cancer deaths decrease among those who stop smoking but still remain higher than among those who never smoked. This graph is on a logarithmic scale and thus the advantages of quitting may not seem evident. If a college student, age 20, quits today and does not smoke for the next 20 years, he or she will have only a slightly greater risk of cancer than would a nonsmoker. But the person who continues to smoke over the next 20 years has an incidence 1000 times greater than today's nonsmoker. Twenty years from now that incidence for the nonsmoker rises to 50 and for the ex-smoker to about 100. Thus the ex-smoker has only one-tenth or less the risk of lung cancer as the continuous smoker has. (From The Report of the Surgeon General on Smoking and Health, 1979.)

important reasons (Figure 22-2). If you have not yet induced a tumor cell you have a very low probability of getting lung cancer once you quit smoking. The chance of an additional mutational change arising spontaneously in a cell already hit by the carcinogens from cigarette smoke is very low. Nor is it likely that an isolated tumor cell, if already present, will multiply and break out if its site of origin unless further damage is induced in the surrounding cells by continued smoking. Ex-smokers have a substantial chance of avoiding lung cancer, and the earlier they quit, the less chance there will be of getting lung cancer.

Smoking Is the Major
Cause of Lung Cancer
Forty percent of all cancers in males are lung cancers. In females the percentage is much smaller (eight percent) be-

cause it is only recently, since 1976, that teen-age smoking in the United States has become equally distributed between both sexes. We can predict from this trend that lung cancer in females will gradually rise and that early in the 21st century, the frequency of lung cancer in females will be equal to that of males. Until World War II, cigarette smoking was considered unfeminine, and since there is a 20–30 year delay in the symptomatic expression of an induced cancer, the bulk of lung tumors already present in females are not large enough yet to be recognizable.

Probably 95 percent of all lung cancers are preventable. Nonsmokers rarely have lung cancer (Table 22-1). In a Norwegian study, there were no lung cancers among some 6000 males who never smoked, but in an age-matched group of males who smoked for 40 years or more, four percent had lung cancer.

Table 22-1 Ratio of Lung Cancer Deaths: Prospective Studies

The 1979 Surgeon General's Report on cigarette smoking and health used eight prospective studies (four U.S. and four foreign) to show that cigarette smokers consistently show higher ratios of death from lung cancer than do nonsmokers. The ratios are established by setting the nonsmokers' incidence of lung cancer deaths at 1.0. These are prospective rather than retrospective studies, which means that, in each study, the smokers and nonsmokers were followed (and in some of these studies they will continue to be followed) before any of them had cancer. In prospective studies large numbers of smokers and nonsmokers are followed, and they are matched for all factors except their smoking habit. The lower ratio of lung cancer among the Japanese reflects the tendency of the Japanese to smoke fewer cigarettes and not to inhale deeply. The British study shows a high ratio because the British prefer strong, unfiltered cigarettes and smoke heavily.

Population	Size	Number of Deaths	Nonsmokers	Cigarette Smokers
British doctors	34,000 males	441	1.00	14.0
Swedish study	27,000 males	55	1.00	8.2
	28,000 females	8	1.00	4.5
Japanese study	122,000 males	590	1.00	3.76
	143,000 females	148	1.00	2.03
A.C.S. 25-state study	490,000 males	1159	1.00	9.20
	562,000 females	183	1.00	2.20
U.S. veterans	239,000 males	1256	1.00	12.14
Canadian veterans	78,000 males	331	1.00	19.2
A.C.S. nine-state study	188,000 males	448	1.00	10.73
California males in nine occupations	68,000 males	368	1.00	7.61

Source: The Report of the Surgeon General on Smoking and Health, 1979.

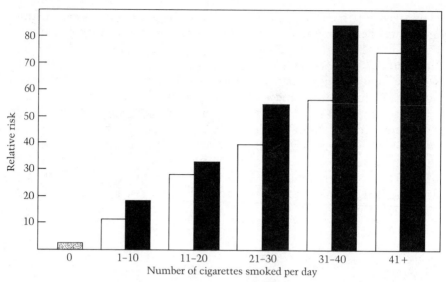

Figure 22-3   Lung Cancer Risk with Filters, Nonfilters, and Total Cigarettes Smoked Daily
The risk of getting lung cancer is directly proportional to the number of cigarettes smoked per day. Note that filters offer very little protection. The unfiltered cigarette (black) and the filtered cigarette (white) both increase lung cancer risks. In this graph the relative risk reveals that the heavy smoker has about 40 times the chance of getting lung cancer as does the nonsmoker. Even light smokers have five to ten times the risk of getting lung cancer as nonsmokers do. (From The Report of the Surgeon General on Smoking and Health, 1979.)

The risk of lung cancer is directly related to the number of cigarettes smoked (Figure 22-3). A nonsmoker has a risk of lung cancer of only 3.4 per 100,000 in the population; a smoker who inhales one-half to one pack (10–20 cigarettes) daily has a risk of 59.3 per 100,000; and a heavy smoker who smokes two or more packs daily faces odds of 217.3 per 100,000. Those who smoke pipes or cigars are less likely to get lung cancer because such smokers usually prefer not to inhale the smoke or they consume less tobacco per day than does a cigarette smoker. They are not without risk, however, of throat and mouth cancer, although the risk is not as high as it is for cigarette smokers.

Three Major Lung Cancer

Tissues Have Been Found   Among the major tissue types recognized in lung cancers, the squamous cells are the most likely to become malignant. About 60 percent of all lung cancers arise in the squamous cells. It should be remembered that squamous cells themselves are produced as a result of long exposure to smoking.

A second major class of cancerous tissues, accounting for about one-sixth of lung cancers, has small, round, loosely-attached cells known as oat cells because of their characteristic appearance. These are also highly malignant and kill their victims quickly once symptoms appear.

The third major category of lung tumors are adenocarcinomas. These tend to arise peripherally, in or near the air sacs, rather than along the trachea or bronchi. Unlike the squamous and oat cell tumors, which are 20–30 times more numerous in smokers than in nonsmokers, adenocarcinomas are only found three times as often in smokers as in nonsmokers. The adenocarcinomas grow slowly but metastasize at an early stage of tumor formation.

The Tobacco Industry Does Not Acknowledge Cigarette Smoking As a Cause of Lung Cancer

The dramatic increase in lung cancer among smokers in the 20th century reflects the parallel increase in consumption of cigarettes. In the United States, annual per capita cigarette consumption rose from 1000 in 1925 to 4000 in 1950. While filters remove some of the *tobacco tar*, a considerable quantity still reaches the lungs. There are more than 4000 compounds produced by the smoking process, and only a few of them have been individually tested for cancer induction. Were it not for the widespread demand for tobacco products and the major role tobacco plays in our agricultural economy (ten billion dollars yearly), cigarette smoking would be more aggressively regulated and criticized than it is today. The tobacco industry, while not acknowledging the strength of the link between smoking and lung cancer, has promoted research into more effective filters and identification of specific compounds in tobacco and smoke condensates that might be the exclusive contributors to lung cancer (Table 22-2). Presently there is no cure for the various types of lung cancer by means of a physiological treatment (e.g., vaccine), and no cancer-free cigarette (with or without a filter). For these reasons only those who do not smoke and those who quit smoking can protect themselves from lung cancer.

Smoking Affects Health in Other Ways Also

While lung cancer is the disease most often associated with tobacco smoking, other disorders are also associated with it. Smoking, according to the 1979 Surgeon General's Report, is "the single most important factor contributing to premature mortality in the United States." For the pack-a-day smoker coronary heart disease is twice as frequent (i.e., 100 percent higher) as it is among nonsmokers (Table 22-3). Other cancers are increased among cigarette smokers as well, including cancers of the esophagus (4.17), the larynx (8.0), the pancreas (2.7), the kidney (1.42), and the bladder (2.0). Each of the numbers in parentheses indicates the relative increase in frequency of cancer among smokers as compared to nonsmokers.

Table 22-2   Known Carcinogens in Tobacco Smoke

When cigarettes are smoked, nicotine (50–2500 μg/cigarette) and tar (500–35,000 μg/cigarette) are inhaled, along with many gases. More than 4000 known compounds are generated from a lighted cigarette. A few of the known carcinogens or tumor-promoting compounds are listed in this table. Some of these agents have higher probabilities of causing cancer of the esophagus, kidney, or urinary bladder than cancer of the throat, trachea, or lungs. The quantities of substance per cigarette are represented in nanograms (ng, billionths of a gram), micrograms (μg, millionths of a gram), and picoCuries (trillionths of a Curie of radioactivity).

Phase	Quantity of Carcinogen per Cigarette
Gas phase	
Dimethyl nitrosoamine	13 ng
Ethyl methyl nitrosoamine	1.8 ng
Diethyl nitrosoamine	1.5 ng
Nitroso pyrrolidine	11 ng
Hydrazine	32 ng
Vinyl chloride	12 ng
Formaldehyde	30 μg
Solid phase (particulates)	
Benzo(a)pyrene	0.05 μg
Dibenz(a)acridine	0.01 μg
Urethane	0.03 μg
N-nitrosonicotine	3.70 μg
Polonium-210	0.07 pCi
β-Naphthylamine	0.02 μg

Source: The Report of the Surgeon General on Smoking and Health, 1979.

Emphysema (a destruction of the air sacs or alveoli) is also a common occurrence among heavy smokers, as is chronic bronchitis. The babies of smoking mothers are about 200 grams lighter, on the average, than nonsmokers' babies. This raises the incidence of newborn babies under 2500 grams. Such babies require intensive care and a longer stay in the hospital, and have a higher risk of mortality or illness.

While most smokers use cigarettes, there are many others who prefer a pipe or cigar. These individuals have lower lung cancer rates than cigarette smokers, but their overall mortality rate is raised about 10–20 percent (Table 22-4). Filters on cigarettes have only a modest effect on reducing the mortality from smoking. If the nonsmokers' rate is set at 1.0, smokers using low-tar and low-nicotine filtered cigarettes have a risk of 1.52, compared to smokers who use high-tar and high-nicotine filtered cigarettes, whose risk is 1.80.

Table 22-3   Coronary Heart Disease and Cigarette Smoking

Although lung cancer has received most of the publicity in the Surgeon General's Reports of 1968 and 1979, the major cause of death in the industrialized nations is coronary heart disease. Heavy smokers (those who smoke more than 20 cigarettes per day) have three times the risk that nonsmokers have of dying of a heart attack. Even if there were no smokers at all, there would still be a lot of coronary heart disease, but there would be virtually no lung cancers. For this reason the sizable reduction in heart attacks that would occur if people stopped smoking is often overlooked.

Category Studied	All CHD Deaths	First Major Coronary Event
Never smoked	1.00	1.00
½ pack/day	1.65	1.65
1 pack/day	1.70	2.08
Over 1 pack/day	3.00	3.28
Ex-smokers	0.80	1.25

Source: American Heart Association Project, United States, 1970.

Note: The frequency of coronary heart disease (CHD) deaths and first coronary attack are set at 1.0 for controls (those who never smoked). The relative increase for various categories of smokers can be seen for both forms of heart involvement.

Table 22-4   Increased Risk of Dying from Different Tobacco Habits

While most smokers are cigarette smokers, a few smoke cigars, pipes, or both. Note that all forms of tobacco smoking reduce life expectancy. Cigarettes are the most hazardous, and pipes the least. The table shows the relative risk, with nonsmoker (control) death rates set at 1.0. All causes of death are included in the mortality ratios. The major threat of pipe smoke and cigar smoke (which are not as frequently inhaled) is cancer of the throat and mouth.

Study	Non-smoker	Cigar Only	Pipe Only	Cigar and Pipe	Cigarette and Cigar or Pipe	Cigarette Only
Males in 9 states	1.00	1.22	1.12	1.10	1.43	1.68
British doctors	1.00	—	—	1.09	1.31	1.73
Canadian veterans	1.00	1.06	1.05	0.98	1.13	1.54
U.S. veterans	1.00	1.16	1.07	1.08	1.51	1.55
Males in 25 states	1.00	1.25	1.19	1.01	1.57	1.86

Source: The Report of the Surgeon General on Smoking and Health, 1979.

Does Involuntary Smoking
Create a Health Hazard?   *Involuntary smoking* occurs when non-smokers occupy the same room as smokers. About five percent of the smoke is inhaled by the nonsmokers. The incidence of bronchitis is in-

creased in children of smoking parents, but there is no convincing evidence that heart disease or cancer is elevated among nonsmoking spouses of a smoker. As is the case for very low doses of x rays, the smoke produced by smokers and inhaled by nonsmokers is difficult to evaluate scientifically for its effects.

The Psychology of Smoking
Is Complex

There are many reasons why 36 percent of all adults in the United States are cigarette smokers. Even if the 250-million-dollar-per-year advertising budget used by the tobacco industry were eliminated, it is not clear that the percent of smokers would diminish significantly. Smoking begins among teen-agers, usually between the ages of 13 and 16. For these teen-agers it marks a rite of passage into adulthood and is a visible symbol of the adolescent's developing independence from parental and school authority. Teen-agers may have only a meager concept of the underlying biology of smoking and its relation to lung cancer. They may not see lung cancer or heart disease as an immediate threat, and for many teen-agers an adult victim in his or her 50s may be too old a person to serve as a model for a biological hazard.

Furthermore, smoking is often habituating and rewarding. The ability to inhale, to form smoke rings, and to daydream through changing wisps of smoke may be personally fulfilling. So too may be the social aspects of smoking—sharing cigarettes and matchbooks, and using the cigarette to fill pauses in conversation, reduce appetite, and stimulate one's thoughts. When these habits are reinforced by peers who also smoke, and repeated thousands of times yearly, they become exceptionally difficult to reject or replace.

What Should Smokers and Nonsmokers
Do to Respect Each Other's Needs?

Certain attitudes, customs, and laws are needed to protect the public from the hazards of smoking and yet still assure smokers opportunities to smoke. Smokers should recognize that in an enclosed room about five percent of their smoke enters the lungs of nonsmokers. Thus a nonsmoker is forced to inhale some smoke regardless of personal values or health needs. In crowded places, such as lecture halls, concert halls, and movie theaters there are bound to be some individuals with allergies, emphysema, asthma, or other respiratory or circulatory disorders. For them public smoking is a hazard. The first rule of smokers should be to confine smoking to their own houses, to the open air, and to rooms and lounges specifically set aside for smokers. Lounges or other space should be provided if smoking is otherwise banned in enclosed public places such as schools, theaters, libraries, and office buildings.

A smoker should routinely ask strangers if it is all right to smoke in a car, taxi, or small meeting place. The nonsmoker should be polite when relaying a negative response, thanking the smoker for being considerate and thus reinforcing an attitude of mutual respect.

Anti-smoking legislation should consider both smokers' and non-smokers' needs. Such legislation should include restriction of vending machines to places rarely frequented by teen-agers; no-smoking ordinances in public places such as auditoriums, meeting halls, retail stores, restaurants, waiting rooms, and so forth; and provision for smoking rooms in public facilities. Other approaches could include a federal tax on tobacco to be used specifically for research on lower-risk tobacco products and for the medical expenses of those incapacitated by long-term smoking (especially victims of emphysema and respiratory tract cancers). Also, long-range planning is needed to replace tobacco with equally profitable crops in those regions of the country where it is raised in large quantities. A similar phasing-out and replacement of the tobacco industry is needed. In the absence of economic compensations, the attempt to reduce smoking among teen-agers and adults is not likely to succeed. Also, it would be valuable if as much psychological and marketing research were invested in influencing individuals not to smoke as is presently spent on the maintainence of smoking as a socially acceptable habit.

Questions to Answer

1. Discuss the changes taking place in a cigarette smoker's bronchial tract (a) one year after starting to smoke and (b) ten years after starting to smoke.
2. Why is lung cancer "95 percent preventable and 95 percent incurable"?
3. About 1975, 40 percent of all cancers in males were lung cancers, but lung cancers accounted for only eight percent of all cancers in females. Discuss (a) the basis for this sex difference and (b) the probable extent of the sex difference about the year 2000.
4. Why is it extremely difficult to prevent teen-age students from adopting a cigarette smoking habit?
5. What is the relation of lung cancers to the number of years as a smoker and the number of cigarettes smoked each day?
6. If you were to design a prevention program for lung cancer, what approaches would you use?
7. Do low-tar or filter cigarettes prevent lung cancer? Discuss the reasons supporting your view.
8. Discuss the psychological reasons why many smokers are more afraid of radiation from operating nuclear plants than of their own smoking habits. In your answer compare the relative risks involved.

9. Do you favor or oppose state laws regulating smoking in public? Cite the reasons for your viewpoint.
10. What is involuntary smoking? If you are a smoker, do you feel you have a "right" to expose nonsmokers to your smoke without their permission? Why or why not? If you are a nonsmoker, do you feel you have a "right" to prevent smokers from smoking in your presence? Why or why not?

Terms to Master

emphysema   tobacco tar
involuntary smoking

Part 6
Molecular
Genetics

It is not easy to appreciate the relationship between an amino $(-NH_2)$ or keto $(C=O)$ group and the color of one's skin, the destruction of one's kidneys at middle-age, the death of a blind and paralyzed infant at the age of four, or ambiguous genitals on an otherwise normal-looking baby. That a change of one nucleotide (less than that—just a side group of one nucleotide!) out of the thousands that compose one gene can make one person bald, another deaf, and still another wracked with agonizing pain seems unbelievable. Yet, through molecular biology, the most complicated organisms can be explained in terms of the simplest biochemical constituents.

Molecular biologists have come a long way toward explaining life. We know the chemical composition of genes (nucleic acids), the way genes work (through a genetic code), what they produce (proteins), how they replicate (from a double helix), and how they mutate (by chemical or physical changes in nucleotides). Even more significant, for us, is how this knowledge can be applied to hereditary disorders. As early as 1949 Linus Pauling was calling sickle cell anemia a "molecular disease" because the symptoms of the disorder and the deformed shape of the red blood cells could all be accounted for by an altered protein of hemoglobin. Geneticists and biochemists demonstrated the exquisite relationship between the altered protein and its mutant gene.

The rapid growth of molecular biology has been immensely helpful in understanding some of the 3300 single gene mutations in humans. While only a few of these mutations can be successfully treated, many more can be diagnosed prenatally or screened among adults. The choices that this technology makes available to us are not easy to make and we sometimes have to wrestle with our conflicting values in order to make a decision. Recently molecular biologists have developed techniques for splicing together genes from different species. Through this splicing technique, several pharmaceutical companies are hoping to make human insulin, growth hormone, interferon, and other biologically active molecules for the treatment of genetic disorders and for cancer therapy. However, the same techniques of gene splicing can be applied to germ warfare and, in the not-too-distant future, to gene therapy (the replacement of mutant genes with normal genes in a malfunctioning tissue), and to eugenics itself. While all scientific discoveries generate potentially useful and harmful applications, few of these applications are carefully debated or analyzed before being introduced. Progress in applying gene splicing techniques was slowed down considerably, not so much because of the fears of its potential abuse, but because the structure of eukaryotic genes is fundamentally different from that of prokaryotic genes, and thus highly ingenious and indirect methods must be used to manufacture human proteins from their nucleic acids.

While the concept of gene splicing has excited the imaginations of those in private industry, the basic molecular biology of development is still unknown. For example, it may turn out to be difficult, if not impossible, to splice genes for nitrogen fixation in the roots of maize, oats, wheat, and rye. It may also be difficult and time-consuming to insert genes that would make the amino acid content of cereal grains as nutritional as that of meat. Further along in the dreams of applied geneticists are genes designed to convert bacteria into efficient producers of hydrocarbons for synthetic petroleum. Whether these new technologies are only a few years away or several decades away cannot be predicted by extension of what we now know.

Chapter 23
The Molecular Basis
of Heredity

At first appearance it seems impossible that a living organism, whether it be a barely visible fruit fly hovering over a bunch of bananas, a cat darting across a street, or a sculptor shaping a terra cotta figure, can be interpreted through its molecular systems. The behavior of organisms is beyond our present understanding of the structure and chemistry of their nervous systems. Some aspects of humanity, including creativity, love, values, and aesthetics, may not ever be reducible to a molecular basis.

Yet consider the routine events that lead to the formation of these organisms. All are products of germ cell fertilization. Each zygote has a single cell origin, and the development that follows results in a fly, a cat, or a person through genetic instructions that are characteristic of the genes of that species. For all organisms, the same basic mechanisms of origin and development are involved: mitosis, growth, differentiation, morphogenetic movements of cells, tissue formation, and organogenesis. In fact, if we were to observe the cat and the person in their single-cell stage, it would be difficult to see any difference between them except at a molecular level. We would be driven to the chemistry of the genes to seek human uniqueness because even the organelles of those two cells would appear indistinguishable.

Muller Advocated the Gene
As the Basis of Life
The belief that our biological uniqueness comes from our genes and that the genes, in turn, constitute the basis for life, was first advocated by H. J. Muller in the 1920s. All other organelles of the cell, as well as the biochemical constituents of the cell, were somehow synthesized by these genes. Mutations of individual genes could lead to loss of organs, defective tissue formation, or failures of

biochemical activity, as in albinos. For these reasons Muller chose the individual gene as his major research interest and worked out novel approaches to mutation rates, the artificial induction of mutations by x rays, and the properties of genes and their mutant alleles. Muller's ideas were gradually accepted by other geneticists and extended to other sciences, such as physics, because the properties of x rays somehow had to be related to the objects, the genes, which they altered so effectively.

A Nobel Prize Physicist Described
Some Problems of Life
Physicists Might Study   During World War II, Erwin Schrödinger, a German physicist who had fled Nazi Germany, resided in Dublin, Ireland. He could not do much research under wartime conditions so he began to read and think about the relation of physics to life. He gave a series of public lectures and published them as a slim book entitled *What Is Life?*

Schrödinger's questions were stimulating. How is it possible that a change in one or a few atoms could be expressed in the gene, in the cell, in the tissue and organ, and finally in the individual person? The number of participating molecules in chemical reactions, even in quantitative chemical analysis, is astronomical. Yet for a disease like achondroplasia only *one* molecule need be altered in a sperm or egg in order for its effects to be immediately manifest at birth in an infant containing about 10^{25} molecules of all kinds.

The Chromosomes or Their Individual Genes
Were Thought to Be Aperiodic Crystals   Schrödinger's insights encouraged other physicists to look upon living matter as fair game for research. Schrödinger believed that the chromosomes were crystals whose genes had an aperiodic structure. Unlike salt and other common crystals which have a simple, repeating (periodic) structure, the genes were thought to be *aperiodic crystals*, each being unique in structure, yet all having a common capacity to replicate. Schrödinger also claimed that genes had a "code script" which permitted the genes to store the metabolic information that they controlled. Such a code, Schrödinger claimed, need not be chemically complex. He used Morse code as an analogy, showing how any message in the English language (which has 26 letters) can be encoded by the use of only three symbols—a dot, a dash, and a space.

The One Gene–One Enzyme Theory
Provided a Model of How Genes Work   During the mid-1930s, George Beadle began a series of biochemical experiments with fruit flies and later with the orange bread mold *Neurospora*. The fruit fly experiments proved that the red eye color is actually the result of two pigments being de-

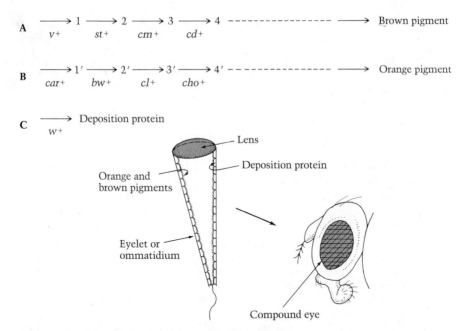

Figure 23-1 Eye Color Inheritance in *Drosophila*
The red eye of a fruit fly is a composite of two pigments. A brown pigment is synthesized by the joint efforts of some 20 different genes in a common biochemical pathway. Thus in A, the gene v⁺ converts a small molecule (the amino acid tyrosine) into a slightly more complex molecule (product 1). The gene st⁺, encountering product 1, converts it into product 2. This synthesis continues until the final product, brown pigment, is produced. Another pathway (B), involving some 20 different genes, synthesizes an orange pigment. In addition to the two pigments, there is a third component, a protein, on which these pigments are deposited. The eye of the fly is compound, containing 750–1000 eyelets or ommatidia. The red eye is the combined effect of two pigments and the interference pattern produced by the eyelets.

posited on a protein which lines the individual eyelets of the compound eye of the fly (Figure 23-1). Each pigment is synthesized by many cumulative steps. One pigment is bright orange; the other is dull brown. The brown pigment is synthesized by about 20 genes, in a series of steps beginning with the amino acid tryptophan. Each step is carried out by an individual enzyme that controls only that function. Each of these enzymes is controlled by a corresponding gene in the biochemical pathway leading to the synthesis of the dull brown pigment (Figure 23-2).

One Character May Be Affected
by Many Mutants Related Through
a Common Biochemical Pathway
By extending this analysis to the bread mold, Beadle and his colleagues proved that any substance, such as a vitamin, an amino acid, a nucleotide, or a sugar, is synthesized by a biochemical pathway from the time it is a simple starting molecule to

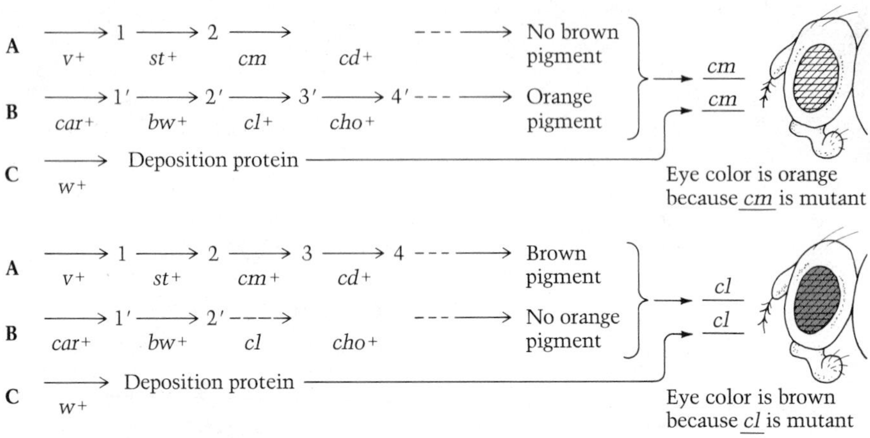

A   —→ 1 —→ 2 —→   - - - —→ No brown
        v+   st+   cm   cd+   pigment
B   —→ 1' —→ 2' —→ 3' —→ 4' - - - —→ Orange
        car+   bw+   cl+   cho+   pigment
C   —→ Deposition protein —————————
        w+

cm / cm

Eye color is orange
because cm is mutant

A   —→ 1 —→ 2 —→ 3 —→ 4 - - - —→ Brown
        v+   st+   cm+   cd+   pigment
B   —→ 1' —→ 2' - - - →   - - - —→ No orange
        car+   bw+   cl   cho+   pigment
C   —→ Deposition protein —————————
        w+

cl / cl

Eye color is brown
because cl is mutant

Figure 23-2 Mutant Eye Colors in *Drosophila*
Although there are some 40 genes involved in producing the red eye color of the normal fly, the homozygous mutation of any one of the genes can result in a departure from red eye color (which is dull red, like ox blood). In pathway A, any one of the 20 genes, if homozygous mutant, prevents brown pigment formation (in this diagram the carmine gene, cm, is homozygous mutant). If all the other genes are normal, then pathway B will make orange pigment, the orange pigment will be deposited, and the eye color will be brilliant orange. Note that a mutant is named for what it looks like rather than for what it synthesizes. The genes for making orange pigment usually have brown names (carnation, brown, clot, chocolate), and the genes for making brown pigment usually have orange names (vermilion, scarlet, carmine, cardinal).

A similar relation exists for a homozygous mutation in the orange pigment pathway. The resulting eye color is brown, because only the brown pathway is fully normal. If you combined the mutants brown and scarlet (bw/bw st/st) both pigments would be missing and the eye color would be white. The white eye color which Morgan found is a mutation of the deposition protein which is missing in X-linked white-eyed flies. Such flies make both brown and orange pigments, but they lack the protein on which these pigments can be deposited. Hence no pigment is seen in the eyelets, and the compound eye is colorless or white.

when it is in its final complex form. The participating enzymes gradually add, shift, delete, and twist the molecule and its additions until it reaches its final form, each event occurring in order of increasing complexity, and each step being carried out by a unique enzyme.

From these studies, Beadle proposed the one gene–one enzyme theory. The function of each gene was to control or synthesize one enzyme. Different sets of genes could be related, not as alleles, but as members of a common biochemical pathway.

Pneumococcal Transformation Suggested a Chemical Basis for Heredity

About the same time that Beadle's one gene–one enzyme theory was developed, a group of microbiologists at the Rockefeller Institute were studying a puzzling phenomenon. It had been known for some time that bacterial pneumonia was caused by certain strains of *Streptococcus pneumoniae.* These strains all produced a gummy coat of carbohydrate that produced glistening smooth colonies on cultured plates. Certain mutant strains of these smooth forms could not synthesize the carbohydrate coat, and these cells formed small

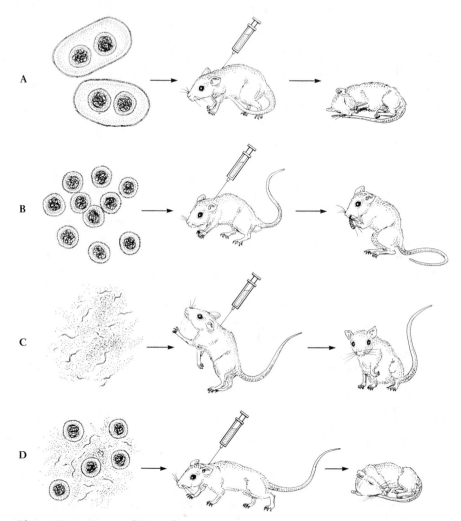

Figure 23-3   Bacterial Transformation
Streptococcus pneumoniae *produces a gummy carbohydrate capsule around each pair of bacterial cells. When injected into mice (A), pneumonia develops and the mice die. Mutant strains of* Streptococcus pneumoniae *arise which lack the ability to make the carbohydrate coat. Such cells, which form rough colonies instead of smooth ones, do not cause pneumonia when the mice are injected (B).*
When the gummy strain is boiled, killing all the cells, the injected extract (C) does not cause pneumonia. But surprisingly, a mixture of rough cells with boiled extract of dead gummy cells (D) produces pneumonia. When the lungs of the dead mice are examined they contain gummy encapsulated cells whose carbohydrate composition is identical to that present in the strain of the donor.

rough-textured colonies on culture plates. The rough forms could not cause pneumonia. The various coated strains could be classified by the chemical composition of the carbohydrates in the coats.

When one strain of one coated form was boiled (which killed the cells), and an extract of it was placed on live bacteria of the rough strain, a transformation took place, either on the culture plate or in the lungs of a mouse infected with the mixture (Figure 23-3). The cells became virulent (capable of producing disease) and the cells once more had a gummy coat, characteristic of the strain that had been killed by boiling. Neither the boiled extract nor the uncoated bacteria could produce pneumonia when administered alone to the mice.

The Transforming Substance
Turned Out to Be DNA   Oswald Avery and his colleagues at
Rockefeller Institute attempted to isolate the chemical basis for the transforming substance. By successfully isolating the carbohydrates, proteins, DNA, and RNA, they proved that transforming activity was present only when DNA was added to the rough strain of *Streptococcus pneumoniae.*

Biologists were surprised to learn that DNA had genetic functions. For many years nucleic acids had been ignored because they were so simple in composition. A nucleic acid is composed of four types of simple units, called nucleotides. Each *nucleotide* consists of three components: phosphoric acid, a sugar containing five carbon atoms (deoxyribose in DNA; ribose in RNA), and a ring-shaped molecule containing carbon and nitrogen. The nitrogenous ring compounds are called *nitrogenous bases.* The nitrogenous bases of DNA exist as *purines* (adenine and guanine) and *pyrimidines* (cytosine and thymine). RNA has the same nitrogenous bases except that thymine is replaced by uracil.

The Versatility and Complexity of Proteins
Offer a Contrast to the Apparent
Simplicity of Nucleic Acids   Proteins had long been
thought to be the genetic material because there were 20 kinds of amino acids which formed thousands of different proteins. Also enzymes were proteins, and they were remarkably versatile in function and uniquely different in structure, each type of protein provoking a different antibody which responded to it alone. Nucleic acids, with only four kinds of nucleotides, were thought to form a repetitive scaffolding around which the proteins were draped.

The transformation of a rough strain of bacterium into a smooth strain is caused by the entry of a piece of DNA from the smooth strain into the bacterium (Figure 23-4). This is an uncommon event. Also, when the DNA of a smooth strain of bacterium is isolated, it consists of hundreds of fragments. Only a fragment bearing the normal gene for coat production will successfully convert the cell so that it can make the coat car-

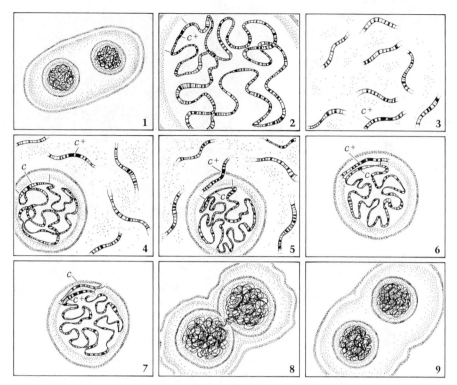

Figure 23-4   The Mechanism of Bacterial Transformation
*In 1944 Avery, MacLeod, and McCarty proved that the substance in the
boiled extract of gummy* Streptococcus pneumoniae *was DNA. Geneticists
later demonstrated how the system worked. The gummy encapsulated cell (1)
contains a gene, c^+, for making the carbohydrate capsule (2). When the DNA
is broken into pieces (3) by boiling, one of the fragments contains c^+. The
DNA fragments are added to the nonencapsulated (rough) strain of bacteria
(4), and the piece bearing the c^+ gene may enter one of the rough cells (5).
The c^+-bearing DNA fragment pairs (6) with its homologous region bearing
the mutant allele, c, which cannot make the gummy carbohydrate. By
crossing-over (7), the c^+ gene changes places with c, and as the cell divides
(8), the c^+ gene produces the coat material resulting in a gummy encapsulated
pair of cells (9).*

bohydrate. The entering fragment pairs with the corresponding allelic
region on the bacterial chromosome and, by a process like crossing-over,
replaces the genes in that vicinity. The bacterium can then multiply,
with the gene for coat carbohydrate production now a part of its chromosome.

The Function of DNA Differs
from That of the Protein in Viruses   Although human parasites are
uncommon today in the United States in well-washed, combed, and freshly
clothed individuals, it was not uncommon about a century ago for a
person to have lice in the hair, fleas in the clothing, bedbugs in the

Figure 23-5   Bacteriophages:
                            Eaters of Bacteria

The virus T4 is a complex organism composed of several proteins and a large, noncircular DNA which has more than 100 genes. In this electron micrograph the viral DNA being injected into a bacterial host cell (E. coli) resembles a series of miniscule syringes. The T4 strain of bacteriophage is widely used to study basic genetics at the molecular level. (© Lee D. Simon/Photo Researchers, Inc.)

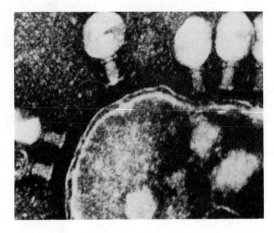

bedding, or worms in the intestines. Parasitism is surprisingly common throughout the world, however. Even bacteria have parasites that live off them. Such bacterial parasites are viruses called bacteriophages (because they eat bacteria).

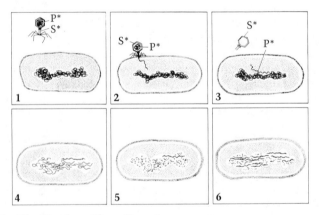

Figure 23-6   The Hershey-Chase Experiment

A bacterial virus contains DNA (labeled with radioactive phosphorus) and protein (labeled with radioactive sulfur). Since DNA does not contain sulfur, and viral protein does not contain phosphorus, the fate of the labeled components can be followed. In (1) the doubly labeled virus is added to a suspension of E. coli cells. The virus attaches to the surface of the bacterial cell (2) and injects its DNA into the host. Note that the viral protein remains on the surface. In fact, that protein "ghost" of the virus can be snapped off the surface (3) by using a milk-shake blender to agitate the infected cells. Almost all of the original, labeled sulfur is recovered. The viral DNA synthesizes enzymes which digest the bacterial DNA (4), but not the viral DNA. Eventually the bacterial DNA is digested and broken down into a pool of nucleotides (5). After biochemical modification by additional viral enzymes the viral DNA (6) replicates. These new viral DNA molecules, in turn, synthesize the head protein (7) into which the viral DNA condenses (8). The noncondensed DNA continues to produce additional viral proteins for the tail

A bacteriophage consists of only one nucleic acid (usually DNA) and proteins (Figure 23-5). The proteins may be numerous, serving functions such as protection, digestion, and host recognition. Some may even function as an injecting mechanism to introduce the viral DNA into the bacterium.

In the early 1950s, A. D. Hershey and his colleagues proved that the life cycle of a virus could be generated by the viral DNA. The viral protein did not enter the host cell. This was demonstrated by radioactive labeling. Nucleic acid contains phosphorus (amino acids do not); and proteins contain sulfur (nucleic acid does not). By growing viruses in a culture supplied with radioactive sulfur and phosphorus, Hershey isolated doubly labeled viruses. These were then used to infect unlabeled bacteria (Figure 23-6). By following the fate of the radioactive phosphorus and sulfur inside and outside the bacterium, Hershey showed that the protein bearing the sulfur never entered the bacterium. Thus the viral DNA (whose radioactive phosphorus is inside the bacterium) was found to provide the instructions leading to the synthesis of both more viral DNA and viral protein so that 100 or more progeny are formed from this injected DNA.

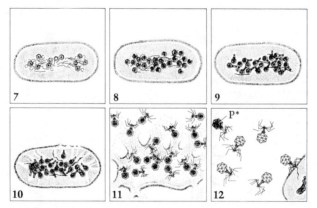

components (9). Up to this point, samples of infected bacteria which are broken open do not release infective viruses. From stage 9 on, however, the viruses within the cells can be used to infect fresh bacteria if the infected cells bearing them are broken open. In the final stages of infection—about 30 minutes from the time the viruses were introduced into the bacterial suspension—the viral DNA not incorporated into mature virus particles synthesizes an enzyme to digest the bacterial cell wall from within (10). The cell then bursts open (11) releasing a shower of virus particles (about 20–100) which can infect other cells in the suspension. Note that these viral particles never contain radioactive sulfur, but some of them contain remnants of the original viral DNA. Since that DNA had the radioactive phosphorus, the experiment demonstrates that (a) viral DNA enters the cell but viral protein does not; (b) the function of the protein is to protect the DNA and permit viral recognition of the host cell and injection of viral DNA; and (c) the function of viral DNA is to synthesize more viral DNA and viral proteins, and carry out the life cycle inside the host by redirecting the metabolism of the host.

The Structure of DNA
Is a Double Helix

The novel idea that nucleic acids might be hereditary material was very influential. When J. D. Watson, a graduate student at Indiana University who did his thesis work on viruses and took courses with H. J. Muller, left for post-doctoral study in Europe, he wanted to work with nucleic acids because he recognized that working out the chemistry or structure of DNA would be equivalent to working out the structure of genes.

While in Cambridge, England, he interested F.H.C. Crick in the same problem. (Crick had been stimulated by reading Schrödinger's *What Is Life?* and was receptive to turning his physical science background to biological problems.) Together they developed a double helical model of DNA (Figure 23-7). It was a difficult structure to work out, especially because neither Watson nor Crick had much biochemical background, and they had to use x-ray diffraction analysis and model-building as the basis for their interpretations.

Their successful model had many features that could be tested experimentally, and this made their model provocative and widely discussed. The model assumes that DNA consists of a crystalline structure with an aperiodic feature, just as Muller and Schrödinger had predicted

Figure 23-7 The Double Helix Model of DNA

A double helix replicates, with each half strand of the duplex generating a new complementary strand. This leads to the semi-conservative pattern of replication.

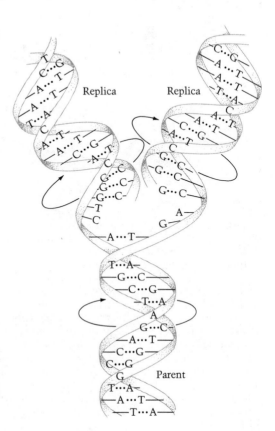

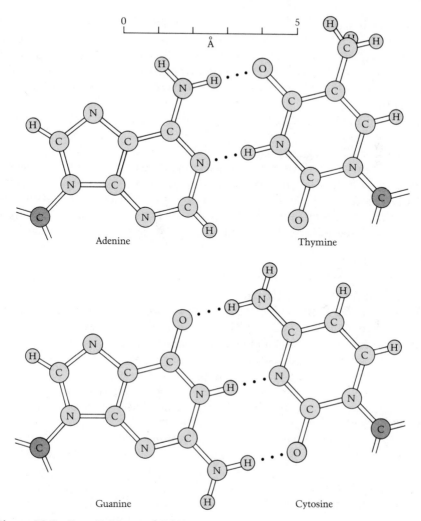

Figure 23-8 Base-Pairing and DNA

The double helix of the DNA molecule arises from the uniform base pairs which are aligned sequentially like steps in a spiral staircase. Shown here are the permissible base pairs: adenine with thymine (held by two hydrogen bonds), and guanine with cytosine (held by three hydrogen bonds). All other pairing arrangements of these four nitrogenous bases lead to oversized or undersized base pairs and an instability from improper hydrogen bonding.

The Watson-Crick model of DNA explained some relations of the nucleotides first observed by Erwin Chargaff. These fixed relations or ratios are called Chargaff's laws. The constant ratios are A/T = 1, G/C = 1, A + G (the purines)/C + T (the pyrimidines) = 1, and A + C/G + T = 1. In this last relation the double helix model predicts that the bases with an amino group (—NH₂) at their apex pair with those having a keto (=o) group at their apex. Since A and C are amino-bearing bases and their complementary pairing bases, T and G, are keto-bearing bases, there will be equal numbers of amino groups paired with keto groups. Chargaff also noted that the A + T/G + C ratio varied with the species used and seldom equalled one. The A + T/G + C ratio, the Watson-Crick model predicted, represented the aperiodic sequences of base pairs.

on purely biological grounds. The crystalline feature consists of paired nucleotides bearing purines and pyrimidines of a constant size—either adenine (A) with thymine (T), or guanine (G) with cytosine (C). Each of these nitrogenous bases is chemically bonded to a sugar (deoxyribose) and to inorganic phosphate. These are hooked to one another like steps in a spiral staircase, each fitting together with its immediate predecessor by shifting a few degrees, and thus generating a helix. Since the base pairs are of constant length, any given pair can have any of the four pairs (AT, TA, GC, or CG) as an adjacent neighbor (Figure 23-8).

If we imagine constructing a spiral staircase using two-by-four wooden beams painted yellow, orange, blue, and green, we could place the two-by-fours in any order desired, in a repeating or nonrepeating pattern. A nonrepeating sequence would be aperiodic. Thus DNA can be a crystal having a paired helical structure and yet have an aperiodic sequence of bases.

Complementary Base-Pairing
Maintains the Double Helix
The two bases in any given permissible *base pair* (AT, TA, GC, or CG) are held together by a weak bonding—*hydrogen bonding*. This bond can be separated by enzymatic activity or by high temperatures near the boiling point of water. When the double helix is unwound by such a process, the single strands of DNA can either join together again into a double helix or they can replicate, if the proper nucleotides, enzymes, and physiological conditions prevail, and form two identical double helixes.

DNA consists of two complementary molecules, like a photographic positive and negative, held together. When the molecules or strands are separated, each strand can generate a complementary half to complete a new double helix. This was the feature that made gene replication, chromosome doubling, and cell division biological realities. It was now possible to talk of a molecular basis for gene replication.

A Genetic Code Is Implied
by an Aperiodic Sequence
of Nucleotides
The Watson-Crick model of DNA also predicted that the sequence of nucleotides was aperiodic, each gene having its own unique sequence of nucleotides, just as words differ from one another by the sequence of their letters. This implied that there was a genetic code, and that information used in metabolism was stored in nucleotide sequences.

Finally, the model predicted that mutations in individual genes could occur if the sequence was altered in the slightest way. For example, a single nucleotide, if replaced in a gene by one of the other three nucleotides, could have a profound effect, just as a change of one letter in a word can completely change the word's meaning.

Proofs of the Watson-Crick Model
Followed Within a Few Years
After the Model Was Developed   Within four years after Watson and Crick developed their model the first tests of the model supported the double helical structure of DNA (Figure 23-9). Bacteria which were fed a heavy isotope of nitrogen produced DNA which conformed to the predictions of the model. The heavy nitrogen was not dispersed indefinitely to the progeny cells, but remained as half-molecules in some of the DNA molecules. Nor did the original heavy nitrogen remain exclusively with the original DNA molecule after one or more rounds of cell division.

Additional agreement was obtained when radioactively labeled DNA was used in plant and animal cells, including human cells. DNA always acted as if it contained two strands which could separate, and either strand could then synthesize another molecule that adhered to it.

Biochemists proved, directly, that DNA synthesis could be done under test-tube conditions, that the molecule was a double helix, and that replication involved the formation of complementary base pairs. Geneticists proved that the structure of genes corresponded to the structure of nucleotide sequences, that each gene was capable of being mapped into a sequence of many dozens of sites, and that each site corresponded to an individual nucleotide.

Protein Synthesis and Nucleic
Acid Coding Are Related   The achievements of geneticists, cell biologists, and biochemists in the 1950s and 1960s were among the most exciting in the history of science. The new field of molecular biology, while having many tributaries leading to its origins, became the dominant focus of attention and research shortly after the publication of the Watson-Crick model.

One of the first problems successfully explored by molecular biologists was the relation of genes to metabolism. It seemed reasonable to assume from the one gene–one enzyme theory that the genetic code somehow converted the nucleotides in the gene to a sequence of amino acids in a protein. Biochemists had noted that another nucleic acid, RNA (differing in its sugar, ribose, and in one of its bases which had thymine replaced by uracil in its nucleotides), was essential for protein synthesis. Cell biologists also noted that RNA was synthesized in the nucleus of a cell and that it then left through the nuclear pores, ending up in the ribosomes of the endoplasmic reticulum of the cytoplasm. From these observations Crick proposed a *genetic dogma* which stated that DNA synthesized RNA, and RNA synthesized protein. This dogma was written symbolically as DNA→RNA→protein.

Crick also proposed, on theoretical grounds, that there were at least two forms of RNA. One type of RNA had to copy the gene's DNA in an RNA form, just as a printed message is a copy of the same message

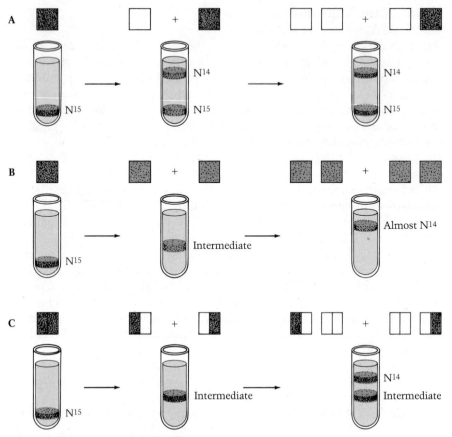

Figure 23-9   Proof of the Semi-Conservative Replication of DNA

Matthew Meselson and Frank Stahl used a heavy isotope of nitrogen in the raw material used as a food source for bacterial cells. Such cells become labeled throughout their DNA after growing 20 or more generations on such heavy nitrogen food. By transferring the cells with the heavy nitrogen in their DNA to food containing the lightweight nitrogen normally present in the environment, Meselson and Stahl were able to test three models of replication.

DNA can be taken from a sample of cells and placed in tubes in a centrifuge. Such molecules, suspended in a heavy metal salt solution, will settle out into a layer as the centrifuge spins at high speed.

In A, B, and C the three models are represented. The squares indicate the amount of heavy nitrogen in the DNA of cells after one or two rounds of cell division. Beneath the squares are the centrifuge tubes which show how many layers form after one or two replications and whether these layers rest at the same position as heavy DNA, light DNA, or at some intermediate position.

The predictions of the three models are uniquely different. Only model C, the semi-conservative replication, produces what is actually found: a single band of DNA of intermediate weight at the end of one cell division, and two layers—one light and one of intermediate weight—after a second cell division. The conservative model (A) is ruled out by the failure to find the predicted two layers in the first replication. The dispersive model (B) is ruled out by the failure to find the predicted single layer at the end of two replications.

written in script. Crick and his colleagues called this type of RNA *messenger RNA* or m-RNA. The second form of RNA had to decode the nucleotides and bring in an amino acid so it could be hooked up to the corresponding sequence of amino acids along the length of the messenger RNA. This amino acid-carrying molecule was later called *transfer RNA* or t-RNA. Both of these forms were confirmed by biochemists, and in 1959 biochemists put these components together in a test tube and showed that protein synthesis could be carried out independently of an intact living cell.

The Genetic Code Was Demonstrated
by Direct Chemical Analysis

The final piece of evidence which proved the molecular basis of biology was the solution of the genetic code. If amino acids were coded by nucleotides, how did the code work? It was clear that 20 amino acids could not be coded by a single nucleotide; only four possible bases could occupy that one site—A, G, C, or T. Nor could a sequence of two sites do the job, because 4 × 4 would give 16 sequences of two consecutive nucleotides, which is four short of 20. By extending this reasoning, geneticists predicted that the ratio of nucleotides to amino acids was three. This sequence of three nucleotides was called a *triplet* or *codon*. An analysis of mutations in bacteriophage suggested that this was the case. However, direct proof of

U		C		A		G		
UUU	phe	UCU	ser	UAU	tyr	UGU	cys	U
UUC	phe	UCC	ser	UAC	tyr	UGC	cys	C
UUA	leu	UCA	ser	UAA	TER	UGA	TER	A
UUG	leu	UCG	ser	UAG	TER	UGG	try	G
CUU	leu	CCU	pro	CAU	his	CGU	arg	U
CUC	leu	CCC	pro	CAC	his	CGC	arg	C
CUA	leu	CCA	pro	CAA	glN	CGA	arg	A
CUG	leu	CCG	pro	CAG	glN	CGG	arg	G
AUU	ile	ACU	thr	AAU	asN	AGU	ser	U
AUC	ile	ACC	thr	AAC	asN	AGC	ser	C
AUA	ile	ACA	thr	AAA	lys	AGA	arg	A
AUG	INIT met	ACG	thr	AAG	lys	AGG	arg	G
GUU	val	GCU	ala	GAU	asp	GGU	gly	U
GUC	val	GCC	ala	GAC	asp	GGC	gly	C
GUA	val	GCA	ala	GAA	glu	GGA	gly	A
GUG	val	GCG	ala	GAG	glu	GGG	gly	G

phe = phenyl-alanine
leu = leucine
ser = serine
tyr = tyrosine
cys = cysteine
try = tryptophan
pro = proline
his = histidine

glu = glutamic acid
arg = argnine
glN = glutamine
ile = isoleucine
thr = threonine
met = methionine
asN = asparagine

lys = lysine
val = valine
ala = alanine
asp = aspartic acid
gly = glycine
TER = end of message
INIT = start of message

Figure 23-10 The Genetic Code

These are the messenger RNA triplets (codons) read by transfer RNA molecules bringing in amino acids to the ribosomes. The code is constructed in this table with the first letter in the leftmost column, the second letter in the top row, and the third letter in the rightmost column. Of the 64 possible codons, one (AUG) initiates a reading with methionine, later removed (at least in the prokaryotes), and three terminate the message (UAA, UAG, and UGA). The 60 remaining triplets code for the 20 amino acids found in proteins. Note that there can be as many as six "synonyms" for one amino acid (the codons AGU, AGC, UCU, UCC, UCA, and UCG can all serve to encode serine).

this was lacking. Nor was there any theoretical way to work out the correct sequence of the three nucleotides for each of the 20 amino acids. In 1960, however, an organic chemist, Marshall Nirenberg, and his colleagues solved the coding problem by making a synthetic messenger RNA consisting of a simple composition. The sequence UUUUUUUUU (poly-U), when placed in the right test-tube mixture for protein synthesis, acted as a synthetic messenger-RNA and yielded a sequence of phe.phe.phe.phe.phe. . . (poly-phenylalanine), where phenylalanine was alone selected from the 20 amino acid types in the test tube. Assuming the triplet model to be correct, Nirenberg then assigned UUU to phenylalanine as the first triplet to be decoded.

Biochemists soon worked out all 64 possible triplet sequences by using different combinations of nucleotides and tested these synthetic

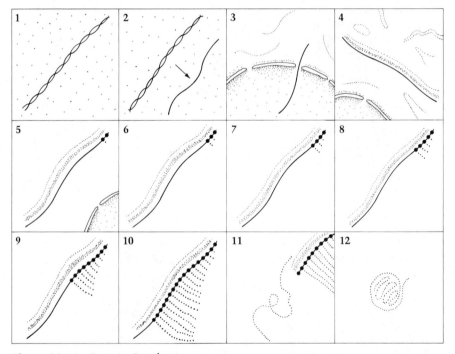

Figure 23-11   Protein Synthesis

A DNA molecule in the nucleus (1) receives a signal permitting an enzyme, RNA polymerase, to make a copy (arrow) of RNA (2) from one of its DNA strands. This messenger RNA (m-RNA) leaves the nucleus through a nuclear membrane pore (3) and enters the cytoplasm (4) where it associates with an endoplasmic reticular membrane. Ribosomes begin to move along the m-RNA (5) and generate amino acids decoded according to the triplet sequences of the m-RNA. Each advancing ribosome adds one new amino acid (6, 7, 8), and as the ribosomes approach the end of the message (9, 10) the length of the protein chain is considerable. The signal to end the message causes a detachment (11) of the protein which then folds into a three-dimensional shape (12). It can then serve as an enzyme, a regulatory signal, or a structural subunit of an organelle.

m-RNAs for protein synthesis (Figure 23-10). The genetic code was solved. Three of the triplets served as a punctuation signal to terminate protein synthesis, equivalent in function to the period at the end of a sentence. The other 61 combinations coded amino acids. Some amino acids had several triplet sequences that recognized them, their transfer-RNAs acting like synonyms in their function.

Protein Synthesis Is a Direct
Consequence of Genetic Decoding
When DNA is converted to m-RNA, the enzyme RNA polymerase reads the nucleotide sequence from a specific starting sequence of nucleotides. The m-RNA increases in length as the enzyme moves along the DNA. When a series of RNA polymerases follow each other along the DNA, a feather-like configuration may be seen (Figures 23-11 and 23-12). The completed m-RNA molecules, in prokaryotic cells, are engaged by ribosomes which enter at one end. These ribosomes continue to move along the length of the m-RNA producing an elongated protein until a stop signal is reached (Figure 23-13) when the decoded protein is released from the ribosome.

The Double Helix Model Permits
Many Genetic Principles
to Be Studied at a Molecular Level
The biochemical analysis of protein synthesis and the successful cracking of the genetic code estab-

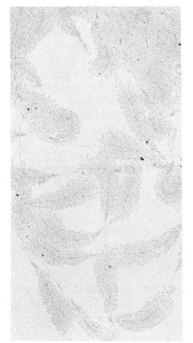

Figure 23-12   RNA Transcription from Chromosomal DNA
These feather-like electron micrographs show RNA copies emerging at right angles from the DNA threads of chromosomes. Note that not all regions of the DNA are producing RNA. The direction of synthesis runs from the narrow tip of the "feather" to its wide base. The genes being decoded in this picture are those that produce ribosomal RNA (r-RNA) which is required in large amounts by the endoplasmic reticulum. The r-RNA gene is repeated many times along the length of the chromosome and when the intact r-RNA is released from the chromosome it enters the endoplasmic reticulum and, in association with ribosomal proteins, forms the ribosomes. The cell used in this photograph is an oocyte of a salamander. As the enzyme transcribing the DNA moves from the apex to the base, the length of the RNA fibril increases. [Extrachromosomal nucleolar (rRNA) genes from an amphibian oocyte. Miller, O. L., Jr., and Barbara R. Beatty, 1969. "Portrait of a Gene." J. Cell. Physiol. 74:Sup 1 225-232.]

Figure 23-13   Protein Synthesis in
Prokaryotes

In the bacterium, E. coli, *two DNA molecules run parallel across the photograph. The lower DNA molecule bears a gene from which messenger RNA is being synthesized. The direction of synthesis runs from top to bottom. Unlike eukaryotes, whose RNA leaves the nucleus and is decoded in the endoplasmic reticulum, the bacterial ribosomes begin decoding the m-RNA while it is still attached to the DNA. The amino acid chains of the growing protein cannot be seen in this electron micrograph. (Unidentified operon of* Escherichia coli *showing coupled mRNA transcription and mRNA translation into protein. Miller, O. L., Jr., Barbara A. Hamkalo, and C. A. Thomas, Jr., 1970. "Visualization of Bacterial Genes in Action."* Science *169:392-395.)*

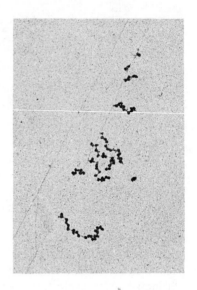

lished the biological worth of the Watson-Crick model. The model predicted how genes replicate, mutate, and function. All three of these vital functions can now be understood in molecular terms. During the rapid flood of experiments and theories which attempted to extend the Watson-Crick model to biology, there were errors, false leads, and oversimplifications. Nevertheless, so many laboratories were checking on each other's work and independently conceiving of similar experiments, that there was no question, by 1962, that the major features of cell metabolism had been identified and the genetic code had been solved.

Questions to Answer

1. Compare the differences between yourself and a dog as (a) adults and (b) zygotes. Why would you expect to find fewer morphological differences in zygotes than in adults? What major difference in (b) would determine which is the dog and which is you?

2. What did Schrödinger mean when he called a gene an "aperiodic crystal"?

3. What is the one gene–one enzyme theory?

4. What is the chemical nature of the "transforming substance" in bacterial pneumonia?

5. What does the labeled sulfur demonstrate in the Hershey-Chase experiment?

6. Why were labeled sulfur and labeled phosphorus chosen in the Hershey-Chase experiment rather than labeled hydrogen and oxygen?

7. What role does hydrogen bonding play in the double helical model of DNA?
8. Why is the term "semi-conservative" appropriate for describing the replication of DNA?
9. Define each of the following terms: bacteriophage, base pairs, messenger RNA.
10. What is meant by the phrase "genetic dogma"?
11. How can you demonstrate that the genetic code must be at least a triplet code?

Terms to Master

aperiodic crystal
base pair
biochemical pathway
codon
complementarity
double helix
genetic dogma
hydrogen bonding
messenger RNA

nitrogenous base
nucleotide
periodic crystal
purine
pyrimidine
semi-conservative replication
transfer RNA
transforming substance

Chapter 24
Mutagens: Monitoring
Our Environment

A mutation is a change in an individual gene. It differs from a chromosome aberration, which requires one or more breaks of the chromosome with a subsequent loss or rearrangement of parts. It also differs from ploidy changes of individual chromosomes or sets of chromosomes, resulting from nondisjunction or from the failure of a meiotic division to take place.

This narrow interpretation of mutation, which separates events within the gene from events associated with chromosome breakage, was proposed in 1921 by H. J. Muller (Figure 24-1). Muller believed such a distinction was useful because a defined category, *gene mutations,* could then be studied for mutation rate and the characteristics of the mutation process.

The Characteristics of Mutation
Were First Described by H. J. Muller
Muller's preoccupation with gene mutation was fruitful. He showed that spontaneous mutation rates could be measured and that they varied with temperature—higher temperatures producing more mutations than lower temperatures. Also, mutation rates varied with the genetic background, making it virtually impossible to establish constant control rates.

Muller also noted that mutations could arise during any stage of development. When we think of a new mutation we usually think of extreme images, like the short-legged Ancon ram which suddenly appears in a flock of sheep, or the child with achondroplastic dwarfism whose parents are of normal height. Such *sporadic complete mutations* probably arose just before the early stages of meiosis, producing a sperm or egg with the new mutant. In fruit flies, however, and also in humans, as we

Figure 24-1 H. J. Muller: Founder of Radiation Genetics

The field of mutagenesis was pioneered by Hermann J. Muller, one of the members of the Drosophila Group at Columbia University. After Muller received his Ph.D. degree in 1915 he left Columbia to establish his own career. He was particularly interested in the properties of the individual gene and the measurement of mutation frequency. In 1926 he designed the experiments which firmly established the induction of mutations by ionizing radiation. Muller's interest in radiation mutagenesis extended beyond theoretical concerns. He actively sought stringent standards of radiation protection and criticized widespread abuses of radiation in medical and industrial practice. His efforts met with bitter opposition from those using radiation who felt that radiation was safe in their hands and that the mutagenic effects of low doses were unproven or inconsequential. (© Fabian Bachrach, 1927)

shall see, mutations are not limited to a defect of a single gene in a mature sperm or egg. Rather, the mutation may be an alteration of a single strand of the DNA duplex of such a sperm or egg; or it may be a mutation which occurred in a mitotic, immature, germ cell (a spermatogonium or oogonium). In fact, the mutation could also occur in the cleavage, embryonic, or later stages of development.

This fact was established by an analysis of the various white eye alleles in fruit flies which was done in the decade following 1910, the year when Morgan found the first case—a solitary male amidst several dozen red-eyed siblings (Figure 24-2). Morgan's white-eyed male was not a mutation of a single egg of its mother because a few of the sisters which mated with him also produced white-eyed offspring, proving that some of those females were heterozygous for white eyes.

Depending on When a Change in a Gene Occurs,
Some Mutations Arise As Complete or
Fully Expressed Traits and Others As Mosaics   Some of the newly-arising white eye alleles were present only in single, isolated males, all of whose sisters were homozygous for the normal red allele. In the case of an X-linked trait like white eyes the existence of a single white-eyed male amidst all homozygous red-eyed brothers and sisters establishes a mutational origin from a single egg.

In some cases the mutant male was white in one eye but red in the other. Some of these mosaic white-eyed mutants produced white-eyed

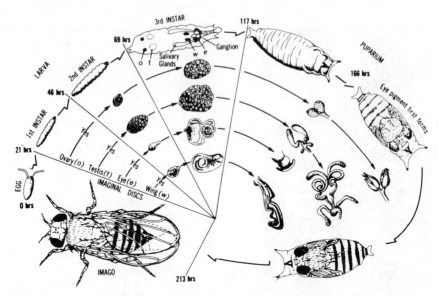

Figure 24-2   The Fruit Fly, *Drosophila melanogaster*
*Although the fruit fly is well-known as the organism used by T. H. Morgan
and his students to establish the chromosome theory of heredity, it is not as
well-known for its other contributions to genetics. H. J. Muller used a
carefully designed stock of flies to detect x-ray-induced gene mutations in
1926. He also used more complex stocks to induce chromosome
rearrangements from chromosomes broken by x rays. Fruit flies were the first
organisms to show mutagenic effects of ultraviolet rays and of chemicals such
as alkylating agents and formaldehyde. Mosaicism and other features of gene
physiology and expression (such as dosage compensation) were first
demonstrated and analyzed in fruit fly experiments. Fruit fly research
continues to flourish today as molecular geneticists study DNA fragments
containing genes of well-known traits in* Drosophila, *including some genes
thought to regulate development.*
    *In the life cycle of the fruit fly, shown here, it takes about 9 days from the
time the eggs are laid until the adult hatches out from the pupa case. The egg
hatches, releasing larvae which go through three moults or instars. In the
larvae are embryonic rudiments, imaginal discs, which form the adult
structures while the larval tissues dissolve inside the pupa case. [Redrawn
from D. Bodenstein, "The postembryonic development of* Drosophila" *in*
Biology of Drosophila *edited by M. Demerec (John Wiley and Sons, 1950) as
adapted in William Baker,* Genetic Analysis *(Houghton Mifflin, 1965).
Copyright © 1965 by William Baker. Used by permission.*

offspring—always with both eyes being white. More often, however, a
mosaic white-eyed mutant did not transmit this new mutation. The gonads only contained the normal allele, giving rise to the red-eyed phenotype. A mosaic of this sort usually arises from a mutation occurring after
the zygote is formed, but it can occur as early as meiosis (after DNA
replication). The later it arises after fertilization, the smaller will be the
percentage of tissue carrying the mutant cells and thus the less likely it
will be that the embryo's gonad will carry that new mutation if the
mosaic also shows the white-eyed tissue in all or part of one eye.

While mutations can arise at any stage of development, they do not do so uniformly at all stages. Mitotic stages are less mutable than stages occurring during meiosis, fertilization, or the first few cleavage divisions. However, it is precisely in these nonmitotic stages that very complex cell processes are occurring with major physiological changes, and this increases the risk that an error of replication will take place.

Radiation Induces Both Gene Mutations and Chromosome Breaks

Muller succeeded in inducing mutations with x rays in 1927. His results were surprising for several reasons. He, and other geneticists, knew that heavy doses of x rays would sterilize fruit flies. He knew that near-sterilizing doses produced a modest increase (about twofold or threefold) in nondisjunction or in crossing-over. Yet all investigators who x-rayed wild flies *en masse* and who allowed these irradiated flies to breed freely with each other reported that the offspring looked normal! They did not realize that almost all mutations are recessive and are not seen in the heterozygous state.

Muller's contribution, which proved that x rays did indeed induce mutations, was the design of a genetic stock that picked up a restricted class of gene mutations—those that kill the male embryo in an early stage of development (Figure 24-3). These *X-linked lethals* were detected by the absence of a class of males that should have appeared when an F_1 female, heterozygous for a newly induced recessive X-linked lethal in the P_1 sperm which generated her at fertilization, produced her two categories of sons. Since a female has two X chromosomes, these two categories of sons can be distinguished by using a different visible gene to mark each chromosome. If, for example, one X has the marker for bar eyes and the other has the marker for yellow body color, then half the sons should be yellow-bodied with normal round eyes and the other half should be of the normal amber body color but with narrow bar eyes. If the father of that heterozygous female had been x-rayed, and if that father was yellow-bodied and round-eyed, then any new X-linked recessive mutant in a sperm would survive in the heterozygous daughter. When such a daughter produced her own progeny, no yellow-bodied sons would appear because all of those X chromosomes would carry the lethal mutation.

Muller's high doses (about 5000 roentgens) showed that x rays did induce mutants, not two to three times the spontaneous rate, but 150 times that rate. The technique that Muller devised, called the *X-linked lethal test*, has been used thousands of times by geneticists since then to study the effects of other radiations as well as chemicals for their genetic effects. Muller succeeded, where others failed, because he used the flies of a special stock of his own design which allowed him to follow the X chromosomes in individual sperms, through the daughters, until they emerged (or failed to emerge) in the grandsons.

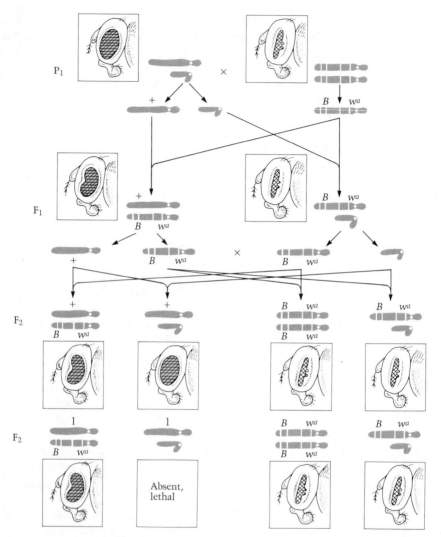

Figure 24-3   The X-Linked Lethal Test

In 1926 H. J. Muller used an X-linked lethal test to detect x-ray-induced mutations. Muller's test is unusual because it does not look for visible mutations among the flies; rather, it looks for the absence of a category of flies normally present.

A normal X chromosome from the male encounters a genetically marked X from the female. The female's X contains the dominant gene, bar eyes (B), and the recessive eye color mutant, apricot (w^a). The X also has several inversions.

In the F$_1$ the females are heterozygous for the bar and apricot alleles and thus appear red-eyed with kidney-bean-shaped eyes. The sons have the bar-shaped and apricot-colored eyes of the P$_1$ mother. When the F$_1$ siblings are crossed they produce F$_2$ bar-apricot-eye males and females, round-eye males, and kidney-bean-red-eye females.

When x rays (or other mutagens) are used, the exposed normal males are mated to the bar-apricot females and each F$_1$ female is individually mated to one or more of the bar-apricot males.

Since each F_1 female arose from one P_1 sperm exposed to the mutagen, the test of such individual females can reveal the presence or absence of a recessive lethal mutation induced on the sperm's X chromosome. Since the genes do not function in a sperm, an induced lethal does not prevent it from fertilizing an egg. The lethal (1) in the F_1 female is recessive and thus its normal allele (in the B w^a chromosome) compensates for it.

In the F_2 those zygotes receiving a Y from the father and the lethal-bearing X from the F_1 mother do not survive, and thus the red-round-eye flies are absent.

The container with the F_2 progeny is simply examined for the presence or absence of red-round-eye males. If a container (usually a small glass vial) has such males in it it is classified as a nonlethal. If that category of males is absent the container is recorded as bearing an X-linked lethal. Note that the lethal is maintained by the red-kidney-bean-eye females and a permanent stock can be kept. The inversions in the B w^a chromosome prevent crossing-over, and thus the integrity of the lethal-bearing X chromosome is also maintained.

Radiation Rarely Produces Monsters or Noticeable Increases in Defects

It is important to realize that a situation like Hiroshima, where the population was irradiated *en masse* and then continued to interbreed, is like the x-ray experiments done on fruit flies before Muller showed how to detect and measure induced mutations. Few induced mutations are dominant. Most are recessive. Lethals and mutations with only slight impairments of a major organ system are induced far more often than are strikingly obvious, visible mutations or gross monstrosities.

It is a fallacy brought about by science fiction writing and movies that an atomic war would produce a civilization of mutant monsters. The offspring of the survivors of Hiroshima, as Muller predicted, showed no noticeable effects of the radiation because recessive mutations in heterozygotes are phenotypically indistinguishable from the homozygous normal. Their only defect would be the hard-to-measure heterozygous effects of induced mutations causing increases in heart attacks, strokes, and arteriosclerosis.

Muller pointed out that a population of wild fruit flies that are given a 4000-roentgen dose of radiation and then bred with one another yields less than one percent of visibly mutant progeny (e.g., progeny with a thickened wing vein, a slightly dusky wing color, or a thinner bristle on the thorax). However, with the proper genetically marked stocks, Muller was able to claim that about four new mutations (all heterozygous) were present in *each* of the offspring from this heavily irradiated population.

The Gene Mutations Induced by the Atomic Bombs in Japan Are Difficult to Detect

In human beings we do not see most of the recessive lethals that are arising spontaneously or that were induced by the Hiroshima and Nagasaki atomic bombs. There are two major reasons for this. First, consanguinity of brother-sister matings is rare, so most of those mutations will remain, for any single generation, in the heterozygous condition. Second, even if the same mutant alleles were to come together (as in marriages of first cousins, which would begin to take place about 2005, if the mutation was induced in 1945 and if the generations were 20 years apart), the most likely thing that would happen is that the embryo would spontaneously abort shortly after implantation. Either the pregnancy would come and go unnoticed or a first-trimester abortion would occur. Only a few of the induced mutations from the atomic bombs have produced visible mutations (e.g., Tay-Sachs, Pompe, or Hurler syndromes, albinism, or other well-defined and recognizable defects) that survived throughout pregnancy. The elimination of these induced mutations will be very gradual, mostly through their effect on the genetic load of the population. No striking changes in sex ratio, new mutant phenotypes, or increases in familiar hereditary childhood diseases will be noted.

Several Molecular Mechanisms of Mutation Have Been Demonstrated

X rays hit DNA directly, displacing several base pairs, so that when the lesion is enzymatically repaired there is a small deletion or rearrangement of those bases in the gene. At high doses of radiation, more than one break may result, producing large rearrangements of chromosome pieces (Figure 24-4 and Table 24-1). X rays also have an indirect effect, converting water and some organic compounds into chemically reactive peroxides. These can react with a DNA base and convert it to an analogue. Such an altered base can mispair during replication, resulting in an amino acid substitution from the altered codon.

The usual *substitution mutation* involves an AT⇌GC replacement. For example, an altered adenine may pair with cytosine rather than thymine. Such a shift in a codon can lead to an amino acid substitution at a specific place in a protein molecule specified by the mutant gene. Thus, a shift from CTG, which specifies valine, to CCG would result in a codon that specifies glycine.

Substitution mutations are produced by many chemicals, including nitrous acid, nitrosamine, formaldehyde, and toxic compounds called alkylating agents. An alkylating agent attaches an organic molecule to the DNA by reacting with one of the amino or keto groups of the bases. The altered base then changes its pairing properties. A protein with this substituted amino acid usually retains its immunological properties, but it may lose its efficiency as an enzyme, structural subunit, or regulator.

A second category of mutations is produced by the loss or gain of one or two bases (or a larger multiple of one or two bases). Such *frame shift mutations* cause the sequence of bases to read as entirely different amino acids from those in the normal allele. Thus, if the first letter were removed from the repeating codons TACTACTAC . . ., the codons would be read as ACTACTACT . . ., specifying an entirely different poly-amino

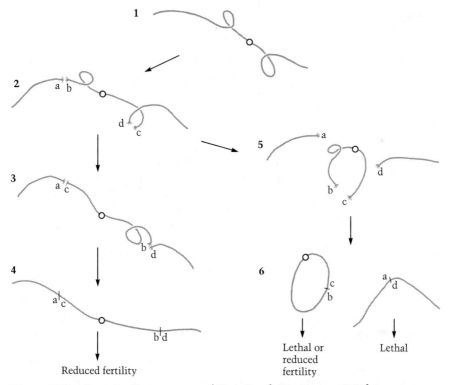

Figure 24-4   Genetic Consequence of Two-Break Events in a Single
 Chromosome

*When two breaks occur in one chromosome the fate of the cell depends on the
way the broken ends are rejoined and the distance between the breaks. The
intact chromosome (1) experiences two breaks (2) producing broken ends a,b
and c,d. If an* inversion *occurs (3) the middle piece rotates 180° so that a and
c as well as b and d are joined together by repair enzymes (4). Such an
inverted region, in the parent heterozygous for it, can cause birth defects or
early fetal wastage among half of the fertilizations that occur. More serious
damage results if the large inner piece (5) joins break points b and c to form a
ring chromosome. The fragments with breakpoints a and d get lost because
they lack the centromere. Even if the ring in 6 is larger, the loss of the genes
causes a malformed baby or an aborted fetus. The ring would have effects
similar to those produced by inversions if the balanced carrier were viable.*

*Small distances between the two breaks can lead to minute deletions or
inversions. The minute deletions might be viable but could show birth
defects, as in certain retinoblastoma cases which are associated with other
physical defects. The minute inversions would be viable and show no
abnormalities in the balanced carrier.*

acid sequence. Such proteins usually lose their immunological properties
as well as their functions. X rays and alkylating agents also induce frame
shift mutations. One class of chemical agents, the acridine compounds,
regularly produces frame shifts as its major mode of mutagenesis. The
acridines are multiple ring structures that slip between adjacent pairs of
nucleotides and distort the helix of the DNA molecule. The acridines

Table 24-1   Two-Break Rearrangements

For all two-break events, the fate of the resulting rearrangement will depend on the number of genes lost or duplicated. If the individuals survive and become mature adults, problems arise during meiosis when they produce gametes. The translocations involve breaks in two nonhomologous chromosomes, and the duplications involve breaks in a pair of homologs (often as a consequence of unequal crossing-over).

Form of Rearrangement	Consequence to Individual
Deletion	Lethal or gross abnormalities
Ring	Like deletion; if viable, causes reduced fertility
Inversion	Usually normal, but reduces fertility and can produce grossly abnormal offspring
Translocation	Usually normal, if balanced, but can produce fetal wastage or grossly abnormal offspring
Duplication	Either normal, if duplication is small, or gross abnormality, if large

are used commercially as antimalarial drugs but it is not known if their therapeutic effect is due to such mutational lesions in the malarial parasites.

The frame shift event is comparable to what happens when you align your fingers incorrectly on your typewriter. Instead of words you obtain unintelligible gibberish. This is quite different from the occasional typographical error you might make by striking the wrong key. Such an occasional error would correspond to a substitution mutation. Also, a typographical error in a word usually slows down your reading and interpretation of that word, but rarely changes it into a word with another meaning.

Since the individual codons for amino acids are sequences of three nucleotides, any loss or gain of nucleotides, other than in groups of three, will cause a frame shift mutation. This also implies, and it has been confirmed experimentally, that frame shift mutations can be reverted from total loss of function to near normal function by another frame shift mutation near it. In such cases the sum of the missing or extra bases of the two mutations is divisible by three and the rest of the nucleotide sequence of the gene is read properly.

Chemical Mutagenesis, Discovered in 1940, Differs from X Rays in Its Effects on Genes and Chromosomes

The first chemical mutagen was successfully analyzed by Charlotte Auerbach in 1940, but the wartime censorship prevented her from publishing her results until 1946. She used the alkylating agent, mustard gas, because a pharmacologist pointed out to her that it caused burns similar to those caused by x rays. In fruit flies the mustard gas caused gene mutations and chromosome breaks. It was *radiomimetic;* that is, its effects were similar to those of heavy doses of x rays.

There were two noticeable differences from the effects of x rays, however. Mustard gas produced more gene mutations than chromosome rearrangements. Also, most of the new visible mutations obtained were mosaics (Figures 24-5 and 24-6). It was not until after the Watson-Crick model came out that a satisfactory explanation of these differences was made. The chemical reaction of the alkylating agent is usually restricted to a single base and thus gene mutations are more likely to occur than chromosome breaks. Furthermore, single base alterations lead to mosaics, especially if mature haploid sperm are exposed to the mustard gas.

Regardless of the agent used, a substitution mutation can lead to the partial or total loss of function of the protein produced by that mutant gene. It may also lead to a "neutral" mutation, one in which the substituted amino acid has little or no detectable effect on the protein function. Substitution mutations are commonly encountered in human metabolic diseases, where the presence of the defective enzymes can be demonstrated by immunological tests. The enzyme is there but it is functionally a "dud." On rare occasions the mutation will change a gene whose protein serves as a major regulator of cell activities. Such changes will disturb the structure or development of the organism. Such regulatory mutations are often dominant rather than recessive.

Humans Exposed to Chemical Mutagens Are More Likely to Have Induced Cancers Than to Produce Mutant Offspring

Mustard gas was used in World War I, and many cases of lung cancer may have been induced by this agent because of the mutations and chromosome aberrations that occurred in the bronchial tissue. Agents that are mutagens are often also *carcinogens.*

The mustard gas was not likely to have reached the testes of the soldiers exposed to the gas. The cell cytoplasm in most bodies serves as an effective barrier; its millions of organic molecules are likely to react with an alkylating compound before it has a chance of reaching the DNA of the nucleus. When x rays are administered, no such barrier exists, because the radiation passes through all the cells in its path. Only if large quantities of an alkylating agent circulate through the body and enter the testes will such germinal mutations take place from the small number of alkylating molecules reaching those reproductive cells. It has also been confirmed experimentally that chemical agents are more effective in inducing mutations in mature sperm, where there is very little cytoplasm present, than in meiotic or premeiotic stages of sperm development or any stage of egg development, where there is a larger cytoplasmic barrier.

This difference in sensitivity of reproductive cells to x rays and chem-

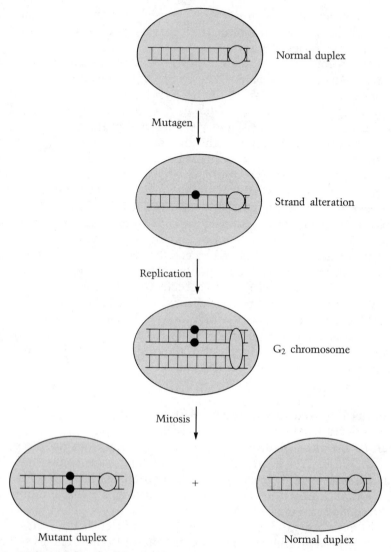

Figure 24-5 Origin of a Mosaic Mutation

If a normal reproductive cell (i.e., a sperm or an egg) contains a DNA duplex which is altered by a chemical mutagen (or by radiation) only one nitrogenous base may be altered in a base pair. The replication of the duplex in a zygote produces two chromatids, but one chromatid has picked up a base which is complementary to the altered base which is different from the base normally at that site (e.g., a G instead of an A may be picked up). When the centromere separates and the chromatids are distributed to the two daughter cells, one cell contains the mutant base pair and the other contains the normal base pair at this site. The resulting embryo is a mosaic, bearing both mutant and normal tissues. When this mosaic origin of the mutation occurs before meiosis begins, the resulting sperm or egg is a complete mutation, and the mosaic origin of the mutation cannot be detected. Most spontaneous mutations, like chemically induced ones, arise from alterations of a single base in a DNA duplex.

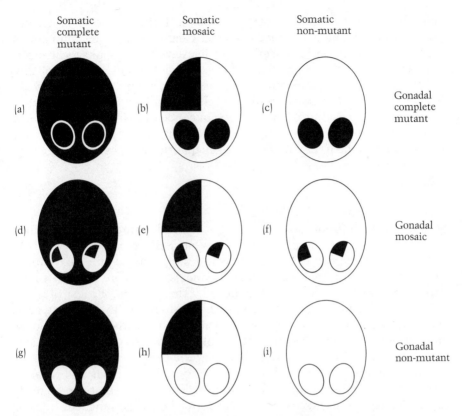

Figure 24-6   Patterns of Mosaicism in the Higher Eukaryotes

In humans (as well as in most vertebrate animals and higher plants) the origin of a gene mutation may occur before meiosis, resulting in a complete mutation [category (a)]. Such individuals produce mutant progeny in accord with Mendel's laws. Nonmutant individuals [category (i)] have no detectable mutant tissue either in their body tissue (soma) or in their offspring. All of the remaining categories in this figure are mosaics, but different genetic techniques are needed to reveal their presence. Categories (b), (e), and (h) are phenotypically mosaic. Categories (d) and (g) look like complete mutations, but (g) does not transmit the trait and (d) does so with less than Mendelian expectation. Categories (c) and (f) are cryptic mosaics because they look normal but transmit the trait (these correspond to normal parents who have two or more children with the dominant mutation, achondroplastic dwarfism). In fruit flies and experimental systems where analysis can be carried out, all eight of the mutant combinations can be observed and counted. In humans the existence of categories (b) through (h) is usually based on indirect evidence.

icals is an important distinction to remember when evaluating hazards of x rays and chemical mutagens. The ionizing radiation of diagnostic x rays, radiation therapy, and atomic bombs penetrates all tissues and there is little in the way of cytoplasmic protection for chromosomal DNA. Thus, when an organism is exposed to radiation, germinal mutations are as much a matter of concern as somatic mutations. In chemical mutagens, however, the site of entry or biological concentration of a compound would be the major focus of genetic damage. Ingested compounds would most likely cause cancers of the digestive tract (esophageal cancer, stomach cancer, colorectal cancer). Inhaled compounds would probably cause lung cancers or leukemias. Chemicals that are not alkylating agents but which are concentrated in marrow could also cause leukemias. Other compounds filtered out of the kidneys in concentrated form could cause bladder cancer. Still other compounds, picked up by red blood cells, could concentrate in the liver when they are dismantled for recycling and cause liver cancer.

Some Compounds Should Be
Regarded with Suspicion
Any chemical that reacts with purines or pyrimidines, that *intercalates* (fits between adjacent base pairs of) DNA, or that breaks the bonds joining nucleotides together, should be regarded as a potential mutagen. Some such chemicals, which are familiar to us, are nitrosoamines, which are produced by the reaction of meat preservatives (sodium nitrites and nitrates) with the stomach's hydrochloric acid. The nitrosoamines are also produced when meat is barbecued and the fat drips down onto the hot charcoals. A volatile spray of nitrosoamines and carcinogenic benzopyrenes will then coat the meat.

It is important to read the fine print on packaged foods and to minimize the consumption of foods containing compounds that can be readily converted to mutagens. If you eat fewer hot dogs, luncheon meats, and canned meats preserved with nitrates, you will consume fewer potential carcinogens. Similarly, if you cover your barbecue grill with aluminum foil and perforate it with a fork, you will get all of the heat needed to cook your meat and enough of the barbecue flavor to enjoy your cookout without incurring a high risk of your meat being coated with carcinogens.

Formaldehyde, which is used as a preservative, is another potentially dangerous compound. It is a volatile substance whose fumes irritate the eyes and respiratory tract. It is also used in the textile industry to produce permanent creases and wrinkle-free fabrics. The workers in textile mills, the garment workers who use the fabrics to manufacture clothing, and the store personnel who sell fabrics for home sewing are all at risk from inhaling these formaldehyde compounds. Clearly this is an area where labor unions and management could negotiate for safer procedures and working conditions, especially adequate ventilation to reduce the indoor atmospheric levels of formaldehyde and its related compounds to minimal levels.

Many Carcinogens Are
Products of Our Technology
Chemical products, such as chlorinated benzenes or complex organic rings with chlorinated side groups (e.g.,

DDT, many other pesticides, and some weed killers), are potentially hazardous because they can cause cancer in mice, rats, trout, and other organisms in the laboratory or in fields and streams. These compounds are fat-soluble and very little of them can be dissolved in water. Thus they are likely to be concentrated in fat-rich tissues or in dairy products. Their major site of carcinogenesis in humans would be in fat-rich tissue such as the breasts or in the colorectal area.

Many other chemical compounds are alkylating agents and many of them have been used widely in the paint, plastics, and automotive industries. Regrettably, it has only been through epidemiological evidence—the increased incidence of cancers—that workers and the surrounding communities have been made aware of the need for proper ventilation and inactivation of these biologically harmful products. Since the fumes of these compounds are always present at nontoxic levels in the companies manufacturing or using them, the carcinogenic or mutagenic hazards are usually not foreseen. Geneticists and cell biologists who suspected this carcinogenic relation many years ago had no direct evidence that cancers would be induced in humans. Thus they presented analogies from affected laboratory organisms and asserted that the same hazards could exist for humans. However, public ignorance of the single cell mutation that marks the beginning of cancer has made and still makes it easy to ignore the hazards to the individuals exposed to these substances and to future generations.

The Prospects for Wide-Scale Mutagen
Testing Are Not Encouraging

About a thousand new compounds are introduced into the environment each year which are ingested, inhaled, or absorbed through our skin, either accidentally or intentionally. Those which enter our foods and drugs are tested for toxic and carcinogenic effects (Table 24-2). Toxicity is easily detected; an organism gets sick or dies, and the reaction appears fairly soon after exposure to the compound. Carcinogenicity is tested by use of special inbred strains of mice or other animals sensitive to carcinogens and results can show up in a matter of a few months. Any compound that causes cancer when used as a food additive or prescription drug is not allowed to enter production for such purposes.

Industrial screening for mutagenic effects is not required. The reasons for this are ample. Unlike toxicity and carcinogenicity, the detection of mutations would involve breeding for two or three generations. Even with small mammals, such as mice, it would be difficult and expensive to raise several thousand progeny from the control and the treated parents. Since most of the compounds would reach the testes or ovaries in very low concentrations, few mutations would occur, and the difference between the control and treated mutation rates might be very slight—recognizable if thousands of progeny are studied, but not if hundreds are used. This would be prohibitively expensive.

Table 24-2   Criteria for Testing Potentially Hazardous Substances

Hazard	Test	Subjects	Testing Frequency
Toxicity	Assays for evidence of sickness, lethargy, weight loss, bleeding, death, and other signs of a reaction to the administered agent.	Usually 10–100 small animals are sufficient	Required by federal law
Carcinogenicity	Assays for tumors. Tests: for skin tumors, agent smeared on shaved part of body; for bladder tumors, dissolved agents are drunk; for intestinal cancers, foods mixed with testing agents are ingested	Usually 10–100 mice or other small animals are sufficient	Required by federal law
Mutagenicity	Bacterial cells cultured in eukaryotic-cell organelle suspensions, such as fragmented endoplasmic reticulum.	Live animals are not used because heavy doses of potent mutagens require 100–1000 animals and weak mutagens (found in foods, drugs, and cosmetics) require at least 10,000 mice.	Not required by federal law, but have been used by government agencies and some private companies
Teratogenicity	Agent mixed with food or water and fed to small mammals soon after onset of pregnancy. Litter size and offspring malformations assayed.	10–100 pregnant animals	Since the thalidomide tragedy, tests for teratogenicity have been added to new-drug testing, if the drug is likely to be used during pregnancy, but most compounds have not been tested.

Microbial Screening Tests Are More Efficient Than Animal Tests

Mutagen screening is still in the process of development. Among the possible models for screening mutagens are microbial tests using bacterial or fungal cells which have been implanted into the peritoneal cavity of a rabbit or a rat. Sometime after a mutagen is injected or ingested by the animal the microbial cells are removed and tested for induced mutations. Such *host-mediated assays* have yielded inconsistent results. In another approach the microbial cells are exposed in test tubes or culture dishes to a mixture of mammalian cell components previously treated with a mutagen. The direct exposure of a microbial cell to a mutagen would be unconvincing to manufacturers because, as we saw, there is a long, devious route between the entry of a compound into our body and its eventual transport to the germ cells. Many compounds may be totally inactivated before they ever reach the gonads.

The ideal screening test would be repeatable, inexpensive, and sensitive enough to represent the approximate conditions of germ cells in our own gonads. It should screen for gene mutations as well as chromosome breaks.

The Ames Test Is Presently
the Most Effective Test Available
for Mutation Screening

Several microbial assay tests have been introduced. Perhaps the best known is the *Ames test* which uses bacteria suspended in fragments of endoplasmic reticulum extracted from beef liver. The tested chemical agent may then be modified in either direction, turning from mutagen to nonmutagen or the reverse, as sometimes happens in mammalian tissues and organs before it reacts with DNA of the bacteria. Most of the time, however, the chemical remains nonmutagenic whether tested directly on microbes or through the Ames test. The Ames test has been used to show that most carcinogens are also mutagens and that many household chemicals, food additives, and pesticides are mutagenic. The Ames test does not test for chromosome breakage, but agents known to break chromosomes almost invariably also produce gene mutations. The reverse, however, is not true, and agents which alter nucleotides or cause frame shift mutations do not necessarily produce much, if any, chromosome breakage. Several refinements of the Ames test have greatly increased its sensitivity for detecting induced mutations.

A serious social problem related to carcinogen and mutagen screening is our indifference to carcinogens and mutagens due to the long period of time between the introduction of a carcinogenic or mutagenic compound and its effect on the body. Cigarette smoking is a good example. The long lag between cancer induction and symptoms creates an illusion of safety. So, too, with the hundreds of synthetic compounds that enter our bodies as food colorings, flavor enhancers, imitation products, thickeners, preservatives, and many other products. The effects may be long-delayed, especially with cancers and germinal mutations. If our individual risk from any one of these substances is very low, the cumulative effect of several thousand of these, from birth on, may be ignored even though some 80–90 percent of all cancers are believed to be produced by environmental agents other than viruses.

Questions to Answer

1. The term "mutation" can be applied to the individual gene or to all classes of heritable changes taking place in the cell. What are the advantages of restricting the term to the individual gene?
2. What evidence demonstrates that mutations can arise at any stage of development?
3. Why is mosaicism more likely to be seen for dominant disorders than for recessive ones?
4. What is an X-linked lethal? How can it be used as a test for induced mutations?
5. What is the movie and comic book version of mutations? How do these fictional mutations differ from mutations which occur in human populations?
6. Why was it unlikely that the children of the survivors of Hiroshima and Nagasaki would show an increase in mutations?
7. What are the indirect ways in which x rays induce mutation?
8. Distinguish between substitution and frame shift mutations.
9. Why do chemicals produce fewer chromosome breaks than x rays, even when they produce the same amount of mutations?
10. Why are most mutagens also carcinogens?
11. Why are animal fats more likely to be mutagenic than animal proteins or vegetable oils?
12. How does mutagenesis or carcinogenesis differ, physiologically, from toxicity?
13. Why would it be difficult to use mice or other animals to test new products for their mutagenic action?

Terms to Master

Ames test	mutagen
carcinogen	radiomimetic
frame shift mutation	substitution mutation
host-mediated assay	teratogen
intercalation	toxicity
mosaicism	X-linked lethal

Chapter 25
Genetic Engineering: Splicing Genes and Manufacturing Their Products to Order

Deeply embedded in popular attitudes about science is the Frankenstein model, derived from Mary Shelley's famous character. The image of Dr. Frankenstein, the scientist whose passion for research led him to create a monster over which he had no control, still haunts us. The fear of new knowledge goes back to mythology. In the story of Pandora's box, human plagues and misfortunes were meted out to punish Pandora for her curiosity and for opening a box which the gods forbade her to open. In another story, Prometheus was bound to a rock and tortured by Zeus for giving a frail humanity the divine fire or knowledge from which technology springs. This fear also appears in our religious traditions, with Adam and Eve banished from Eden for disobeying the divine order not to eat the forbidden fruit of knowledge.

New knowledge is suspect because it can be abused. Yet our sympathies really lie with Prometheus, Pandora, Adam, and Eve, because it is only when humanity acquires new knowledge that our likelihood of survival increases and our appreciation of the universe becomes enriched. For all the hazards that new knowledge brings, neither the fear of abuse nor the fear of divine retribution deters most of us from seeking it.

Genetic engineering is the most current of the Frankenstein models. It is tainted by the shadow of the military use of atomic energy, and the guilt many physicists felt when their scientific triumph became the instrument for the destruction of Hiroshima and Nagasaki. Unlike the development of the atomic bomb, however, there is no national emergency to spur current research in genetic engineering. Instead, scientists are thrilled by the basic knowledge this technology offers and its potential for medicine.

Genetic Engineering Is a Collection
of Techniques to Alter Heredity
Although the term *genetic engineering* can be used in a broad sense to cover all applications of genetics to humans, it is more selectively applied to biochemical efforts for altering, synthesizing, designing, and creating cells whose new heredity can be used for many scientific purposes. Some of these purposes are medical, including the treatment and prevention of genetic disorders. Some are agricultural, for increasing yields or improving the nutritional content of foods. Some uses still exist at a pure research level in pursuit of the molecular genetic mechanisms for the regulation of metabolism, for the control of development, and for the organization of the prokaryotic and eukaryotic cell.

Genes Can Be Spliced into Chromosomes
by Recombinant DNA Techniques
In 1973, Stanley Cohen, Paul Berg, and other molecular biologists worked out a technique which they called *recombinant DNA synthesis*. They learned how to splice a piece of DNA from one organism into an unrelated organism (Figure 25-1). The chimeric host harboring such foreign DNA could then use its own metabolic machinery to produce the products of the inserted genes. Thus a bacterial cell could produce the globin protein of a mammal if the gene for that protein were inserted into its own chromosome. Theoretically, the genes from any of our chromosomes could be inserted into viruses or bacteria by these techniques and their protein products could be isolated and analyzed for function or used for the production of important medical products, such as human insulin. As we shall see, this is actually more difficult to achieve than the techniques imply.

How Recombinant
DNA Works
There are several methods that can be used to insert foreign genes into a chromosome. The most widely used and discussed method for genetic engineering uses enzymes to cut DNA at specific places along the chromosome. These regions are usually *palindromic* or symmetrical sequences of nucleotides. ATTTTA is a palindrome because it has mirror-image symmetry. In our language we have palindromic names (OTTO), words (RADAR), and sentences (SLAP NO GAG ON PALS). The enzymes that recognize specific palindrome sequences are called *restriction enzymes*. When a given restriction enzyme is used, the chromosomes of a cell are cut into sections terminating at each end with a half strand of DNA that contains complementary sequences (e.g., TAA and ATT). If such strips are then mixed with a small virus or a *plasmid* (a small cytoplasmic ring of DNA often containing one or more genes useful to the bacterium) containing only one such palindrome in its circular chromosome, that open virus or plasmid can accept any of the introduced fragments. Another enzyme, called a *ligase*, can then syn-

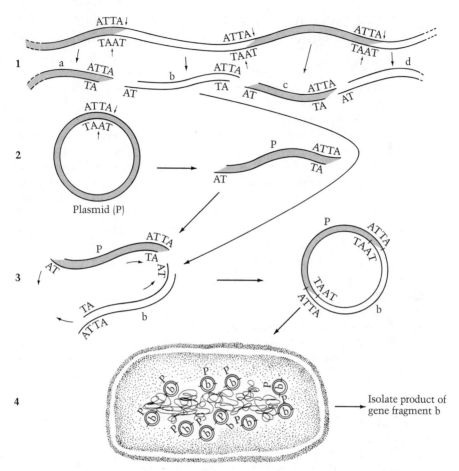

Figure 25-1   Recombinant DNA: How Genes Can Be Spliced

At certain sites in a DNA molecule there are small sequences, usually palindromic (readable the same way whether read from left or right). Restriction enzymes can cut an intact DNA at those sites into several pieces (1), each usually containing at least one complete gene. The enzyme cuts the sequence so that a small segment of single-stranded nucleotides remains, each end being complementary to the other end of a fragment. The same enzyme is used to cut a small circular virus (plasmid) that has only one such sequence in its DNA. The opened plasmid (2) is introduced to the fragments of the DNA (which can be from a prokaryotic or eukaryotic cell) and some of the plasmids and fragments join together in a new circular form with the donor DNA spliced into the plasmid (3). The splicing involves the complementary pairing of plasmid and donor DNA ends and an enzyme (called a ligase) which binds them.

In the diagram, fragment number 2 is inserted into the plasmid. The plasmid is then introduced into a suitable host cell (a bacterium or fungus) and the plasmid will multiply rapidly in it (4). The host cell will also divide, and within a day or two a large quantity of cells bearing a donor gene (and its protein product) may be present. This technique, slightly modified, has been used to make human growth hormone from its isolated gene. The technique is not yet a routine one and there are many problems, technical and social, in exploiting this for commercial use of human gene products.

thesize the bond to tie these paired sections together, and the virus or plasmid then has a spliced insertion of one or more genes derived from the cell.

This technique usually produces many types of recombinant viruses. These can then be used to infect bacterial cells (and by proper dilution only a single virus or plasmid will infect a bacterium) where they will multiply, producing a *molecular clone* of the inserted genes. Since each cell has a different chimeric virus or plasmid whose DNA sequence and function are unknown, the geneticist has to design a selective environment to isolate a particular gene that might be of interest. This can be achieved, for example, by using a bacterial strain that lacks an essential enzyme. If a strip of DNA inserted into one of those viruses or plasmids has the gene for an enzyme the cells need, they will survive on a food medium lacking the essential product that enzyme usually makes. Such an experiment, where special environmental and biochemical techniques are used to detect a desired inserted gene from many hundreds of other inserted genes in other bacterial cells, is called a *shotgun experiment* because somewhere in the profusion of cells bearing foreign DNA there is likely to be one bearing the lucky hit, the inserted gene for a particular function desired by the geneticist (Figure 25-2).

Shotgun experiments require a special selective scheme to isolate one gene from many unknown genetic combinations in a large population of cells. It is a needle-in-a-haystack situation. A more precise method makes use of a specific messenger-RNA that may be particularly abundant. Immature red blood cells in the marrow would be actively synthesizing hemoglobin, and their ribosomes would be rich in the m-RNA for the globin protein message. The m-RNA, after isolation, can be made to make DNA by adding an enzyme, *reverse transcriptase*, which certain viruses manufacture, permitting the conversion of their RNA to DNA inside cells. The DNA made from the RNA can then be chemically altered so its ends can be spliced into a bacterial chromosome. Once inside the bacterial cell the functioning gene can manufacture globin and a molecular clone of the gene can be made, limited only by the quantity of bacterial progeny desired by the scientist.

The Hazards of Recombinant DNA
Research May Be Real or Unlikely
In 1973 Paul Berg and several of his colleagues questioned the safety of the techniques they had developed. They feared that abuse and accidents could lead to chimeric bacteria that were potentially dangerous. Many scenarios have been discussed. Here are a few:

Case 1. A terrorist group has as one of its members a microbiologist who removes the toxin-forming genes of the botulin-producing bacterium. These genes are then inserted and spliced into *E. coli*, a harmless

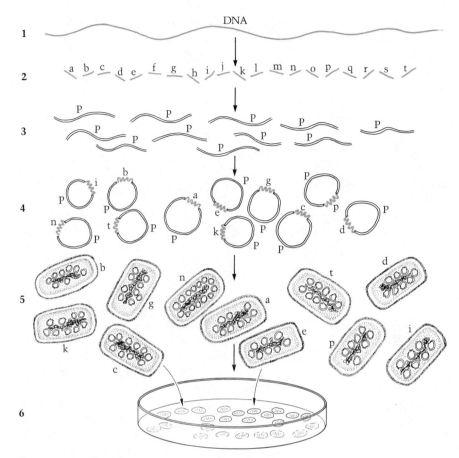

Figure 25-2 Shotgun Experiments Using Recombinant DNA

Before the structure of eukaryotic genes was known to differ from that of prokaryotic genes, concern was raised about the safety of recombinant DNA obtained from human tissue. A possible threat would arise if the following events occurred. The DNA of human cells (1), after extraction, is digested by restriction enzymes (2). The fragments are added to plasmids (3) similarly treated with the restriction enzymes. These recombine and some of the plasmids contain the human DNA (4). These are diluted and added to appropriate host cells (5) so that no cell contains more than one plasmid (P). The cells clone the inserted genes as they form colonies (6) on a Petri dish. If the investigators are looking for a specific enzyme, such as the letter c fragment in the diagram, they may select for it by an appropriate chemical test. However, the original source of the DNA may have had a gene which promotes colorectal cancer (fragment e in the diagram). This might be inhaled, ingested, or introduced into the body of an investigator by other accidental means. If the oncogene (cancer-inducing chief gene) is multiplied, the person may come down with colorectal cancer 20 or more years later. The possibility of such unwanted disasters made some of the early workers in recombinant DNA technology call for a moratorium on such experiments until the hazards could be evaluated and appropriate safety measures could be put into practice to prevent accidents. Such a catastrophe is not likely to occur because the genes of eukaryotes produce a different, nonfunctional product in a prokaryotic host cell.

bacterium that grows in our gut providing the bulk of our feces. The terrorists threaten to release the chimeric bacteria at unknown sites in several cities unless their demands are satisfied.

Case 2. A scientist uses a shotgun method to prepare a chimeric *E. coli* cell containing a gene for the human enzyme hexoseaminidase A. He hopes, after isolating it, to clone it so that he can use enzyme replacement therapy *in utero* for fetuses known to have Tay-Sachs disease. While carrying a test tube with some of the chimeric bacteria he stumbles and splashes some of the broth on his hand. He casually rinses his hands but some of the bacteria remain under his fingernails and in the creases of his fingers. When he later eats a sandwich at lunch he ingests some of these bacteria. One of them is an *E. coli* cell bearing a defective regulator gene whose product alters cell function to the tumor state for colorectal cancer. The bacterium multiplies and is spread to the scientist's family, friends, and neighbors over the next 20 years before the cancer first appears in an apparent epidemic form.

Case 3. The director of microbial warfare in a small country threatened by its neighbors and by the vacillating political pressures of the Soviet Union, the United States, and China decides to develop protective weapons. He believes that the big powers secretly are doing such research and that one of his neighboring countries may be supplied with such weapons, not to harm the population directly, but to destroy its livestock and agriculture. To counteract this threat he authorizes the chimeric production of antibiotic-resistant strains of the organisms that cause typhoid fever, cholera, and pneumonic plague.

At issue are many factual and ethical problems. Is it possible to produce chimeric bacteria with the genes for lethal toxins and antibiotic resistance? Very likely this could be done, either by a shotgun and selection technique or by isolation of the desired genes and their direct insertion. The major technical problem is not the isolation and insertion itself but the functional capacity of the inserted genes. Only if they contain a responsive regulatory system can they work. They either have to be read along with an adjacent gene that is turned on or they have to have their own regulator in a turned-on state to make them continuously functioning. Most genes, it should be remembered, are turned off in a cell. They are turned on by regulatory signals in response to specific environmental conditions. Even if the toxin genes for diphtheria or botulism are properly isolated, they may not be readable in *E. coli* in the absence of a proper regulatory signal.

Nevertheless, this probably represents only a temporary delay in the making of such pathogenic chimeras. Precisely the same research, applied to innocent regulatory genes, can be used to work out techniques for turning on any desired gene. This would then permit the abuse of these techniques for malevolent purposes. All knowledge can be abused, of course. It is not the invention of levers or gears or the discovery of oxidative chemical reactions that deserves our criticism, but their misuse when they are applied to machine guns or cannons.

The creation, through accidents, of chimeric organisms that become more virulent, acquire new resistance, or cause cancer epidemics is far less likely than the creation of such organisms through deliberate abuse. Such an unusual event would require many adaptive features which are rarely observed in nature. The extra genetic piece would either have to be neutral or beneficial in its physiological effect on the bacterium. We know that spontaneous or induced insertions or duplications of genetic material are usually harmful to the organisms in which they arise. Such alterations tend to disrupt normal gene activity because it is easier for random change to diminish a function than to enhance it. Although a bacterium may contain an inserted gene for cancer production the bacterium may die in the gut rather than multiply. Even if transcribed, the product may remain inside the bacterial cell and may only be effective when it is inside a human cell in large quantities. If the bacterial cell dies and its DNA fragments enter human cells they will probably be digested by lysosomes long before they can enter the cell nucleus and provide a source for transcription.

The Political Aspects of Genetic Engineering
Reflect Both Caution and Approval
At the Gordon Conference in New Hampshire in June 1973, Paul Berg went public with his concern about the hazards of recombinant DNA research. With the help of the National Academy of Sciences and other agencies, molecular biologists held an international conference at Asilomar, California, in February 1975. Temporary guidelines for research were suggested with certain types of experiments temporarily banned until their safety could be assured. At the same time the National Institutes of Health began a number of studies to work out safe procedures and guidelines for research. These guidelines were finally approved in July 1976 (Table 25-1). They are not laws, however; they are a means of exerting moral pressure on scientists, and they serve as guidelines in the awarding of federal grants for research.

The guidelines identify three levels of biological containment: special strains of bacteria for experiments that transfer genes from primates, including humans (EK3); bacterial strains that use genes from other mammals or toxins from any source (EK2); and bacterial strains that use genes from plants, lower animals, and nonpathogenic bacteria and viruses (EK1). Only the EK1 level is considered a safe system for shotgun experiments.

Additionally, the guidelines define four facilities for laboratory research. P1 laboratories have open access and need no special precautions. They can be used for EK1 experiments. P2 laboratories require an autoclave to sterilize materials, safety cabinets to handle more hazardous EK2

Table 25-1   NIH Guidelines (1976): Stringent Safeguards Initially Imposed

After consulting with committees of scientists working in recombinant DNA research as well as those concerned about the hazards of such research, the National Institutes of Health issued guidelines for recombinant DNA research, rather than legal regulations, because not enough was known about the safety or risks of such research at the time the guidelines were formulated. As evidence of safety has become available, the physical and biological containment conditions have been relaxed (or made more stringent, if experiments indicated greater hazards than predicted). These 1976 guidelines represent the caution urged by scientists who preferred to begin research with stringent rules for safety, gradually relaxing them on the basis of experiments and experience.

Classification	Conditions to Be Used
Shotgun experiments with E. coli	
Primate tissue, adult	P3 + EK3 or P4 + EK2
Primate tissue, embryonic	P3 + EK2
Other mammals	P3 + EK2
Birds	P3 + EK2
Adult tissue, other vertebrates	P2 + EK2
Embryonic tissue, other vertebrates	P2 + EK1
Lower eukaryotes	P2 + EK1
Plants	P2 + EK1
Prokaryotes that normally exchange genes with *E. coli:*	
nonpathogenic prokaryotes	P1 + EK1
low-risk pathogenic prokaryotes	P1 + EK1
medium-risk pathogenic prokaryotes	P2 + EK2
high-risk pathogenic prokaryotes	banned
Prokaryotes that do not exchange genes with *E. coli:*	
nonpathogenic	P2 + EK2 or P3 + EK1
moderate-risk	P2 + EK2
high-risk	banned
Cloning plasmids in E. coli *with:*	
Animal viruses	P4 + EK2 or P3 + EK3
Plant viruses	P3 + EK1 or P2 + EK2
Animal viruses as cloning vectors	
Defective SV 40 with nonpathogenic DNA source	P3
Defective SV 40 with pathogenic DNA source	P4

Prohibited experiments
Cloning cancer viruses
Cloning toxin-producing genes (e.g., botulism, diphtheria, snake or spider venoms)
Cloning genes that extend host range or increase virulence
Transferring drug (antibiotic) resistance to those normally lacking resistance

Note: *Physical containment*

P1   No special design. A typical microbiology laboratory.

P2   A typical microbiology laboratory with autoclave (steam sterilizer) and safety cabinet (where one's hands enter rubber gloves and do not come in direct contact with the organisms); restricted access to the laboratory; no food storage (coffee, snacks).

P3   Special engineering safety features, including a separate access so that only authorized persons can enter or leave; separate corridors; safety cabinets and autoclave in the laboratory that sterilize the air and water leaving them.

P4   All of the P3 features plus a separate building (isolated from other areas); special safety cabinets designed to prevent escape of microorganisms; personnel enter and leave through decontamination showers and wear special clothing upon entering; air flow is engineered so that it is sterilzed upon leaving P4 unit.

Biological containment

EK1   *Escherichia coli*, strain K12 (the one used in most microbiology laboratories).

EK2   A modified *E. coli*, K12, with numerous mutations that make it virtually impossible for it to survive outside the laboratory (less than one survivor among 10^8 cells). *E. coli* strain × 1776 was certified as an EK2 organism.

EK3   An EK2 strain which has been tested in animals (especially primates) without evidence of cloned genes entering the animal cells after inhaling, ingesting, or being injected with the EK2 cells.

experiments, and a more secured means of access to the room. The P3 facilities are very elaborate, with air flow being carefully monitored and none of the air in the chemical hoods or cabinets being allowed to mingle freely with the vented air of the building. A separate corridor has to be built so that only those authorized to enter can do so. Both EK2 and certain EK3 experiments can be done in a P3 facility. Most universities have P1 and P2 laboratories, but considerable architectural and engineering skills are involved in P3 facility construction. Finally, very risky experiments, such as shotgun human cell experiments, are done in a P4 facility. This is an isolated building with safety features comparable to those used at Fort Detrich, Maryland, where germ warfare research was done. Totally banned from any facility are cloning experiments with virulent pathogens, cloning of toxin genes, experiments that would increase the host range of pathogens, the deliberate release of recombinant cells into sinks or sewers, and the introduction of antibiotic-resistant genes into human pathogens.

Only some ten percent of current university research using gene splicing techniques is regulated by the guidelines. Most of the research projects in this field have been shown to be safe, and special precautions for such exempted research are not needed. As progress in recombinant DNA technology has been made, the knowledge gained has been used to modify the regulations.

There Are Many Benefits Predicted
from Genetic Engineering

Those who favor all but the banned experiments in P1 or P2 facilities do so for two reasons. First, they reject the risk of biological hazards as remote, either on evolutionary grounds or based on prior experience with processes like bacterial transformation or the viral transmittal of bacterial genes. Secondly, they believe that the benefits from the recombinant DNA research are enormous. They believe human gene mapping can become as detailed as that of *Drosophila*, *E. coli*, or the bacteriophages. Molecular clones of a par-

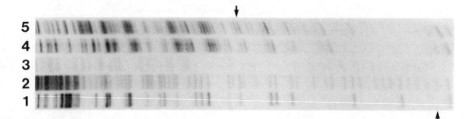

Figure 25-3   The Direct Reading of Nucleotide Sequences
Maxam and Gilbert developed a biochemical method of digesting DNA under special conditions so that the location of A, T, G, or C becomes evident in the gel used to separate the treated fragments from the length of DNA. The sequence of this DNA from a kangaroo rat, reading from right arrow to left arrow, is AAA AGA AAT AGA CAA TAT CAG TCG GTT AC. The gels for the numbers 1–5 represent: 1 = G, 2 = G + A, 3 = A > C, 4 = C, 5 = C + T. Thus a bar in 1 and 2 is G, in 2 but not 1 it is A, in 4 and 5 it is C, and in 5 but not 4 it is T. The third row is a control check for A in case row 2 is not sufficiently clear. (Courtesy of Andrew Thliveris)

ticular DNA fragment (or its m-RNA product) can be labeled and used to form paired complexes with the allelic region they came from in a culture of human cells. The molecular clone of a given DNA fragment can be used to yield its sequence or its protein product (Figure 25-3). Additional biochemical tests can be used to identify its function as a specific enzyme, organelle subunit, or regulator. A viral encyclopedia of human gene functions can be developed by a systematic study of molecular clones (Figure 25-4).

In medicine such studies would provide human insulin and other hormones and essential metabolites that are often lacking in children or adults with birth defects and genetic disorders. The clonally produced enzymes can be inserted into lipids or empty red blood cells and find their way past the body's immune system and into target organs where they might function. *Enzyme replacement therapy* could be used to treat autosomal recessive diseases in humans, even to the point of injecting such protected enzymes into the fetal blood system during pregnancy.

There are many biologists involved in cancer research who are hoping for an exemption from the research moratorium so that they can isolate those gene fragments in human tumor tissue that can induce tumor formation in tissue culture or primates. Once such cancer genes are isolated, their DNA sequences can be worked out, their proteins grown in abundant quantities for scientific study, and their functions tested in detail. Of course such cloning of oncogenes from human tissue would have to be done in P4 facilities with extreme precautions.

The field of agriculture would benefit from the creation of strains of rice, wheat, and other grains rich in all of the 20 amino acids needed for protein synthesis. Genes for the increased synthesis of a specific amino acid normally scarce in the plant proteins could be grafted onto one or more of the chromosomes of the cereal grain cells. Strains of the nutri-

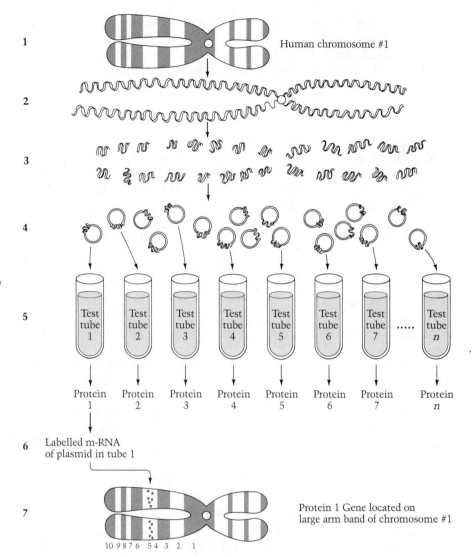

1   Human chromosome #1

2

3

4

5   Test tube 1   Test tube 2   Test tube 3   Test tube 4   Test tube 5   Test tube 6   Test tube 7   .....   Test tube n

Protein 1   Protein 2   Protein 3   Protein 4   Protein 5   Protein 6   Protein 7   Protein n

6   Labelled m-RNA of plasmid in tube 1

7   Protein 1 Gene located on large arm band of chromosome #1

10 9 8 7 6   5 4   3   2   1

Figure 25-4   Constructing a Library of Genes: A Possible Approach

Human chromosomes (1) can be multiplied by growing their cells in tissue culture. The DNA (2) can be digested by restriction enzymes and these pieces (3) can be inserted into plasmids (4). These can be cloned in an appropriate host cell. If the right dilutions are used during the infecting process each plasmid will enter a single host cell. The cloned genes will thus be present in large amounts in each test tube (5). These genes will be different from the plasmid or the host cell. If test tube number 1 has such a cloned DNA, the messenger RNA for it will be found in large amounts also. This RNA can be isolated, copied so that it is made into a radioactive form (6), and then made to adhere to the DNA complementary to it on the intact chromosome (7). Thus the site for DNA fragment number 1 would be identified. In some cases the m-RNA for a known human protein can be synthesized and used as a probe (radioactively labeled) to identify its map location, and this, in turn, can be used to identify the tube in which it is cloned.

tionally enriched crops could be used to greatly diminish malnutrition in populations where animal protein is rarely eaten for religious or economic reasons. Artificial photosynthesis might even be achieved, with immense quantities of sugar being produced directly from the atmosphere, water, and sunlight, bypassing any living organism for its production.

There Are Also Potential Risks
and Abuses of Genetic Engineering
The optimistic view must be tempered by the risks—both hidden and apparent—of recombinant DNA research. There is something called "the ice-nine effect," based on Kurt Vonnegut's novel *Cat's Cradle*. The hero of the book, a scientist who has little interest in anything but his work, designs and produces a form of water that crystallizes at room temperature. Military strategists are pleased because it can be used to harden mud, making spring campaigns and tropical war tolerable. Unfortunately, though, most people have limited vision of the applications and implications of their activities. Not only does ice-nine congeal mud; it also extends to the ground water, streams, ponds, rivers, and oceans. It can also congeal blood. Like the Midas touch of classical legend, technology can be a nightmare when it touches things or beings that it was not intended to touch. Of what use is a glue that is so potent that it requires surgery to separate one's fingers after using it to make a model airplane? Of what use is a recombinant DNA bacterium that digests old rubber boots and plastic wastes in our polluted environments if it causes our overshoes to fall off our feet when we walk in the rain or snow? Can a chimeric bacterium be safely used to clean up oil spills if it turns the gasoline in our automobile tanks to cheese? How often do scientists reflect on the unexpected, the catastrophic, and the potential abuse of their inventions?

The Controversy over Recombinant DNA
Has Subsided Because of New Findings
Much of the debate over recombinant DNA research has subsided. The major reason for a change in concern was the unexpected discovery that the structure of genes and the way proteins are synthesized in eukaryotes differ profoundly from the structure of genes and the protein synthesis process in prokaryotes (Figure 25-5). Prokaryotes are *colinear* in their production of proteins from genes; eukaryotes are not. In the eukaryote the informational sequence for a protein is interrupted by *intervening sequences* (*introns*) whose functions are not understood. Enzymes snip out the appropriate segments of transcribed RNA and assemble these into a messenger RNA. Thus a eukaryotic gene, taken out of a chromosome and spliced into a bacterial plasmid, will not produce the protein for that gene because the enzymes needed to assemble a messenger RNA from the *informational sequences* (*exons*) are missing. This makes it virtually impossible for a shotgun experiment to introduce a human gene capable of producing a

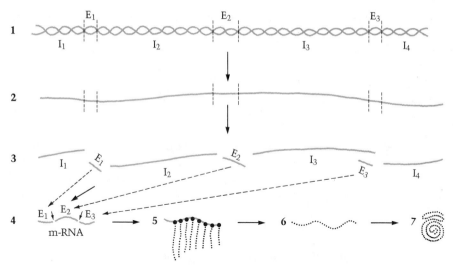

Figure 25-5   The Discontinuous Organization of Eukaryotic Genes
Unlike prokaryotes, where the flow of information from DNA to m-RNA to protein is colinear, the eukaryotes have genes which are not colinear with their proteins. The eukaryotic gene has informational segments, called exons, and intervening sequences, called introns, of yet unknown function. In the diagram these are represented by E for the exons and I for the intervening segments. The intact DNA (1) shown here has three informational and four intervening sequences.

The entire DNA sequence bearing the collection of informational and intervening sequences is transcribed as a single RNA molecule (2). Special enzymes (probably in the nucleus) cut the RNA (3) and the informational segments are assembled into a messenger RNA (4) which is translated (5) into protein normally by ribosomes in the endoplasmic reticulum (6,7).

Besides the discontinuous nature of the eukaryotic gene, there are other differences in protein synthesis between prokaryotes and eukaryotes. The site for the enzyme that transcribes DNA into m-RNA is located within the gene in eukaryotes, but it is adjacent to the gene in prokaryotes.

The absence of the enzymes in stage (3) in prokaryotes means that scientists cannot insert human DNA fragments into prokaryotes or their plasmids and expect gene translation to yield functional proteins. To use this prokaryotic system, geneticists have had to use a "back door" approach, isolating m-RNA, using a reverse transcriptase to convert it into DNA, and then using that DNA, appropriately spliced into a plasmid with an appropriate regulator, to generate a protein.

protein. It also diminishes the chances of a rapid exploitation of human genes useful for medical treatment or pharmacological sale. Such genes would have to be constructed in more complex ways (by direct synthesis, or by copying DNA from the m-RNA).

A second assurance of safety comes from experiments carried out in P4 facilities. There is no introduction of tumor genes from plasmid carriers to the gut cells of recipient mammals when these mammals are fed *E. coli* bearing these genes. Accidental ingestion of the *E. coli* would not be worrisome if massive amounts of bacteria are incapable of transmitting the clonal DNA they carry in their plasmids (Table 25-2). Also of significance is the discovery that our gut *E. coli* naturally harbor plasmids and exchange DNA with other bacteria and plasmids. Untold billions of natural recombinant DNA experiments have taken place for millions of years. This makes it unlikely that shotgun experiments are a serious hazard.

The problems of germ warfare, paranoid behavior, and the shifting of the host cell from *E. coli* to fungi and other eukaryotic cells aside, some of the controversial aspects of recombinant DNA technology are unresolved. Not all of the critics of this new science believe the procedures to be safe, and many of those engaged in the research are aware of the potentially embarrassing criticism they face if they become overconfident and ignore the NIH guidelines.

As in the story of Pandora's box, the lid has already been opened and whatever potential hazards exist in recombinant DNA research have already been circulated throughout the scientific world. After Pandora opened the box and human miseries flew out, she slammed the box shut, but

Table 25-2   NIH Guidelines: Major Changes in 1979

The ongoing debate over recombinant DNA continues. Congress has so far declined to make the NIH guidelines federal law, although certain aspects are likely to be written into law (e.g., regulation of patent rights based on research funded by federal agencies). The main issue, safety, seems likely to be resolved by the experience of scientists using the guidelines.

Among the events favoring continued caution is the accidental death of a person (working in an unrelated laboratory) in England from smallpox. The laboratory director of the pox laboratory subsequently committed suicide. His smallpox work had been permitted on the basis of his experience and reputation, and the stringent safety demands of the World Health Organization had not been followed. Also, at Plum Island, off the tip of Long Island, New York, an outbreak of hoof and mouth disease occurred among some cattle. Plum Island is a Department of Agriculture Research Station, closed to public access, and using P3 and P4 facilities for studying agricultural pests.

Offsetting these accidents are the many indications of safety in recombinant DNA research projects. Foremost among these is the presence of intervening sequences in eukaryotic DNA which are not excised during prokaryotic protein synthesis (and thus no functional protein emerges).

The relaxation of physical and biological containment requirements marks the recent actions of the NIH following its study of guideline modifications. Of particular interest in this table are the shifts from *E. coli* to *Neurospora* and yeast as host cells for recombinant DNA cloning.

1. Approval of *Neurospora crassa* (orange bread mold) and *Saccharomyces cerevisiae* (yeast) as hosts for plasmid clones. The strains to be used for P1 or P2 containment have mutations that prevent their survival outside a laboratory or in cases of accidental mating with wild strains. These are not pathogenic to humans, animals, or plants, and are thus of no risk to the natural ecology. If wild strains are used, P3 facilities must be used.

2. The containment restrictions on primate shotgun experiments were lowered to P2 + EK2 or P3 + EK1.

3. Containment restrictions on shotgun experiments for mammals other than primates were lowered to P2 + EK2 or P3 + EK1.

the only thing left for humanity was hope. In our time, we are more apt to rely upon our own reason, experiments, and technology than upon hope to provide the benefits and eliminate the risks. However, the history of occasional error, accident, and oversight in the application of our technology should make us realize that for us, too, hope and faith, not only the assurances of human reason and good will, govern the potential of new technologies.

Questions to Answer

1. What is meant by the term "genetic engineering"?
2. What is the natural role of restriction enzymes in cells?
3. Why are restriction enzymes used in recombinant DNA technology?
4. What is recombined in recombinant DNA technology? How does it differ from the recombination of crossing-over or independent assortment?
5. What is a ligase? How is it used in recombinant DNA technology?
6. What is meant by a shotgun experiment?
7. What is the difference in safety philosophy between physical containment and biological containment?
8. What is a molecular clone?
9. What were the risks imagined by critics of recombinant DNA technology in its early years of existence?
10. What were the applied uses of recombinant DNA technology imagined by its supporters during the early years of its existence?
11. What discovery made both the fears and the applications of recombinant DNA technology less likely to be an immediate reality?
12. What techniques can be used to clone human genes in (a) a bacterium and (b) a yeast cell?
13. If a new technology arises, who should regulate it: (a) the scientists who developed it, (b) the industries that apply it, or (c) the government? Discuss the advantages and the risks involved for each of these choices.

Terms to Master

enzyme replacement therapy	molecular clone
exon	recombinant DNA synthesis
intron	restriction enzyme
ligase	reverse transcriptase
palindromic sequence	shotgun experiment
plasmid	

Part 7
Population
Genetics

The complex relationship between the characters or features of organisms and their genes in different environments is often difficult to demonstrate experimentally. Yet many geneticists would argue that the overall health and life expectancy of individuals depends upon this relationship.
When individuals have combinations of genes that diminish organ function, those persons are more likely to experience organ failure and to
experience it earlier than other people. Geneticists have made estimates
of this effect of the individual genotype, and roughly one-fifth of those
born are fated to die ten or more years before their mean life expectancy
because of their combination of less advantageous alleles.

There are several implications of this theory of disadvantageous genotypes. In the population as a whole it is measured as a change in the
percent of gene frequencies passed on to the next generation. The potential increase or decrease of genotypes leading to changes in mean life
expectancy for the population is called the genetic load. To interpret
genetic load we have to acknowledge, first, the vast reservoir of genetic
variation in human populations, not all of it consisting of genes for neutral or beneficial traits. Second, we are faced with a constant influx of
newly arising gene mutations, usually resulting in impaired function of
the altered genes. This may be as high as one new mutation in every ten
sperm or eggs. Third, we can accept the balance of newly arising mutations because they are often compensated for by the elimination of detrimental genotypes, usually through disease and malnutrition. This was
true for most of humanity before the 20th century. Fourth, we must
recognize that modern living (with public health programs, high standards
of living, and substantial medical care) enables many who would have
died young because of their disadvantageous genotypes to survive to maturity and pass on part of their constellation of genes. Thus we have a

paradox: curative medicine (which we surely wish to practice) in the long run leads to an increased genetic load for the population and forces us to think about alternatives such as differential breeding, artificial insemination, and other techniques of eugenics.

The implications we considered for ourselves are characteristically found in natural populations of plants and animals. Darwin called attention to the consequences of this variation in organisms and used it for his theory of evolution through natural selection. However, the theory of evolution disturbs many people. It is also widely misunderstood. Evolution theory neither affirms nor rejects the existence of a god. It does reject the idea of a purposeful direction in which organisms (especially humans) are headed. It also rejects the idea of an inner driving force within individuals and the concept of any mystical guidance of the process. The theory of natural selection assumes that variations arise from chance mutations of genes and rearrangements of chromosomes. It assumes that species arise through the gradual adaptation of character traits to changing environments, mostly by differential survival of individuals with those constellations of genes that provide advantages.

These evolutionary concepts are hard to accept. We may believe in our favored uniqueness among all species. For many people that is too important a concept to give up. Those who retain a belief in this special relation of humanity to a creator may attempt to reconcile religious needs and credos with evolution theory (sometimes modifying Darwinism to a divinely directed process). Others may reject evolution theory altogether because there is no way that a literal interpretation of biblical creation could be fully in accord with the time involved, the sequence of events; and the lack of purpose associated with Darwinism.

The implications of population genetics and evolution theory touch upon two controversies—what to do about our genetic health in the generations to come, and what an acceptance or rejection of evolution theory does to our sense of values and meaning. Very clearly, whatever our values and religious beliefs might be, we cannot ignore our genetic health indefinitely. Nor can we long avoid the overwhelming mass of evidence demonstrating the antiquity of life on earth and the relatedness of life which evolution theory has pieced together from molecular to organismal levels.

Chapter 26
Genetic Load: An Undesired Shift in Gene Frequencies

New mutations arise every generation. This has been observed for many dominant disorders in humans, such as retinoblastoma, a cancer of the eye in children; achondroplasia, a disproportionate dwarfism affecting the limbs and skull; and Apert syndrome, which produces a severely malformed head and the coalescent fusion of the fingers and toes. The sudden origins of some X-linked mutations have been carefully documented, such as a line of hemophilia (bleeder's disease) in Queen Victoria (or an immediate female ancestor of hers not more than two or three generations removed). Most mutations are recessive and their origins are difficult to trace because the likelihood is overwhelming that an affected sperm or egg, after fertilization, would be heterozygous.

Out of all those mutations, past and present, comes the variability that makes each of us, except monozygotic twins, genetically unique. Since we all differ from one another in appearance we can infer that we also differ in genotype for our other bodily traits, including our vital organs. Even more variability may be revealed in our cells through the use of immunological and biochemical techniques applied to our proteins or DNA.

Elimination of Mutations Must Occur to Compensate for Newly Arising Ones Some of this variability probably arose hundreds or thousands of generations ago. Since individuals harboring certain mutations had a greater chance for survival, these mutations may have been passed on. Yet most mutations do not increase one's chance of survival and, in fact, only a small number of all mutational changes have been retained. If the average frequency of spontaneous mutation for a gene is about 1 in 250,000, and if a human gamete contains, conservatively, 25,000 genes, then, as Muller estimated, one new mutation occurs in every ten human sperm or egg cells. This implies that the present world population should have 100 million more mutations than the world population had in 1945. Humankind has existed for tens of thousands of generations and thus immense numbers of harmful genes have arisen, and if they were all present in the population they should be revealed by inbreeding, as in cousin marriages. Cousins have $\frac{1}{16}$ of their genes in common and only about one marriage in a thousand involves first cousins. Yet most of these couples produce normal children, and the risk of death or genetic defect is only slightly higher (about five percent) for their children than for children of noncousin couples. If each of us carried hundreds or thousands of detrimental mutations from the past then virtually no child of a cousin marriage would survive; but we do not carry all past mutations because they are being eliminated every generation. If inbreeding is uncommon, so are random marriages between heterozygotes for rare genetic disorders. How then does a harmful gene eventually become extinct?

Danforth and Muller Proposed a Quantitative Measure of Mutations in a Population

In 1921 Charles Danforth worked out a few arithmetical relations between mutation frequency, the fitness or impairment of the mutant gene when homozygous, and its frequency of extinction when it became homozygous. In 1950 Muller applied these equations in a new way and proposed the concept of *genetic load* (Figure 26-1). According to Muller's theory, a recessive mutation is seldom 100 percent recessive. Instead, it expresses a small, but measurable, impairment on the heterozygote because two doses of a normal allele in an individual have a slightly better metabolic efficiency than a single dose of that normal allele. This small impairment can be demonstrated experimentally in fruit flies heterozygous for recessive mutants.

The Genetic Load Is Expressed Through the Partial Dominance of Recessive Mutations

Muller's evidence for the slightly disadvantageous effect of a recessive mutant in a heterozygote came from several sources. He noted that humans heterozygous for many recessive disorders show partial impairments. He also showed experimentally in fruit flies, as did other geneticists, that newly arising recessive lethal mutations reduce the viability of the heterozygotes carrying them by about five percent. If a new mutation produces an impairment of this magnitude, then 1 out of 20 heterozygotes carrying that mutant gene will die before maturity or remain sterile. Muller called this extinction of a gene through the homozygote or, more frequently, through the heterozygote, a *genetic death*.

Figure 26-1   Genetic Load in a Population

An average genetic load consists of 8 detrimental genes in the heterozygous condition. Such an average is represented in white. Ten unrelated individuals in the first generation are followed for five generations. Those with a mean load less than 8 are represented in black; those with a mean load higher than 8 are shaded. A genetic death (failure to leave progeny) is assigned to an individual with a genetic load of 12 or more. The average of the population remains constant (it is represented as 80 for these ten kindred lines). In general, high genetic load individuals have a tendency to pass that load on to one or more offspring. Similarly, low genetic load individuals tend to pass on a low genetic load. In the diagram a replacement with a mean of 8 is introduced from an unrelated family. It is important, in discussing genetic load, to recognize the heterozygous accumulated damage to an individual. The extinctions are not caused by homozygosity for a lethal gene, but by the more subtle damage to health and fertility of the high genetic load. This includes such factors as smaller family size, delayed marriage, and premature death, all of which diminish overall capacity to survive.

Which of the 20 possible carriers will die prematurely? Muller claimed that there are enough past detrimental mutations in existence to account for the equivalent of a genetic load in the population of four lethal mutations per person. This does not mean that each of us carries that equivalent, because Mendel's laws predict that we shuffle our genes every time we make eggs or sperm so that some of our children have more and others less than the average amount of this heterozygously expressed detrimental genetic material. If a zygote has the misfortune of receiving an above-average number of heterozygous mutations then that individual may be aborted or die in childhood when stressed by environmental hazards, such as pneumonia or other infectious diseases, malnutrition, or hostile weather. Since the greater part of humankind faced these hazards throughout history, most of those individuals with higher frequencies of heterozygous mutations risked a premature death. Since any combination of heterozygous mutations contributing to a genetic death could occur in a zygote, it is a matter of chance which particular mutant gene is eliminated.

The biological effect of genetic load in a population is the formation of an equilibrium between new mutations arising spontaneously and mutations eliminated by genetic deaths in the population. This equilibrium preserves our genetic load at an average of 4 lethal equivalents per person, rather than 40 or 400, which might have been the case if all eliminations arose only from homozygosity for a harmful gene.

There Is More Than One Theory
to Interpret the Effects of Heterozygosity
Whenever two related populations, each relatively homogeneous, are brought together, they produce hybrid offspring. Those hybrids often show a dramatic improvement over the parental populations in such qualities as height, resistance to disease, and efficiency in surviving. This is called *hybrid vigor* or *heterosis* and it is particularly striking in crop plants and farm animals.

A lot of human effort is required to cultivate a domesticated plant, to increase its yield, to select its most nutritious and aesthetic features, and to make it survive the conditions on the farm. Each crop has to be divided into two portions, one for the next year's planting and one for family consumption or sale to the community. Since a single plant may produce a large amount of seed, the selection of a few plants with highly favorable features, such as the number of rows of kernels on an ear of corn, establishes the new trait but at the same time diminishes the total variability in the population. This causes many more genes, each generation, to become homozygous because the only available plants with pollen or ovules are descendants of the few individuals selected from the previous season's crop.

If two strains of maize from distant farms or distant countries are crossed in a field, their progeny will show hybrid vigor in ear size, kernel quantity, number of rows, and ears per bushel. Similarly two highly inbred breeds of dogs, when mated, will often produce mongrels that may lose their economic value but will be more robust than either of the parental strains.

A heterozygous individual does not have any special quality that makes it more vigorous than the homozygous normal or mutant individual. Indeed, at a molecular level, it is not readily apparent why one dose of a normal allele should be more effective than the normal two doses.

East and Jones Proposed a Theory
of Compensation to Account
for Hybrid Vigor
When hybrid corn was first developed by E. M. East and D. F. Jones early in the 20th century, they proposed a model of compensatory dominance in which homozygous recessive genes from each strain are made heterozygous when crossed. Each strain they used brought in normal alleles that rendered heterozygous certain harmful recessive alleles received from the other strain. Furthermore, each strain often had numerous traits linked on a chromosome. Those genes which were more closely linked would not as readily segregate at random in the F_2 generation. This made it almost impossible to construct stocks of homozygous dominant maize by selection in a few generations. Instead, East and Jones, working in their university gardens, developed hybrids which received their dominant genes from four inbred strains. If strains A and B produced hybrid AB, and strains C and D produced hybrid CD, then a cross of AB × CD yielded a combined ABCD hybrid with many more of its recessive genes made heterozygous and not expressed by the dominant traits selected in strains A, B, C, and D.

The Theory of Overdominance Assumes
an Advantage for the Heterozygote
A different interpretation of hybrid vigor is based on the theory of *overdominance*. The heterozygote in this model is more fit than either of the two alleles in their homozygous state. This model was developed most effectively for a human disorder, sickle cell anemia (Figure 26-2). The victims of this disease suffer episodes of pain, frequently get pneumonia, and are stunted in their growth from the poor oxygen-carrying capacity of their blood. They usually die before maturity unless they have effective hospitalization and blood transfusions. The disease is caused by a homozygous recessive gene. Heterozygotes show none of the anemic symptoms although their red blood cells, when cultured in low-oxygen conditions, will collapse into the sickle cell shape of the homozygous sickle cell blood (Figure 26-3).

Sickle cell anemia is a *molecular disease.* The gene mutation produces a defective hemoglobin molecule. Yet despite the severity of sickle cell anemia, the frequency of the gene is high in certain parts of Africa, the probable place of its origin some 10,000 or more years ago. The distribution of high sickle cell anemia areas in Africa closely paralleled the distribution of areas of high malarial infections (Figure 26-4).

Sickle Cell Anemia Is Used
to Demonstrate Overdominance
Studies of sickle cell gene frequencies by A. C. Allison and others soon revealed an unusual genetic situation. Individuals who were sickle cell heterozygotes had less severe

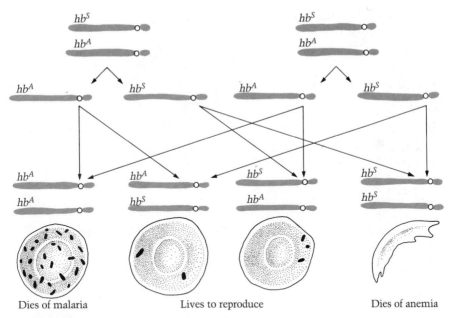

Dies of malaria   Lives to reproduce   Dies of anemia

Figure 26-2   Balanced Polymorphism and the Sickle Cell Trait

If two heterozygotes for the sickle cell gene, hb^S hb^A, marry and have children, each heterozygote will segregate the hb^S from the hb^A allele during meiosis. The resulting fertilizations will yield 25 percent homozygous hb^A hb^A, 50 percent hb^A hb^S, and 25 percent homozygous hb^S hb^S. The sickle cell homozygotes die of the anemia. The homozygous hb^A hb^A individuals are more likely to die of malaria than the carriers, hb^S hb^A, whose infected cells provide less oxygen and have a tendency to form crystals of hemoglobin when the malarial parasites are present. Thus the higher survival of the heterozygotes forces the perpetuation of homozygotes of both types in future generations. If there is no malaria, however, the homozygous hb^A hb^A frequency will increase because there will be no compensation for the loss of the hb^S hb^S homozygotes.

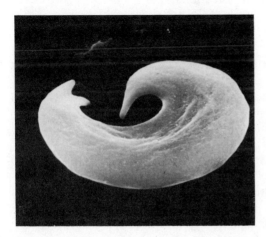

Figure 26-3   Sickle Cells

The electron micrograph of sickle cells from an individual homozygous for the hb^S gene shows the characteristically altered shape of the cell. This arises from the formation of liquid crystals (tactoids) of hemoglobin and the subsequent collapse of the cell membrane around the crystallized material. (Photograph courtesy of Ruth Blumershine, Indiana University School of Dentistry)

Figure 26-4 Distribution of Sickle Cell Anemia and Malaria

Map (a) shows the frequencies of the hemoglobin S gene in southern Europe, Africa, the Middle East, and India. The highest concentrations are in the tropical regions of Africa. Map (b) shows the distribution of the falciparum parasite which causes a severe life-threatening malaria in individuals lacking the sickle cell gene. (From A. C. Allison, Article 3, Annals of the New York Academy of Sciences, *Vol. 91,* © *1961.)*

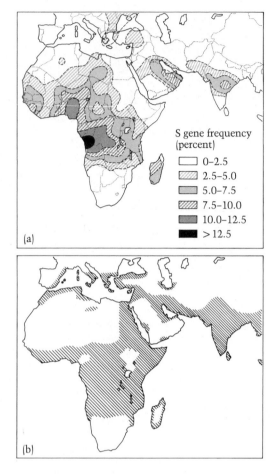

S gene frequency (percent)

□ 0–2.5
▨ 2.5–5.0
▦ 5.0–7.5
▨ 7.5–10.0
▩ 10.0–12.5
■ >12.5

(a)

(b)

attacks from malarial plasmodium infections than did homozygotes who were free of the sickle cell mutation. If we designate sickle cell anemics as $hb^S hb^S$ and normal individuals as $hb^A hb^A$, then heterozygotes, $hb^A hb^S$, would have a better chance of surviving *in that environment* (where malaria is abundant) than either of the two homozygotes. The heterozygote $hb^A hb^S$ shows overdominance or better fitness than the homozygote $hb^A hb^A$. Individuals who are $hb^S hb^S$ would die of sickle cell anemia whether they lived in malarial or nonmalarial areas.

If there are more $hb^A hb^S$ survivors than $hb^A hb^A$ survivors, there will be more $hb^A hb^S \times hb^A hb^S$ matings. Mendel's law of segregation predicts that in each such marriage there is a 25 percent chance of having $hb^A hb^A$ children, a 25 percent chance of having $hb^S hb^S$ children, and a 50 percent chance of reestablishing the heterozygote, $hb^A hb^S$. Although heterozygotes are favored by the selection against homozygous normal individuals, they can never "breed true," that is, they always generate some offspring with the homozygous genotypes. An overdominant condition thus leads to *genetic polymorphism,* the maintenance of two or more allelic varieties that persist in a population.

Sometimes the Heterozygote Formed by Two Different Mutant Alleles Is Normal

An additional source of overdominance is the activity of different alleles in a population. In the fruit fly there is a mutant allele for the dumpy condition, with obliquely cut wings (*o*) and thoracic bumps or pits surrounded by whorls or bristles. This thoracic feature is called vortex (*v*). The dumpy mutant allele (*ov*) is one of many alleles, one of which only causes the wing defect, oblique (*o*), and one of which causes only the vortex (*v*). If dumpy is crossed with oblique their progeny are (*o*); if dumpy is crossed with vortex their progeny are (*v*). When oblique is crossed with vortex, however, their progeny are normal, with neither the wing nor the thorax showing the mutant defect. The expression of the normal trait by two different, but mutually heterozygous, mutant alleles is called *complementation* (Figure 26-5).

There are many allelic series in humans. The blood groups and tissue types have numerous alleles, but there is no strong evidence that specific combinations of these alleles confer overdominance (or hybrid vigor) traits for longevity, fertility, health, disease resistance, or other beneficial effects. It is possible that these human genetic polymorphisms were established in the not too distant past as an adaptation to diseases that later became extinct and that the alleles have persisted since then because there is no strong detrimental effect for any of the homozygous forms of these alleles. It is also possible that many, if not most, of these allelic polymorphisms are neutral and that they arose in the distant past and are still being passed along.

Genetic Load Has Medical Implications

Muller believed that genetic polymorphisms similar to the sickle cell case were temporary adaptations to an environmental crisis. Given a long enough time to evolve, more effective mechanisms for disease resistance would replace the inefficient means by which the sickle cell gene achieves its increase in the population. He believed that most new mutations would exert a harmful, rather than a beneficial, effect on the heterozygotes through genetic load.

This was especially true for conditions of modern civilization which resulted in reduction or elimination of malnutrition, famine, droughts, exposure to severe temperature extremes, and infectious diseases. Public health programs, urbanization, and the rise of technology have enabled people with higher genetic loads to reach maturity. Muller believed that such people would be less vigorous than those with average or less than average loads of mutations. They would be more likely to have major surgery or to depend on medication for survival. They would experience more absences from their work and activities. Their illnesses might make

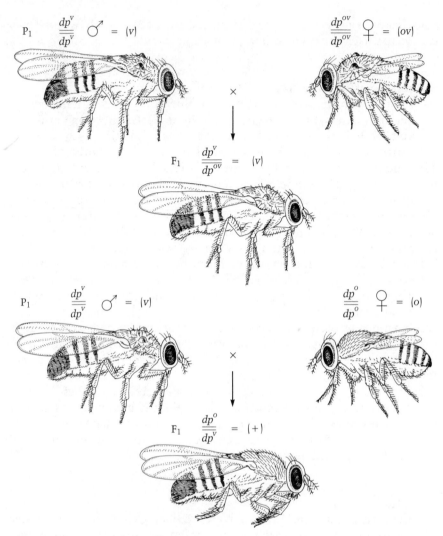

Figure 26-5 Genetic Complementation

Recessive alleles are usually expressed when homozygous; they are not expressed when the individual is heterozygous (except for difficult-to-detect effects on genetic load). In the dumpy series of alleles, an oblique wing and a thoracic bristle disturbance (producing a vortex of small bristles around two pits or raised mounds) are the major phenotypes affected by these alleles. When the vortex allele is crossed to dumpy (which has both the oblique and vortex characters), the F₁ are vortex in appearance. Similarly, if the oblique wing allele is crossed to dumpy, the F₁ are oblique-winged. When the oblique wing is crossed with the vortex fly, however, the F₁ are wild-type, showing neither wing nor thorax disturbances. The normal appearance of heteroalleles (a diploid compound of two different alleles in mutually heterozygous condition) is called complementation. Many mechanisms of complementation exist. One common method involves the assembly of molecules from subunits. If the subunits are defective in different places the assembled membrane or complex protein may compensate these defects and the normal function of the complex will prevail.

them more irritable and less productive. Since they would live to maturity they would be likely to have children, thereby increasing the risk of passing on their greater load to the next generation. Such individuals, despite medical care, would tend to have a lower life expectancy than others of their sex and nationality. They would have increased risks of strokes, heart attacks, cancer, and other common causes of death.

The Changes in Genetic Load May Not Be Noticeable for Several Generations

Since effective public health measures have only existed in the 20th century, most of humanity has the same genetic load today as it had at the beginning of this century. Gene frequencies change very slowly in a population. Thus increases or decreases in genetic load may not be measurable for many generations.

In the industrialized western countries not more than one or two generations, as we saw in Chapter 1, have experienced low infant mortality rates and protection from major medical problems such as infectious diseases. In the past individuals who had high genetic loads would have been eliminated by natural means, but they are now living into maturity. If one gamete in ten contains a new mutant and if elimination of an equivalent amount of mutants is prevented by modern living conditions and modern medicine, then one in five zygotes per generation will have a new mutation added to the average genetic load. Doubling the present load from four to eight lethal equivalents would then take $4 \times 5 = 20$ generations (about 400 to 500 years). This increase, we saw, would make the average family member likely to experience major surgery and to have a greater dependence on medical services. The calculation assumes that most of the newly arising mutations, in combination with previous mutations, will result in treatable disorders. We don't know what medicine will be like in the centuries to come. If it is very effective, the estimate of 400 to 500 years will be accurate. If it is not much better than today's medicine, the doubling of the genetic load might require a millennium or more.

There Is No Agreement on What Constitutes a High Genetic Load

No one knows what the upper limit of the human genetic load is. Nor is there any direct way that individuals with higher risk can be identified. Only rough guidelines can be used by those who seek a reduction in genetic risk for their own families or for humanity itself. Those guidelines involve the concept of compensating for high-risk conceptions by considering eugenic options in our reproduction habits. We would have to reconsider our values about the purpose of becoming parents, the size of family in keeping with our genetic as well as our economic condition, and the quality of life we would like to give our children.

Questions to Answer

1. What is the frequency of spontaneously arising mutations in human gametes?
2. What is the fate of newly arising mutations in a population?
3. What is the most likely mechanism for the extinction of a newly arising recessive mutation?
4. What is meant by the term "genetic load"?
5. What evidence supports the existence of the genetic load in humans?
6. What is hybrid vigor? When does it usually arise?
7. What is genetic polymorphism?
8. Is genetic polymorphism the same as hybrid vigor? What differences and similarities do you see in these two phenomena?
9. What is the difference between dominance and overdominance?
10. What mechanisms before the mid-19th century are believed to have kept the genetic load of humankind constant?
11. What effect, if any, do you believe public health programs, antibiotics, a balanced diet, and modern medicine will have on genetic load? Discuss the reasons for your answer.

Terms to Master

complementation	heterosis
genetic death	hybrid vigor
genetic load	molecular disease
genetic polymorphism	overdominance

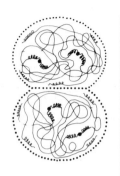

Chapter 27
Darwinism,
Neo-Darwinism,
and Their Critics

When I walk the two and a half miles from my home to my laboratory in Stony Brook, New York, especially in the early morning, I feel a kinship with Walt Whitman, who, a century ago, loved to walk in his native Long Island. I often notice and think about the diversity of the life I see on campus. There are wild raspberries, forsythia, poison ivy, goldenrod, Queen Anne's lace, and stands of oak, dogwood, maple, chestnut, and locust trees. A vast number of weeds hold tenaciously to the meager, sandy soil. Lichens encrust some of the tree bark; Virginia creeper ivy is making a tentative effort to climb the new buildings; and a plague of dandelions has conquered the grass.

Sometimes a startled rabbit will dart across my path. Grey squirrels and occasional chipmunks hide in the shrubs or scurry out of sight along the limbs of the trees. If I look closely I can see ants, earthworms, fireflies, daddy longlegs, and June bugs. When I look at a leaf, it may be distorted with the galls and tumors of insects unknown to me. Fungal colonies can be seen on some of the leaves like buff paint drops. Bracket fungi adorn an old stump. Crows, robins, pigeons, sparrows, and blue jays make known their favorite trees and territories.

No doubt if I were to take a shovel and dig into the soil there would be much more life to see. This would also be true in the campus pond and under the decomposing leaves of the natural woods that have not yet been disturbed by bulldozers and chain saws. Probably 100 different species of higher plants could be collected by a diligent student, and about 50 mammals, birds, reptiles, amphibians, and invertebrates could be catalogued. If a dissecting or compound microscope were used, several hundred more species could be identified.

A Diversity of Life
Exists in Different Locales
The plants and animals on the campus at Stony Brook differ somewhat (or perhaps a lot) from those on other campuses. They would, for the most part, seem strange to a student from the UCLA campus who was used to eucalyptus, palm, and jacaranda trees. A student from Queen's University in Kingston, Ontario, would wonder why there were no black squirrels around. The student from the University of Oregon at Eugene would miss the tall Douglas fir trees that fill the Oregon air with a resinous fragrance. The U.C. Santa Cruz student would be deprived of the ring of giant redwoods emerging like Medusan clones from the abandoned stumps where lumberjacks sawed the trees close to the ground years ago.

Each campus has hundreds of species of plants and animals and no doubt if each school published its own handbook of "campus life" the particular kinds and quantities of plants and animals found there would be as unique and distinctive as the school's buildings, student body, and faculty. Throughout the world some two million different species of plants and animals live. Yet most people see only a small percentage of this immense variety.

Although Diversity Exists
the Basic Components
of All Life Are the Same
All this diversity, however, is based on variations of common components. Except for the viruses these numerous species are cellular, mostly eukaryotic organisms. The same organelles are universally present: chromosomes, nuclei, mitochondria, lysosomes, and the vast, spongy layers and crypts of the endoplasmic reticulum. All the organisms use the same four nucleotides to make their DNA, and the genes from any one of them could probably be translated by the ribosomes and other decoding machinery isolated from any of the other two million species. This universality of the chemical basis of life stands in sharp contrast to the diversity of organs, functions, and life cycles characteristic of living species.

If there is an underlying universality among all living things, does this imply a kinship among them? Such is the view of evolutionists who claim that today's diversity originated from the accumulation of changes gradually occurring over hundreds or thousands of generations. Evolutionists claim that the plants and animals that are familiar to us today were virtually nonexistent 50 or 100 million years ago. Yet there was abundant life then. If we were able to go back even further—200 or 300 million years—life would not seem as diverse, especially animal life on land. About a half billion years ago, there would have been few visible signs of life, with perhaps nothing more than bacterial and algal scums near the seashores to mark the existence of life that was already ancient.

**Evolutionists and Creationists
Have Different Views
on the Origin of Life**   In contrast to the evolutionists'
model, with its imperceptible changes requiring staggering numbers of
generations for noticeable differences between species, there is another
model to account for the diversity of life. It is not a scientific model, in
the sense that we could, ourselves, bring about similar diversity by applying its underlying mechanisms. Rather, like holism, it implies some
supernatural process which, at best, might be observed but never imitated. This is the view of *creationism* which derives its major evidence
from religious traditions which account for diversity by attributing it to
the unique, episodic, or continual activity of a creator. Just as there are
holists who do not attribute their outlook to a mystical or religious basis,
there are creationists who believe that species arise all at once from
preexisting species by processes unknown to biologists.

**Charles Darwin Proposed a Theory
of Natural Selection to Account for Evolution**   Few scientific concepts have had as much influence on culture as the theory of *evolution*.
Although the idea itself is ancient, the link between evolution theory
and Charles Darwin is appropriate because he was the first to develop an
explanation for it using known principles of science (Figure 27-1). Darwin

Figure 27-1   Charles Darwin
The author of The Origin of
Species *ranks as one of the
major scientists of all time.
The theory of evolution
through natural selection
caused a storm of controversy
on human origins that has
continued to the present day.
(Courtesy of the American
Museum of Natural History)*

was born February 12, 1809. He came from a distinguished and wealthy family. His father, Robert, was a prominent physician and his grandfather, Erasmus, a many-talented physician who was also a poet, philosopher, and popularizer of science. Darwin's uncle, Josiah Wedgwood, founded a porcelain and china factory which became internationally reknowned for its high quality.

Darwin's Early Schooling Did Not Suggest an Unusual Talent

The young Darwin was not particularly fond of school, preferring hunting and hobbies like collecting insects. At best he had curiosity and a quick mind, but to his teachers he seemed undisciplined and unscholarly. Darwin's father sent him to Edinburgh to learn medicine and thus follow the family tradition, but Darwin had no stomach for surgery or medical rounds, and he was bored by the dull lectures he had to sit through. His passion for riding horses and hunting seemed frivolous to his father and his professors. Darwin dropped out of his medical training and transferred to Cambridge where he (and his father) believed he would profit more from his education and learn to be a clergyman. Instead, Darwin found more pleasure taking walks with colleagues and professors who, like himself, enjoyed the diversity of life and made a hobby of collecting butterflies, beetles, rocks, and other examples of nature's wonders.

Darwin's Voyage Around the World Provided Him with Material to Study Diversity

In 1835, still far from being committed to the clergy, Darwin was recommended by one of his professors for a position as naturalist on the H.M.S. *Beagle*. After his father reluctantly gave him permission, Darwin left England for a round-the-world voyage that changed both his life and the life sciences.

Darwin prepared for the voyage by reading extensively in geology, geography, and natural history. He had a small library of science books which must have been read thoroughly during the long voyage to South America. Darwin's remarkably keen observation and his careful note gathering supplemented his skill in preserving, dissecting, and describing the anatomy of the hundreds of species he collected. He was struck by the many relations he found among the plants and animals. The South American animals were, for the most part, quite different from European and North American organisms. Sloths, monkeys, armadilloes, giant rodents (capybaras), and llamas were some of the more visible of the thousands of unique South American animals. Fossil bones that he unearthed resembled the prevailing South American animal types, but differed in size or had other slight modifications of their general features. When he observed the animals on the two sides of the Amazon River, he noted that, unless they could swim or fly, these constituted two distinctly different populations.

Wherever the *Beagle* stopped, Darwin amassed information on various familiar and unfamiliar species. The most striking of all his observations was made on a group of volcanic islands about 750 miles west of Ecuador, the Galapagos Islands. In contrast to the South American continent, these islands had very few species on them. These resembled some of the South American animals living in Ecuador, but they were unique and could only be considered characteristic of the Galapagos. Yet this placid group of islands was similar in climate and equatorial location to the Canary Islands in the Atlantic Ocean. None of the Canary Island species could be confused with those on the Galapagos. The Ecuadorian species, while more similar to those on the Galapagos than those on the Canary Islands, were in an environment quite distinct from either of the oceanic islands.

After Studying the Species on the Galapagos Islands Darwin Doubted the Fixity of Species

This was a puzzle. Why did unique species exist in such numbers on the Galapagos (Figure 27-2)? Why did some species live only on one or a few of the islands? Each island in the Galapagos seemed to have its own collection of common and unique species.

Darwin departed on the *Beagle* with the confident belief that species were fixed, each individual belonging to a species whose attributes distinguished it from all other species. Towards the end of the voyage, however, this view was shaken. He found that species varied so much that a fixed set of characteristics of any species was no longer appropriate or accurate.

In 1839 Darwin read Thomas Malthus's essay on population. Malthus believed that the growth of human population outpaced the rate of food supply and available land for farming and living. The natural checks on overpopulation included chronic disease, starvation, war, and epidemic plagues. Malthus encouraged population control by abstinence from sexual activity rather than by the natural checks which caused so much human misery. Darwin applied Malthus's analysis to all species. Unlike humans, all other species have no way of increasing their own food supply, and thus the means of existence for them would be relatively constant each generation. Thus, in a certain spot, he argued, there may be eight pairs of birds. If half of them each year produce four eggs each and rear the offspring successfully, there would be some 2048 birds at the end of only seven years, barring accidents and predators. Yet local birds do not increase in such monstrous numbers. Something must check their population growth.

Darwin's Theory of Natural Selection Explained the Origin of Species

The *Beagle's* voyage demonstrated vividly that there was an enormous amount of variation in the population of any species. The tendency to overpopulate, coupled with

Figure 27-2   Darwin's Finches and the Origin of Species

In 1835 Darwin visited the Galapagos Islands, a volcanic archipelago some 600 miles west of the coast of Ecuador. These islands were inhabited initially by 2 species of mammals, 5 species of reptiles, and 11 species of birds. While some of these colonized land animals resemble Ecuadorian forms, others are strikingly different and appear to be so profoundly modified that they are unique to these islands. During the estimated million years when these volcanic islands were emerging from the Pacific Ocean, the founding species became varied in shape and function. Darwin was much shaken in his initial belief that species were fixed in their collected traits after noting the iguanas and tortoises on these islands. A similar diversity in 14 species of finches unique to these islands was later used to support his theory that an evolution of life had taken place, probably from one founding species which had settled there.

In this diagram the tree finches differ mainly in their beak structure and feeding habits: 1, 2, and 3 are ground finches feeding on seeds; 4 is a sharp-beaked ground finch; 5 and 6 feed on cactuses; 7 is a vegetarian tree finch; 8, 9, and 10 are tree-dwelling insect eaters; 11 is a woodpecker that uses a cactus spine to dig out insects from bark; 12 is a mangrove finch; 13 is a warbler feeding on insects in bushes; and 14 is a Cocos Island finch. The body color of all 14 species is a drab grey-brown, characteristic of the poorly vegetated lava rock background of the islands.

These finches represent the effects of geographical isolation. As each colony of birds on an island developed a particular feeding habit, later migrations permitted a coexistence (on the larger islands) of three or more species, each adapted to a different mode of living (e.g., coastal, desert, bush, tree, or swamp). (Adapted from David Lack, Darwin's Finches, *Cambridge Univ Pr, 1947.)*

subtle variations in each individual, would lead to a *natural selection*, in which those individuals who were best adapted to the conditions they lived in would survive to reproduce, and those whose characteristics made them less effective in coping with their environmental circumstances would die before maturity or fail to mate. Since the surviving variations would be passed on in greater quantity than those that lost out, the next generation would be slightly different in its characteristics. Over thousands of generations the original population would appear quite different from the contemporary population—so different that if naturalists were to compare a specimen from the old and the new populations, they would classify them as two different species.

Darwin's *The Origin of Species* was published 20 years after he developed his theory of natural selection. In those intervening years he did not publish his ideas on evolution and confided them only to a few trusted friends. Instead he worked daily on the collection of organisms he had sent back to England, read copiously in all fields of natural history, corresponded with naturalists and breeders, and visited numerous breeders' shows and agricultural fairs to gather notes on the origin of domestic varieties of plants and animals.

Independently of Darwin, A. R. Wallace developed an almost identical theory and sent it to him for his comments. Although Darwin felt he was deprived of the glory of being the first to develop the theory of natural selection, his friends arranged for the joint publication of their essays. Soon after that, on November 24, 1859, Darwin's *The Origin of Species* appeared in book form.

Darwin Could Not Account Satisfactorily for the Origin of Variation in a Population

The major difficulty with Darwin's theory of evolution was the source of variation. Darwin was ambivalent about the idea that the environment directly altered heredity. Certainly the differences between the plants and animals on the Galapagos and the Canary Islands indicated that something more than climate, latitude, and volcanic soil was involved. Despite this weakness, however, Darwinism succeeded as a scientific theory because no other concept made sense out of the relatedness of life and the relation of life to time and geographic location.

Neo-Darwinism Is the Fusion of Natural Selection with Modern Genetics

After the rediscovery of Mendelism in 1900, a bitter feud developed between those who sought statistical models for explaining variation (biometricians) and those who used Mendel's laws as evidence that variation was discontinuous rather than continuous. Most Darwinists supported the biometrical approach to heredity and most of the Mendelians supported alternatives to natural selection. Thus Hugo de Vries, one of the rediscoverers of Mendelism, believed that new species arose from their parents at once without intermediate ancestors. T. H. Morgan, much impressed by de Vries's work, tried to find new species in the laboratory, eventually choosing to study fruit flies which yield large numbers of offspring each generation as well as many generations per year (about 30).

The fruit fly studies, as Morgan found out after his first three years of work with them, did not show new species arising suddenly. Rather, the variations that arose were usually modest and tended to support Darwin's original view. This was particularly the case for quantitative traits, as Muller demonstrated in several cases. The subtle modifier genes and the relation of character expression to environment showed that the range Darwin had seen in nature was also observable in the laboratory. The new variation, which arose by mutation, recombined with the already existing variations carried over in the population. Meiotic events sorted out the chromosomes from a random alignment and shuffled the paternal and maternal genes within any given pair of homologs by crossing-over.

The Early Controversy of Darwinism and Mendelism Was Resolved Through Population Genetics

At the same time that geneticists were taking a second look at Mendelism and applying it to more complex situations, biometricians were finding that Mendel's laws were not incompatible with Darwinism after all. The British mathematician G. A. Hardy and the German physician W. Weinberg showed that the frequency of a gene in a population remains constant provided selection does not favor or eliminate it (Table 27-1). The *Hardy-Weinberg law* then became the basis of mathematical population genetics. Very subtle factors such as selection, mutation frequency, interaction of genes, and the fragmentation of large populations into small colonies could be used to generate models of the evolution of traits in a species.

This contribution of classical genetics, spontaneous mutation, and population genetics supplied the mechanisms for the origin of the continuous variations seen in a species. When applied to natural selection the supplemented theory of evolution was called *neo-Darwinism*. Additional mechanisms for neo-Darwinism were discovered shortly afterward through studies of variations in chromosome number, resulting in polyploidy or nondisjunctional forms. Some of these variations have become established as new varieties or species, especially in plant evolution. Also, chromosome rearrangements provide a basis for the evolution of plants and animals, either as obstacles to reproduction between close species or as a source for new genes. Foremost in the development of neo-Darwinism were Julian Huxley, R. A. Fisher, and H. J. Muller.

Table 27-1   The Hardy-Weinberg Law

In a population the sum of the frequency of allele A and allele a is 1.0 (100 percent). Let us call the frequency of A, p and the frequency of a, q. Thus $p + q = 1$. This frequency is the same for the sperm as for the eggs produced. When sperm and eggs form zygotes we see that $p + q = 1.0$ sperm fertilize $p + q = 1.0$ eggs, or p^2 zygotes will be AA, $2pq$ zygotes will be Aa, and q^2 zygotes will be aa.

If we know the frequency of homozygotes for a recessive trait (albinism, Tay-Sachs syndrome, Rh-negative blood type) we can determine the frequency of p^2, $2pq$, p, and q. Thus if Rh-negative blood is found in 16 percent of the U.S. population, $q^2 = 16$ percent (0.16), $q = \sqrt{0.16} = 0.4$, $p = 1.0 - 0.4 = 0.6$ (60 percent), $p^2 = (0.6 \times 0.6) = 0.36$ (36 percent), and $2pq = 2 \times 0.4 \times 0.6 = 0.48$ (48 percent). The Hardy-Weinberg law must be modified when used with small populations, partial dominance, a selective advantage for either A or a, and high mutation frequencies of either A to a or a to A.

Regardless of how the population shifts in size, the frequencies of p and q will remain constant given the following ideal conditions: there is random mating and no selection for either allele; the population remains large; and there is a very low mutation frequency.

Frequency of allele $A = p$
Frequency of allele $a = 1.0 - p = q$
Generation$_1$: $p + q = 1.0$
Generation$_2$: $(p + q)^2 = 1$ or,
$$p^2 + 2pq + q^2 = 1$$
$p^2 =$ frequency of AA individuals
$2pq =$ frequency of Aa individuals
$q^2 =$ frequency of aa individuals

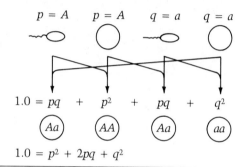

$$1.0 = p^2 + 2pq + q^2$$

The Mechanisms of Evolution
Can Be Seen in Moth Wing Color

Evidence for the effectiveness of natural selection in determining survival and character formation was most convincingly reported by Kettlewell (Figure 27-3). About 1850 in England most of the moths of the species *Biston betularia* were white with a few faint spots and streaks of grey and black. This gave it the popular name peppered moth. With the rise of coal-fueled industries in areas like Manchester, a deposit of soot began to cling to the birch trees. By the end of the 19th century many of the peppered moths were noticeably darker, especially near the cities, where they were almost a charcoal grey-black. This was also true for other species of moths, and the phenomenon was called *industrial melanization*. The moths were not covered with soot; they were born with their dark color.

How did the melanic forms arise? Did the coal dust directly cause their pigment cells to respond with a greater output of melanin? Did some inner will stimulate those cells during development to multiply in greater number? Kettlewell disproved these environmentalist theories and showed that predator selection was involved. He placed equal numbers of rural moths (mostly white) and urban moths (mostly black) on urban birch bark. Birds quickly ate the easy-to-see white peppered moths.

(a)   (b)

Figure 27-3   Natural Selection in Action

The peppered moth, Biston betularia, *inhabits the lichen-encrusted, woody areas of England. In the mid-19th century some naturalists noted a new form,* Biston betularia carbonaria, *which appeared in the industrial cities and surrounding suburbs. Experiments in the early 1950s by H. B. D. Kettlewell demonstrated that birds readily ate peppered moths on the sooty trees of the industrial neighborhoods and missed seeing the melanized (carbonaria) moths. The reverse was true in the unpolluted rural regions where these moths lived. (a) The peppered moth demonstrates its camouflage on an unpolluted tree. (b) A soot-blackened tree makes the melanized moth blend into the background while the peppered moth is sharply outlined. The differential survival of the two forms is characteristic of natural selection. The mutations causing increases of melanin occur in both strains, but the color of the trunks of the trees on which the moths rest during the day determines whether such increases are beneficial or detrimental. (Photographs from the experiments of Dr. H. B. D. Kettlewell, University of Oxford, England. Courtesy of Sabin, Bacon, White & Co., Birmingham, England.)*

In a reciprocal experiment, where members of the two populations were placed on the bark of a rural birch tree, the birds ate many more black moths than white moths. The variations in melanin, which arise by mutation and recombination, are selected by the predators, and as the trees gradually darkened, so did the population of moths. In this way, a population of moths was maintained each generation. Natural selection favored melanization in the industrial areas and the nonmelanized forms in the nonindustrial areas.

There Are Many Factors
That Affect Natural Selection
Many other factors in evolution have been observed. Some populations exist in multiple varieties, each variety

having adapted to a particular shift in temperature, moisture, or seasonal foliage. As we learned earlier, such polymorphisms are often maintained by heterozygosity, with the homozygous forms filling in other environments so that all three forms can coexist within the range of the species.

Many populations get their start from one or more small groups of individuals isolated from their original population. They may arise as a consequence of earthquakes, accidental settlement of a group of islands, or other natural means of isolation. The random distribution of allele frequencies among such isolated small groups is called *genetic drift*. If certain rare recessive or dominant genes happen to be present in one isolated group of the original population, then these may be multiplied as the population flourishes in the less competitive environment of the new location. This process is known as the *founder effect*. In isolated human populations, such as the Amish, the Dunkers, or the inhabitants of Tristan da Cunha, the founder effect has established a high incidence of rare recessive disorders.

Isolating mechanisms need not be geographical. If two populations develop chromosome inversions or translocations at remote regions of their range, they may eventually become reproductively isolated even when brought together again (Figure 27-4). The gradual selection of behavioral differences, changes in color, and other components of courtship could also lead to specialization.

The Hardy-Weinberg Law Is Used
to Calculate Gene Frequencies
It is even possible to calculate the frequencies of genes in a population. This knowledge can be used to determine the frequency of homozygotes and heterozygotes for any human gene. The basic rule for population genetics, we have seen, is the Hardy-Weinberg rule. This rule states that the gene frequency remains constant provided the population is large, there is no selection against either allele, and the mutation frequency from the normal to the mutant allele is very low (Table 27-1). Departures from this rule occur as a result of new migrations into the population, the breakup of the population into several small isolated populations, assortative (nonrandom preferential) mating with respect to the trait, or environmental changes that favor one allele over another. The formula for the Hardy-Weinberg rule is a simple algebraic expression which Hardy considered so trivial that he didn't bother to mention it in his autobiography, *A Mathematician's Apology*.

All of these mechanisms of evolution support the belief that new species are the consequence of the gradual modifications of preexisting species. This makes all life related through descent from simpler and now extinct forms. Despite the loss of these ancestors (except for the occasional fossil remains of a few species), zoologists and botanists familiar with the comparative anatomy of living things can reconstruct the paths by which the two million species of plants and animals evolved

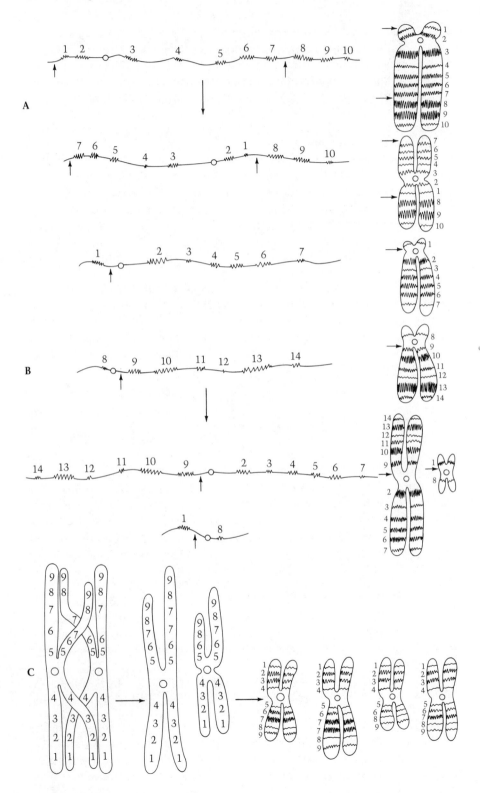

A

B

C

Figure 27-4   Chromosome Evolution
Chromosomes evolve in length and in the location of their centromeres as a result of chromosome rearrangement. The inversion of a segment containing a centromere (A) can shift its location to change an acrocentric (off-center) chromosome into a metacentric (dead-center) chromosome. If two chromosomes (B) each have a break, two rod chromosomes of equal size can be converted into two V-shaped chromosomes of unequal size. Most important, however, for increasing the number of genes in a chromosome are occasional instances of unequal crossing-over (C) between homologs during meiosis. This leads to the duplication (or deletion) of a segment of genes. The duplicated genetic material, if viable, can then undergo gradual mutational change leading to new gene functions.

(Figure 27-5). The phylogenetic trees of the major groups of plants and animals can be refined for more specialized segments at any level. The branching of the phylogenetic trees has been confirmed by the divergence in amino acid sequences of the proteins having a common function in these species.

The Rate and Mechanism
of Speciation Is Controversial   Palaeontologists have long claimed that the fossil record of major groups of organisms (e.g., shellfish, echinoderms) is rather stable. Certain living bony fish have fossil ancestors that lived 100 million years ago that look startlingly like the skeletons of their contemporaries. Changes in the fossil record are more often abrupt and then continue in a new stable phase. Such observations have led to a theory of *punctuated equilibrium* (the sudden appearance of new species which then persist with little change) proposed in the 1970's by S. J. Gould and N. Eldredge. Opponents of this theory believe neo-Darwinism includes punctuated equilibrium through isolating mechanisms created by natural disasters, geological changes, and ecological shifts (e.g., new rivers, mountain formation, desert formation, new competitors). These isolating mechanisms lead to genetic isolation and the populations of one species may diversify into several species as Darwin established for many organisms when he travelled around the world in H.M.S. *Beagle*.

Neo-Darwinists reject punctuated equilibrium as a replacement of the more gradual change emphasized by Darwin and his later interpreters. The episodes of change may take place in a few tens of thousands of years, still involving thousands of generations, and neo-Darwinians reject the view that punctuated equilibrium implies a *macromutation* or single generation reorganization of genetic patterns leading to new species. All attempts to demonstrate such macromutations have failed and cases thought to be such instances have turned out to be produced by conventional genetic mechanisms embraced by neo-Darwinism.

While the current debate remains unresolved, both Neo-Darwinists (sometimes called *gradualists*) and those supporting punctuated equilib-

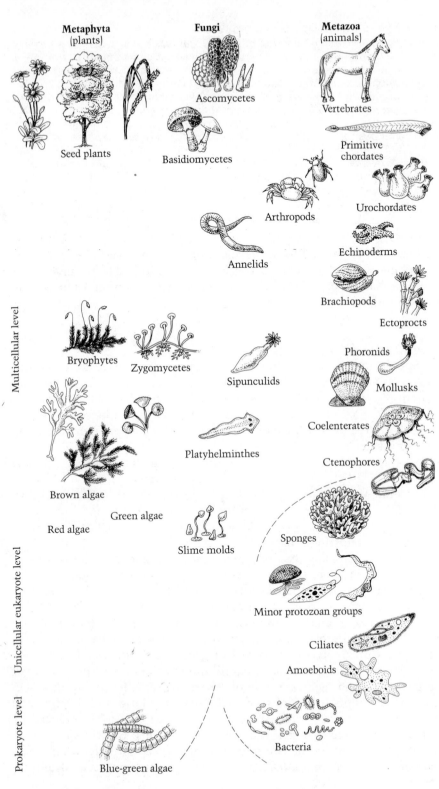

Metaphyta
(plants)

Fungi

Metazoa
(animals)

Ascomycetes

Vertebrates

Seed plants

Basidiomycetes

Primitive
chordates

Urochordates

Arthropods

Echinoderms

Annelids

Brachiopods

Ectoprocts

Bryophytes

Zygomycetes

Phoronids

Sipunculids

Mollusks

Coelenterates

Platyhelminthes

Ctenophores

Brown algae

Green algae

Red algae

Slime molds

Sponges

Minor protozoan groups

Ciliates

Amoeboids

Bacteria

Blue-green algae

Multicellular level

Unicellular eukaryote level

Prokaryote level

Figure 27-5   The Five-Kingdom Classification

It was traditional until fairly recently to assign living things to one of two kingdoms, plants or animals. More contemporary evaluations of taxonomy have favored splitting the classification into three or more kingdoms. Robert H. Whittaker has proposed a five-kingdom classification based on energy production and acquisition. The prokaryotes, which include bacteria and blue-green algae, are in the new kingdom, Monera. The unicellular protozoa are in the kingdom Protista. Multicellular organisms include the kingdom Fungi and the more familiar kingdoms, Metazoa (animals) and Metaphyta (plants). Further subdivisions may occur when the metabolism of the prokaryotes and eukaryotes is more fully analyzed at the molecular level. For example, there may have been an earlier prokaryote-like organization of cells in which there was a eukaryotic method of transcribing and translating genes. No living prokaryotes, however, have been found with intervening sequences in their genes.

rium are unanimous in their rejection of creationism as an explanation of the abrupt disappearance of old species and the sudden appearance of new ones. It should be noted, as Darwin pointed out, that the fossil record is like the ruins of an ancient library, almost all of whose volumes have been completely destroyed, with rarely a book left intact and for most of these survivors only a page or fragment of a page remains to guide us in interpreting the book's contents.

The Philosophic Aspects of Evolution May Conflict with Our Sense of Uniqueness

A person who studies the phylogenetic trees of living organisms may experience the same thrill of historical kinship that the genealogist finds when studying his or her family history. However, unlike the genealogist, who uses actual records of birth, marriage, and death, the evolutionist must rely on a theory of evolution that connects primates to one another in a way distinctly different from their relation to rodents and other groups of animals and plants. Non-evolutionists reject the idea of kinship to other organisms because they believe in the unique creation of species. These two schools of thought are mutually exclusive interpretations of diversity.

It is no surprise, then, that Darwin's theory of evolution has been criticized, rejected, and resisted by those religious groups who interpret the book of *Genesis* as a creationist history rather than as allegory, metaphor, mythology, or the attempt by our ancestors, thousands of years ago, to make sense of the world as they knew it (Figure 27-6 and Table 27-2). Most religious sects, at least in the Judeo-Christian tradition, have accepted evolution just as they eventually accepted the Copernican model of the solar system.

Creationists Believe Evolution Theory Weakens the Religious Beliefs of Students

The creationist-evolutionist controversy is nevertheless painful in many parts of the United

EICHENBERG

Figure 27-6   An Artist's Reconciliation
The wood engraver Fritz Eichenberg here portrays a Biblical view of creation with an improbable gathering of extinct and contemporary life forms. ("The First Seven Days," wood engraving by Fritz Eichenberg © 1955. Reprinted by permission of the artist.)

Table 27-2   Comparison of Biblical and Evolutionary Sequences of the Origin of Life on Earth

There is a controversy between those who accept a literal interpretation of Biblical scripture (often called fundamentalists) and most evolutionists, whose theories of the astronomical, geological, and biological origins of the universe and life on earth are based on technical scientific studies. On the left are the events described in *Genesis* concerning the seven days of creation. On the right are the major events as most scientists would reconstruct them. Note the creation of the sun, moon, and stars *after* the creation of angiosperm plants (grass, seed plants, fruit-bearing plants). Other irreconcilable differences may be seen by comparing the two columns. For this reason fundamentalists have fought against the teaching of evolution theory in public schools. Nonfundamentalists (most Protestants, virtually all Roman Catholics, and most Jews) would have no major difficulties accepting the sequence of events proposed by evolutionists because they regard the Bible as an inspirational or religious history rather than as a scientific text. They regard the account of creation in *Genesis* as symbolic or allegorical, and they attribute the discrepancies between current scientific theory and that of some 5000 years ago to the meager state of knowledge of scientific disciplines in ancient times.

Sequence in Genesis	*Sequence in Evolution Theory*
Day 1. Light and dark, "day" and "night"	1. Steady state or "big bang" origin of the universe
Day 2. Sky and water	2. Galaxies condensing from nebulae
Day 3. Oceans and land, grass, seed plants, fruits	3. Stars (including our sun)
Day 4. Sun and moon, stars	4. Planets (and moon)
Day 5. Fish and birds	5. Microbial life (unicellular prokaryotes and eukaryotes)
Day 6. Cattle, reptiles, and wildlife; man and woman	6. Invertebrate animals, early chordates, fish
Day 7. Rest, holy day	7. Flowering plants
	8. Amphibia and reptiles
	9. Mammals and birds
	10. Humans
	11. Cattle domesticated

States where creationist groups apply political pressure on school boards to declare creationism as an equivalent scientific theory of life which should be taught in the public schools. Many school boards, fearing equal political pressure from scientists who accept an evolutionary model of life, choose, instead, to delete the study of evolution altogether or to use euphemistic phrases in place of terms like evolution or natural selection. In some midwestern states about two-thirds of high school students attend schools in which their biology courses do not discuss or mention evolution by natural selection.

Why should the evolutionary theory of life be so threatening to the religious beliefs of creationists? Why is neo-Darwinism, in particular, so often shunned in the biology courses of secondary schools? When examined closely, neo-Darwinism is seen to rely only on known physical and chemical laws. The mutations that change nucleotides are directed neither by the environment nor by supernatural forces. They occur randomly from self-produced chemical errors and physical events, such as background radiation. If a given sequence that is altered leads to an improvement (a very rare event) it may be preserved by natural selection. Note, however, that the right environment and the right genetic combination are coincidences. Once selected, of course, the mutation will usually increase in frequency because of the likelihood that the surviving organisms carrying it will be more fertile. It may seem unbelievable to those not familiar with the life sciences that chance events and atomic properties can lead to complex molecules that become self-replicating and eventually lead to cellularity, multicellularity, organ formation, and intelligent life, yet some of these processes have been demonstrated experimentally in the simplest form of life, the viruses.

Purpose and Meaning Are Needed by Humans But May Not Be Exclusively Obtained from Religious Traditions

Quite at odds with this scientific view is a religious tradition that ennobles human life by giving it purpose and meaning. The creationists believe that by substituting Darwinism for this tradition, humans lose an imposed or historical purpose or direction to life and people are forced to create their own purpose and meaning of life. Critics of evolutionary theory believe that unless that purpose was given by a creator, life is meaningless and all values are relative. Evolutionists deny these implications. They feel that, although humans are descended from nonhuman primates and share an ultimate origin with nonliving matter, purpose and values are part of human culture and are necessary for both our survival and our sense of self-worth. Human values had changed dramatically long before Darwin was born. We no longer believe, as we once did, universally, that government is the divine right of kings. Nor do we condone capital punishment for witchcraft, blasphemy, or heresy. We no longer believe that disease is a punishment for individual or collective lapses of morality or piety.

Meaning and purpose in human life are found among theists, atheists, and agnostics. It is the burden and the responsibility of all human beings to seek those values, whether from the tenets of a specific religion or from the experience and wisdom of our past and present cultures. Neo-Darwinism will neither destroy nor provide the need for values, but it does force us to expand our views of life, time, and the human condition.

Questions to Answer

1. What arguments favor a common origin of all plants and animals?
2. How do creationists' views differ from those of evolutionists?
3. What observations led Darwin to the belief that a process of evolution had taken place?
4. What theory did Darwin propose to account for the origin of species? What are its major assumptions?
5. What major feature of evolution theory was not available to Darwin in his lifetime?
6. What is the Hardy-Weinberg law?
7. How does neo-Darwinism today differ from Darwinism?
8. What does industrial melanization contribute to the understanding of evolutionary mechanisms?
9. What is genetic drift?
10. Why has evolution theory been criticized by some religious groups?
11. What are the assumptions used by evolutionists to explain the origin of life?

Terms to Master

creationism

evolution

genetic drift

Hardy-Weinberg law

industrial melanization

isolating mechanisms

natural selection

neo-Darwinism

Chapter 28
Human Evolution

About 45,000 years ago, in Shanidar Cave in northern Iraq, a 40-year-old man who lived in the cave was crushed by an avalanche of rocks let loose by an earthquake (Figure 28-1). His fossil skeletal remains indicate that he was arthritic and had suffered since birth from handicaps and injuries. He was born with a deformed and poorly developed right shoulder blade, collarbone, and arm. The deformed arm was later amputated above the elbow and the surgical wound had healed. To compensate for his lost arm, he grasped objects with his jaws and this wore down his teeth. Early in his life he suffered a severe accident, in which he lost his left eye and damaged the bones on the left side of his face. These bones eventually healed. He was unable to hunt, but was apparently cared for by his family and may have contributed to their welfare by doing household duties.

Also found in Shanidar Cave were the graves of three individuals who were buried by their relatives (Figure 28-2). From a microscopic analysis of the soil around their bones, a surprising fact was revealed. They had been buried with flowers, as shown by the eight species of fossilized pollen present. Soil in other parts of the cave and near the burial site did not contain such pollen concentrations. Both the careful burial of the dead and the use of flowers some 40,000 years ago in a funeral ceremony were not expected. The *Neanderthals* had often been portrayed as brutish hunters in popular depictions of our fossil ancestors.

Fossil Human Skeletons Differ Somewhat from Contemporary Human Skeletons   The skeletons of all these prehistoric beings are clearly human, but their skulls differ from those of contemporary humans. They have a thick-ridged beetle brow and a much heavier receding chin. Because of this brutish appearance the

Figure 28-1 Shanidar Cave
The Neanderthals resided in many parts of Europe, Africa, and Asia. In Shanidar, Iraq, archaeologists have found cave dwellings containing the hearths and burial sites of Neanderthal people who lived about 40,000 years ago. [From R. Solecki, Shanidar: The First Flower People *(New York: Knopf, 1971), p. 66.]*

Figure 28-2 Burial of the Dead
Human beings are the only animals that carry out a formal ceremony of burial (or cremation) of the dead. The Neanderthal ceremony included floral tributes, as demonstrated by pollen concentrations in the soil of graves such as this one. The bones in such a burial site are shown in this photograph. Whether such funeral rites imply a belief in an afterlife or represent a symbolic tribute to the person cannot be determined by the burial rite itself. It does, however, prove that there was a concern for the individual and a community tradition among the Neanderthals. It also modifies the image we may have of Neanderthals as brutish, an image brought about by their beetle brows. [From R. Solecki, Shanidar: The First Flower People *(New York: Knopf, 1971), p. 231.]*

Neanderthals have often been thought of as slow-witted and unfeeling. That view is unjustified. The archaeological evidence of Shanidar Cave suggests that Neanderthals, who ranged mostly throughout Europe and the Near East, were compassionate, loyal, and sensitive persons with a religious or philosophic appreciation of the rites of mourning.

Neanderthals (*Homo sapiens neanderthalis*) became extinct about 40,000 years ago and were replaced by modern humans (*Homo sapiens sapiens*). No one knows what caused the disappearance of Neanderthals. They may have been killed by our direct ancestors in a competition for shelter and hunting territories. They may have starved while modern humans, with better tools, depleted the scarce game and edible plants that the Neanderthals had hunted and gathered for thousands of years. Most likely, the Neanderthal bred freely with our ancestors and thus merged with them.

Palaeontologists Use Many Techniques to Date and Analyze Fossil Remains

Physical anthropologists study the anatomy of humans and other primates so that they can apply their knowledge to the study of the bones of prehistoric beings. Healed fractures, birth defect malformations, scar tissue, and disease-induced deformities provide useful clues to the life history of a person. So, too, do the position of the bones and their condition. If they are cleanly fractured in numerous places the person met a violent death, possibly by avalanche or by falling. If the bones are scattered the person may have been dragged off by animals. If the person was a victim of cannibalism, that too can be determined, since the bones may have been broken open for marrow. Each bone and tooth provides clues to the age, sex and activities of that former inhabitant of the world.

The remains may be dated by the accompanying bones of extinct animals that served as food for the cave-dweller. The shape and variety of stone tools present in the soil around the bones, as well as the workmanship that went into the tools, can help determine the person's cultural age. Radioactivity in the carbon of the hearth or in the bones (if the original material has not been replaced by minerals) can be used to date more recent skeletal remains. The composition of the soil can be used to tell whether the climate was glacial or temperate.

The Fossil Record of Early Human Ancestors Is Incomplete

Reconstructing the life and times of our human ancestors is a difficult task. The older those ancestors are, the fewer remains exist. There are fewer than 200 known Neanderthal skeletons, yet the Neanderthals were some of our most recent ancestors. As we go back 150,000 or more years, the total fossil remains of humanlike beings consist of only a few dozen very incomplete skeletons. Much more abundant, because they are so indestructible, are the stone weapons and tools which bear witness to the intelligence and skills of our ancestors. From these tools we gain knowledge of these people as food preparers, hunters, and founders of communal lifestyles.

Humans Are Included Among the Primates
But Differ in Unique Ways

Humans differ markedly in several ways from all other living primates. We have a spoken language by which we can convey a rich description of our universe and our emotions. We can shape permanent and reusable tools out of stone, metal, and wood. An ape or a chimpanzee, at best, can only strip the twigs off a limb to make a temporary tool to fetch an object or capture prey. Our brains are about two to three times larger than the largest brain of any other primate.

There are 211 species of primates alive today. Of these, only twelve (the apes and humans) are more complex than monkeys: seven species of gibbons, two species of orangutans, and one each of chimpanzees, gorillas, and human beings. The monkeys are more numerous with 88 species living in Africa and Asia (the Old World) and 74 species, divided almost equally into marmosets and ceboid monkeys, living in South America (the New World). The remaining primates, less complex than monkeys, are called prosimians. Among them are the lemurs, tarsiers, and tree shrews. Apes and humans do not have tails, but all the monkeys and prosimians do.

The Chromosomes of Apes
and Humans Are Similar

Our chromosome number is 46, but chimpanzees, gorillas, and orangutans have 48 chromosomes. Gibbons have 44. The different species of monkeys have more variable chromosome numbers (42–72). Tarsiers have the largest number—80. The banding pattern of some of the chromosomes in other primates is similar to that of human chromosomes.

The fossil remains of gibbon-like primates can be dated to 39 million years ago. Chimpanzees, orangutans, and gorillas can be traced back about 28 million years. Human-like remains are much more recent, and the genus *Homo* is only about two to four million years old. Human-like fossils older than that have more ape-like features, and such ancestors are given different generic names.

The oldest members of our genus, *Homo habilis*, lived in East Africa and South Africa (Figure 28-3). Their brain volume, about 700 cc, was larger than a gorilla's (550 cc) but about half that of a contemporary human (1350 cc).

Other hominids (fossil primates with some human features) about the same age as *H. habilis* are the ape-like individuals who walked erect, such as *Zinjanthropus* or *Australopithecus*. Yet, even though they walked upright as do humans, their skulls were much more ape-like, with cranial volumes less than 600 cc. Apparently the capacity to walk erect preceded (or was independent of) a large brain volume. The teeth of the Australopithecines were more human-like than ape-like, suggesting that they were meat-eating animals, like humans. (Ape teeth are not adapted for

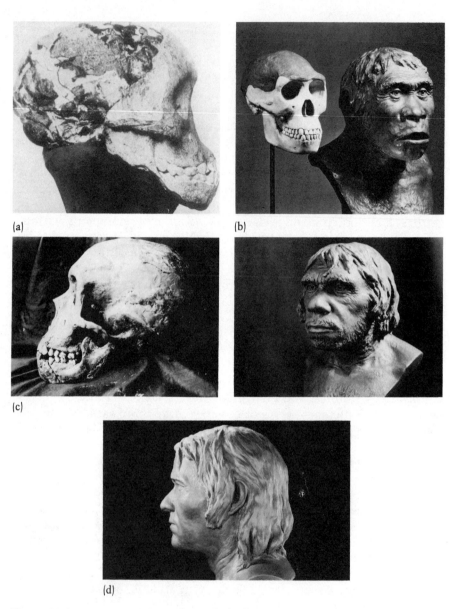

(a)

(b)

(c)

(d)

Figure 28-3 Reconstructions of Fossil Skulls
The four ancestors of modern humans depicted here illustrate the increase in brain capacity, reduction of a forward protrusion of the jaw, smaller jaw size, and diminished brow ridge. Guesses on how hairy our ancestors were are based more on artistic license and analogy to primate features than on known facts about hominid integuments. The ancestors include: (a) Australopithecus africanus, *(b)* Homo erectus (Pithecanthropus erectus), *(c)* Homo sapiens neanderthalis, *and (d)* Homo sapiens sapiens (Cro-Magnon). *[(a) Neg. no. 320282 (Photo: Alex J. Rota). (b) Neg. no. 333193. (c) Left: Neg. no. 125964. Right: Neg. no. 412359. (d) Neg. no. 313683 (Photo: H. S. Rice). All photos courtesy of The American Museum of Natural History.]*

eating meat.) Crushed baboon skulls found near Australopithecine sites support the view that walking erect and hunting occurred earlier than the massive development of brain size which characterizes the genus *Homo*. The Australopithecines used some very primitive stone tools, possibly for scraping hides into mats or coverings for lodgings.

Food Gathering and Hunting
Require Different Degrees of Skill   The evolution of hunting by prehistoric humans involved three phases. The first animals to be captured were smaller, slow-moving animals like tortoises and easily-caught catfish and aquatic birds. In the second phase the immature offspring of larger animals were stalked or trapped. The final phase was the capture of larger animals. Each successive phase was more difficult than the previous one, requiring more sophisticated reasoning and more effective

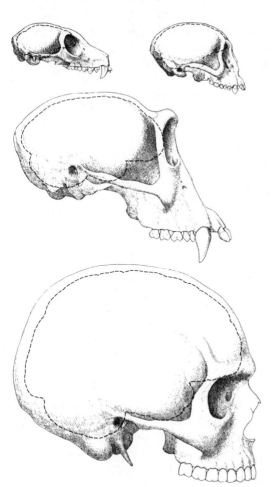

Figure 28-4   Relative Brain Capacity of Primates

The brain of a prosimian, such as a lemur or tarsier, is contrasted with the brains of the monkey, chimpanzee, and human. Note the enormous increase in volume in the human brain. Most of the upper skull in humans is set aside for the brain, while in the prosimians, the brain occupies somewhat less than half the volume of the skull. (From "The Casts of Fossil Hominid Brains" by Ralph L. Holloway. Copyright © 1974 by Scientific American, Inc. All rights reserved.)

tools. Meat-eating also reduces the need for massive canine teeth and large grinding molars, as well as the need for a heavy musculature and a large oral capacity. These changes in teeth and jaw structure have accompanied the fossil record as specimens of more recent deposit have shown.

A brain capacity of about 775–940 cc was found in the skulls of people who once lived in Java and Africa (Figure 28-4). This species, *Homo erectus*, lived about a half million years ago. Several subspecies of *H. erectus* have been found. The African *H. erectus* hunted larger animals, bringing them down with bolas, stones tied to leather strips and thrown against the legs of escaping prey. They also made more completely chipped hand axes and cutting tools. The Peking subspecies, *H. erectus pekinensis*, had about 20 percent more brain than the Javanese or African varieties.

A form intermediate in shape, sharing characteristics of *H. erectus* and the Neanderthals (*H. sapiens neanderthalis*) was found in Java near the Solo River. The Solo people had a brain capacity of 1100 cc. They made tools and spearheads from bone and antlers. The remains of several other fossil ancestors sharing characteristics of *H. erectus* and *H. sapiens neanderthalis* have been found, including a skull with a 1280 cc brain volume. This prehistoric variety of humans lived in Rhodesia and southern Africa.

The Neanderthals Had a Larger Brain
Capacity Than Modern Humans

It is somewhat surprising that the Neanderthals had a larger brain capacity (1550 cc) than modern humans (1350 cc). In the 100,000 years that Neanderthals roamed the earth, they established a relatively uniform culture. Their stone implements consisted of spear-shaped flakes, side scrapers, borers, knives, and spheres. Whether the Neanderthals had the same intellectual capacity as *H. sapiens sapiens* (ourselves) is not known, but certainly our smaller-sized craniums suggest that more than brain volume is involved in the evolution of human intelligence. Although Neanderthals and modern humans differ in brain size, hybrid crosses between *H. sapiens neanderthalis* and *H. sapiens sapiens* have been found in the Middle East and have been dated at 40,000 years old.

A highly cultured prehistoric subspecies of *H. sapiens sapiens*, the Cro-Magnons, lived in southwest Europe about 20,000 years ago. Their wall paintings are anatomically meticulous and aesthetically attractive. Cro-Magnon brains were huge—about 1660 cc. Nothing is known about the origin of the Cro-Magnons. Many other fossil forms of *H. sapiens sapiens* have been unearthed. The skull of one, found in Swanscombe, England, had a 1325 cc brain capacity. Another, in France, preceded *H. sapiens neanderthalis*, thus proving that both subspecies of *H. sapiens* existed for a long time before the disappearance of the Neanderthals.

Varieties and Subspecies
Make a Fixed Definition
of Species Impossible
The term subspecies is often used to distinguish varieties or races within a species. In general, a variety usually differs by one or a few minor character traits (like breeds of cats or dogs). Subspecies have several differences that suggest many mutational differences but not enough to prevent successful fertile hybridization. Races are intermediate between varieties and subspecies. There is thus only one subspecies of *H. sapiens sapiens* alive today, our past subspecies having merged into us or become extinct. As populations of *H. sapiens sapiens* improved their skills for survival they migrated more effectively and prevented the formation of new subspecies by hybridization with other populations they met. Thus human beings today differ in slight ways, none of their differences in skeletal features being as pronounced as the differences between *H. sapiens sapiens* and *H. sapiens neanderthalis*. In general the vast continents have served as partial isolating mechanisms for modern humans, enough for racial differentiation but not enough for subspecies formation. The black (Negroid), white (Caucasoid), and yellow (Oriental) races bear the impermanent differences that marked their continental adaptations, but none have been isolated long enough to develop into new subspecies, and all three have hybridized with each other to multiply the varieties of humans.

The Evolution of Human
Intelligence Can Be Inferred from
the Tools of Fossil Ancestors
When we examine the implements of prehistoric humans, especially those of the Neanderthals and Cro-Magnons, we realize how many skills these people must have had. It is not easy to chip rocks. Few of us could locate the right type of stone to strike so that small pieces could be flaked off without shattering the rock into pulverized bits or irregular fragments. Even if we did, it would take some effort and experience to judge the amount of force needed to fashion a triangular-shaped tool or to flatten an edge so that it could cut through hide and reveal the meat beneath it. It is also difficult to strip the integument from a large mammal. Unlike carnivorous animals who ate their kill on site or dragged a limb up a tree to chew it at a more leisurely pace, human hunters often dismantled the carcass, wrapped the meat, and carried it back to a shelter. Cooking the meat, a process that dates back to the Neanderthals, requires additional sophisticated knowledge. In order to be combustible a material must be thin, dry, and crumbly, and the heat necessary to start a fire requires friction (by elaborate rubbing) or sparks (by striking the right materials together).

Gathering food is also a sophisticated task. Some foods are poisonous. Berries, roots, leaves, insects, and seeds must be identified correctly, and that information has to be passed on from parent to child if the diet is varied and new environments are occupied.

Human Intelligence May Not Have Changed Much over the Past 50,000 Years

The existence of magnificent prehistoric art and skillfully crafted tools suggests that no major changes have occurred in human intelligence over the past 50,000 years. A comparable human intelligence was certainly not true for *H. erectus* or *H. habilis* whose brain size was more limited and whose tools lacked the variety and aesthetic shaping of *H. sapiens* tools. Yet it seems strange that the intelligence that could fell a bison or a wild boar could some 40,000 years later also create an algebraic equation, a violin sonata, or a computer program, or send a person to the moon and back. If the contemporary human brain can entertain such complex and abstract thoughts, what functions in prehistoric society gave it that capacity?

Perhaps we can answer this question by considering essentials for human survival in the absence of civilization. One would need protection from predators, rock slides and other natural hazards, drought and flooding, fires, and disease. Since humans are without a natural layer of mammalian fur, clothing and shelter would be essential to prevent death from exposure. Memory and language would be important in order to communicate experiences and knowledge. Relational thinking would be important so that one could envision how to convert an animal's integument into a bed, a rooftop, a raincoat, a pair of moccasins, a camouflaged decoy, a weapon, a tote bag, or a snare. Problem-solving ability would be necessary to relate distance traveled to a safe journey home, to provide a sense of time and an anticipation of seasons, to approximate the range of a spear, arrow, or bola, and to estimate how much food would have to be collected to meet the community's needs.

The hallmark of humanity is the replacement of immediacy by planning. We do not have to wait until we are hungry to hunt. Rather, we plan ahead by buying and storing food for future use. We have the capacity to create environments that free us from the necessity of an immediate response. Our shelters (caves, Cape Cod houses, apartments, or mansions) enable us to work in 100°F or −30°F (38°C to −35°C) weather. We can travel by foot in cold weather because we know how to clothe ourselves. We can smoke, preserve, and store foods for seasons when fresh plants and animals are not available.

Humans Require Fewer Innate Responses to the Environment Than Do Other Animals

Because humans have been able to adapt mentally to new situations and to circumvent environmental hazards we are less dependent on genetically built-in responses in our nervous system. If there is an unfamiliar sound we do not scatter on impulse like frightened deer. We process the noise and determine what our best response should be: to ignore it, to hide, to run away, to light a fire, to pick up an axe, to signal our neighbors, or to dial a telephone number.

The variations in our responses make it unlikely that we have an inherent sense of territoriality, or that our genes carry some ancient response that makes us drive away intruders. Nor is it likely that our genes make us innately aggressive, cooperative, monogamous, or polygamous. There are so many varieties of human responses to stress or to the abundance of resources that no single behavioral attribute can be considered characteristic of all humans. The history of human cultures demonstrates that values and temperaments are more often inconstant than constant. A nation can be run by a despot and exist as a democracy within the span of one or two generations. People may regard marital infidelity as a normal activity (as during the Restoration period of Samuel Pepys and his contemporaries), or as a personal scandal (as it was in the Victorian era). A country may wage war for centuries and then spend more than a century at peace, as happened in Sweden.

Innate Behavior or Biological Determinism in Humans Is Mostly Unproven

If there are ancient traces of animal instincts in us, they are usually sublimated and redirected by our culture. Our biological signals are minimal: hunger and thirst are physiologically unambiguous, as is our need to urinate and defecate; sexual arousal involves subtle emotional cues turned into physiological passions. There are, however, no such definable cues or physiological changes associated with a need for property and aggressive behavior. Instead they are more likely to reflect our upbringing. It is usually our experiences that lead most of us to be mean or gentle, cruel or compassionate, solitary or gregarious, selfish or sharing—in short, to be human.

Questions to Answer

1. How did Neanderthals differ in appearance from contemporary humans?
2. What are the implications of the discovery that the Neanderthals buried their dead with flowers?
3. How can the age of fossil remains be determined?
4. What major physical differences exist between humans and other primates?
5. How far back can human-like remains be dated?
6. Who were the Cro-Magnons and what is their relation to modern humans?
7. What evidence would you cite to support the view that no major change in human intelligence has occurred in the past 50,000 years?

8. What features of nontechnical societies would require a brain that is capable of learning calculus, writing sonnets, and playing a violin sonata?
9. What is meant by the term "biological determinism"?
10. Present the arguments for and against a biological determinism of human aggression.

Terms to Master

biological determinism

Homo habilis

Cro-Magnons

Neanderthals

Homo erectus

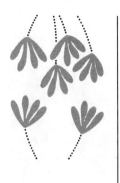

Chapter 29
What Should We Do
with Our Genes?

The history of eugenics, from Galton's idealism at the turn of the century to the mass execution of Jews in Nazi concentration camps during World War II, provides a warning to humanity that any planned program of differential reproduction must be carefully questioned, debated, and examined for potential abuses of racism and spurious elitism. Yet the reality of human spontaneous mutation must also be faced. If an equilibrium in the past depended on infectious diseases, malnutrition, and exposure to harsh environments, and the resultant deaths of infants and children with high genetic loads, then some thought must be given to the consequences of the altered conditions of civilization.

Public Health and Modern Medicine
Have Eliminated Infectious Diseases
As a Major Selective Force
The developments of modern medicine are impressive. Antibiotics have been mass-produced, beginning with penicillin about 1945. Many millions of today's adults would not have survived childhood without penicillin, streptomycin, erythromycin, tetracycline, and other antibiotics. Many of these were probably individuals with high genetic loads. A tour of a pediatric ward in a hospital will also demonstrate how effective medicine is in preserving the lives of sick children. In the intensive care unit will be premature babies breathing a regulated mixture of oxygen that is sufficient to keep life processes going but not so rich that eye damage will occur. The infants may be bathed in fluorescent light to break down hemoglobin pigments before they can cause damage to the nervous system.

Some of the infants will have birth defects (Figure 29-1). A child with hydrocephalus will have its excess cerebral fluid drained by a catheter

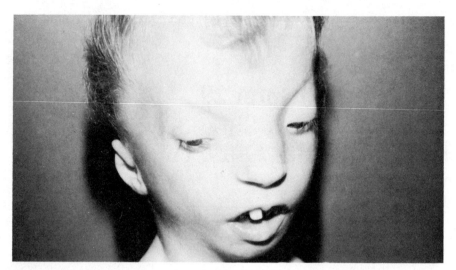

Figure 29-1   Birth Defects and Eugenics

Individual families, as well as societies, differ in their responses to infants born with severe birth defects. A very profoundly malformed child (a cyclops, for example) is often not even shown to the parents. Those with gross physical defects but normal intelligence are often given heroic surgery and medical attention, whether the parents want the child or not, even though the resulting treatment may be more cosmetic than functional, as in the early treatment of thalidomide babies. These issues are usually more immediate than eugenic considerations. In most instances the parents' basic question is: Will this happen again? Unfortunately the physician may not know the genetic or developmental basis for the defect, and either too pessimistic or too optimistic an answer may be given. Rarely do we ask ourselves if we have any responsibility to the future in reproducing our own genotypes. Eugenic practices today are largely carried out at an unconscious level. (Courtesy of R. J. Gorlin, D.D.S.)

that snakes along the body under the skin, eventually draining into the digestive system. An infant with spina bifida will have a fist-sized balloon of membrane surgically removed and replaced with skin to prevent meningitis. Another child will have a compressed bandage around a large umbilical hernia, and the pressure will be slowly increased to gradually push the intestines back into the abdominal cavity. A child with an imperforate anus, a cleft palate, or a cardiac defect has an excellent chance of successful corrective surgery. Only in a few cases are the situations hopeless—a child with an extra chromosome 13 or 18, a hydrocephalic child with insufficient brain development, or a baby born without kidneys.

Prenatal Diagnosis May Maintain or Increase the Frequency of Mutant Alleles in a Population

In other rooms are children with metabolic defects: children with kidney defects arising from an inherited vitamin deficiency, babies with phenylketonuria on a special diet, and hemophiliacs receiving transfusions. Children with other problems are also treated, such as the retinoblastoma patient who comes in for a yearly check-up, and children with defective bladder drainage or bowels that will not pass the digested food along the rectum. Most of these children can live to maturity and reproduce if they so wish. Their disadvantageous genotypes, which may have contributed to their birth defects, will then be passed on.

However, medicine does more than treat infectious diseases and birth defects. Pregnant women may request amniocentesis to detect lethal metabolic genetic defects, like Tay-Sachs disease or the Hurler syndrome. By sparing themselves the tragedy of having to watch their child be devoured by a progressive illness for two years, they can plan to have two or three normal children, each such child having a two out of three chance of being heterozygous. One might think that amniocentesis would reduce the genetic load by aborting a fetus that would have died before maturity if born. Yet amniocentesis may actually increase the mutant gene frequency if the family size is at or above the mean family size in the population. This would be particularly true for diseases like Tay-Sachs where genetic screening programs can identify carrier parents before they marry. There are more than 1000 such single-gene recessive conditions, and if prenatal diagnosis becomes available for most of them, the genetic load will increase.

Environmental Agents May Also
Increase Our Load of Mutations   Improvements in medical diagnosis and treatment are not the only aspects of modern civilization that increase our load of mutations. We are exposed to ionizing radiation in routine check-ups, and not all physicians and technicians provide adequate lead shielding of the gonads or use the minimal dose effective for the particular exposure. Industrial wastes, agricultural fertilizers, insecticides, food additives and preservatives, automobile pollutants, and many more chemical hazards have been introduced into our lives. Very little is actually known about the effects these have on germinal material and how much of an increase they represent over the spontaneous mutation rate.

Several Factors Still Act
to Decrease Genetic Load   The situation would be alarming if the conditions of contemporary society *all* led to an increased genetic load for future generations. There are other features of life, however, which counteract this trend (Table 29-1). Those individuals with a severe medical problem, such as an uncorrectible heart defect, a chronic kidney infection not readily prevented by antibiotics, chronic emphysema, or paraplegia, or those who are retarded or who depart too much from average height, weight, or physical appearance, either cannot or do not marry as readily as their normal peers. If they do marry, many of them

Table 29-1   Assortative (Nonrandom) Mating

Nonrandom or *assortative mating* can lead to genetic differences in populations. It is also a factor in human evolution. If mating were random there would be no correlation (0.0), between any two categories of people and if opposites (e.g., thin men and plump women) attracted each other, the correlation would be negative. The nine traits shown in this table are examples of *positive correlation*. The mean age difference between married couples is 2.7 years. Most couples are of the same religious and racial background, but scarcity can break down such preferences. The highest miscegenation rate in the United States is in Hawaii (12 percent). As values change, so too will these correlations. Mating is still random for traits that are not apparent to courting couples, such as blood groups and fingerprint patterns.

Trait	Correlation	Trait	Correlation
Eye color	0.26	Ethnic background	0.45–0.82
Height	0.30	Age	0.76
Hair color	0.34	Religion	0.79
Intelligence	0.50	Race	0.88–0.97
Residence	0.71		

do so at an older age than the average. They may also elect to have fewer children, either because it taxes their health to raise them or because they do not wish to risk passing their defect on to their children. In some cases, the incidence of the defect is genetically self-regulating because a severely retarded individual, if institutionalized, is sexually isolated for life. In some cases the condition may inhibit sexual desire. People who are chronically or seriously ill often have a diminished sexual drive.

Genetic Counseling, Although Nondirective, May Lead to a Reduction in Genetic Load   Genetic counseling can also be beneficial. When parents learn that their child is sick or deformed because of a mutation both of them carried from distant ancestors, they may choose to restrict their family size. Usually when the defect is chronic, expensive, debilitating, or lethal, the parents do elect to have fewer children if there is a hereditary basis for it. Muscular dystrophy, cystic fibrosis, Lesch-Nyhan syndrome, and Tay-Sachs disease are examples of such defects which usually result in smaller family size because of the emotional devastation. Amniocentesis for the detection of Lesch-Nyhan and Tay-Sachs disorders may, however, reverse the tendency to not breed when a high-risk birth is at stake.

A Good Environment Promotes Assortative Mating, Which Is Eugenic   A prosperous environment is also eugenic because it permits healthier and brighter individuals to select similar mates (Table 29-2). This arises from the near

Table 29-2   Average Number of Offspring per Individual in Relation to IQ

Family planning in industrial nations extends from the poor to the rich, with the two-child family usually being the ideal. This was not true before Margaret Sanger's birth control movement in the 1920s when those with more education (or higher IQ) had fewer children than those with less education (or lower IQ). Eugenists in that period warned that the mean intelligence would decline if the trends were not reversed. This fear was groundless, however. Even if there is a major genetic component to IQ (a controversial viewpoint) the family planning of the last 20 years demonstrates that families with higher IQs have more children. Those with the lowest IQs have the fewest children. These trends suggest that those with very low IQs have difficulty marrying (many of those with IQs below 70 are institutionalized) and those with very high IQs are likely to be college-educated and can afford the houses or larger apartments for three or more children.

IQ Range	Kalamazoo Fertility Study*		Minnesota Mental Retardation Study†		Third Harvard Growth Study‡	
	Number Reporting	Mean No. Offspring per Individual	Number Reporting	Mean No. Offspring per Individual	Number Reporting	Mean No. Offspring per Individual
More than 130	23	3.00	18	4.06	78	2.40
116–130	107	2.51	133	3.33	201	1.99
101–115	344	2.08	296	3.77	510	2.18
86–100	427	2.30	209	3.42	506	2.21
71–85	75	2.05	41	3.12	216	2.00
56–70	3	0.00	13	2.92	19	1.21
0–55	0	—	3	0.00	3	2.00
Total sample	979	2.24	713	3.53	1533	2.14

*Bajema, 1963. Fertility ascertained at age 45 or time of death, if before age 45.

†Waller, 1971. Fertility ascertained at age 45 or time of death, if before age 45.

‡Bajema, ms.

universal literacy of developed nations, the existence of coeducational colleges and universities, and the mobility afforded to those who are neither physically nor mentally severely handicapped. *Assortative mating*—that is, like marrying like—occurs mostly unconsciously, but when, for example, the population of potential mates for college students consists mostly of fellow college students then mating choice is no longer random (e.g., few college graduates marry high school dropouts). In undeveloped nations an illiterate and impoverished individual would have less opportunity for an assortative mating based on intelligence or general health, and a very bright person would have very little chance of finding another person of the same intelligence in a small, impoverished village. Assortative mating becomes eugenic when the family size for such individuals is larger than for those with severe medical or mental handicaps. This is what has actually occurred in the United States and Europe.

Both eugenic and *dysgenic* (genetically harmful) trends are occurring in our times. We do not know the net effect of these two processes but there are some factors which make the dysgenic trends seem more prevalent than the eugenic ones. Foremost among these factors is the dramatic reduction of infant and child mortality from infectious diseases and malnutrition throughout most of the world. Even in most underdeveloped countries public health measures have brought life expectancy to levels that are not too different from those of well-developed nations. Yet these same underdeveloped countries have many years to go before their literacy, class mobility, and living standards approach those of European or North American countries. Only a miniscule percentage of the world's married couples know anything about the heritability of birth defects. Genetic counseling would have to be done on a worldwide scale in order for it to be of any consequence. Even in affluent nations like the United States, only a small percentage of couples receive adequate genetic counseling.

Strategies for Reducing Genetic Load
Involve More Conscious Decisions
About Becoming Parents
Each of us, by becoming educated, can do something to keep the genetic load from increasing. In order to do so, of course, we have to be morally committed to the view that increases in genetic load are undesirable and that the preservation of the *gene pool* (the population's total sum of genotypes) and the prevention of its deterioration require individual effort. On the other hand, if we are convinced that there are compensatory laws of nature that protect us, even though we have not yet discovered them, then we should reject the advice that follows.

For the fertile couple likely to have disadvantageous genotypes, the best strategy is to have fewer than the average number of children. That is, they should have one or at the most two children. Among those who are at risk are those newly married individuals who had no surviving grandparent by the time they were 20 years old. Since most 20-year-olds have two or more surviving grandparents (in their mid-60s or older), the absence of all grandparents would indicate that they died before reaching their mean life expectancy (obviously, natural causes, not accidents, would be used in calculating high or low risk). This would not necessarily be true for a 20-year-old who was the youngest of a large family or whose parents married late.

Also in this risky category is the married person who does not have a surviving parent. Since most newly married individuals are in their mid-20s, their parents are, on the average, in their 50s or early 60s. The failure of both parents to reach their normal life expectancy (about 68 for male and 72 for a female) may indicate that the parents had disadvantageous genotypes.

Any person who, at the time of marriage, has had a prior history of a major medical problem (congenital heart defect, chronic kidney disease, ulcers) requiring major surgery, repeated hospitalization, or lifelong medication, should consider his or her genotype to be at risk. So, too, should a person with a major genetic defect that has proved to be disabling. While it is true that many individuals compensate for their disabilities and achieve happiness and success in their personal lives and careers, far more of those who are burdened with physical problems find their joy in life diminished and their range of goals narrowed. The child of such a genetically impaired parent, although unaffected, should consider the risks of being heterozygous for that defect or of having a higher probability for the expression of that defect in a later generation as a polygenic defect. This would also apply to the sibling of a child with such a disorder.

**Low Genetic Load Couples May Wish
to Have Larger-Than-Average Families** Married couples with low risks may wish to have more than two children (three or four in most cases) if the values associated with eugenics outweigh those associated with reducing world population size by setting a personal example. This would improve the chances for lowering the genetic load in the population. The married couples with three or four surviving grandparents on each side, the couples who have been fortunate not to have experienced any major surgery or major medical problems, and the couples whose family histories show no unusual genetic defects in the past three generations are likely to enjoy low genetic loads.

In some cases a married couple may have to balance a less-than-optimal family history against compensatory talents in art, music, creative writing, science, or other areas of unusual achievement. While there is no guarantee that outstandingly successful people owe some credit to their genes, it is nevertheless true that their children have higher probability of achieving success than the average person. Even if a genetic component is a major factor in eminence, Galton's regression to the mean (and the meiotic basis for it) greatly reduces the odds that a child will be as eminent as a successful parent. A polygenic mechanism would predict an increase of potentially eminent persons in the entire population of children of such parents because the parents would necessarily have more of those genetic factors than the average person.

Sterile males (who constitute about three to five percent of all males) have the option of eugenic family planning by selection of a sperm donor who has been checked out for genetic defects, family history, and special talents. If the sterile couple has the choice (which they can probably obtain by asking for it and presenting reasonable arguments to their physician), they can specify the qualities they want and have the physician locate a suitable person for donation. This might be more sound, from a eugenic perspective, than the present policy where the donor's family history is kept secret and the physician attempts to match blood groups and other personal features of the father to the potential donor.

Whether or not artificial insemination is morally right depends upon the couple's values. If the couple has a strong desire to raise a family, and if the child's health and prospects for living an enriched life outweigh the parents' need for a child who resembles the sterile father then the decision may not be a difficult one.

In both high and low genetic load assessments we have, at present, only an inferred load of a vague nature based on family history. Our knowledge at both the molecular and specific gene level is virtually non-existent for human couples. Yet it is precisely the same family history that guides insurance companies in assigning premium payments. Actuarial studies support the basic predictions of high or low genetic load.

The Philosophic Implications of Differential Breeding

Need Careful Study   Would the genetic strategy suggested here lead to a loss of variability in humanity? Few values are more cherished by geneticists than the variability that accounts for individual uniqueness. Yet there have always been violations of this variability. Before the 20th century few persons in the world married outside their villages. The high degree of inbreeding in isolated populations, such as the Amish or Dunkers in the United States, the decendants of shipwrecked sailors and slaves on Tristan da Cunha in the Atlantic Ocean, and the Tahitian-English descendants of the mutineers from H.M.S. *Bounty* on Pitcairn's Island, is proof that even with a limited number of founding families a population can increase in numbers over the centuries without importing new populations to restore variations (Table 29-3). Furthermore most of the protein variability (polymorphism) detected in human populations has no known relation to natural selection or survival.

Table 29-3   Ethnic Distribution of Genetic Disorders

Over thousands of years, isolated populations have accumulated different frequencies of recessive alleles for monogenic traits and clusters of alleles for polygenic or multifactorial traits. The full extent of ethnic disorders is not known because only a few ethnic groups have been studied extensively (e.g., the various Jewish communities). Very small, isolated populations, such as the Amish or Dunkers in the United States, the inhabitants of Tristan da Cuhna, or the descendants of the H.M.S. *Bounty* mutineers, are not listed here.

Ethnic Group	*High Incidence Disorders*
Mediterraneans (Greeks, Italians, Albanians)	Thalassemia; glucose-6-phosphate dehydrogenase (G6PD) deficiency; familial Mediterranean fever
Africans	HbS (sickle cell) anemia; HbC anemia; β-thalassemia; G6PD deficiency; lactase deficiency (adult); polydactyly*; systemic lupus erythematosus*; hypertension*
Chinese	α-Thalassemia; G6PD deficiency; lactase deficiency (adult); nasopharyngeal cancer*
French-Canadians	Morquio syndrome; tyrosinemia; agenesis of corpus callosum
Irish	Neural tube defects*
Japanese	Cleft lip or palate*; gastric carcinoma*
American Indians	Congenital dislocation of hip*
Finns	Amyloidosis type V; aspartyl glycosaminuria
Ashkenazic Jews	Bloom syndrome; Gaucher disease; familial dysautonomia; Tay-Sachs disease; torsion dystonia Gilles de la Tourette disease; ulcerative colitis*; Burger disease*
Sephardic Jews	Ataxia-telangiectasia (Morocco); cystinosis; familial Mediterranean fever; lactase deficiency; cystinuria II and III (Libya)
Oriental Jews	α-Thalassemia (Yeman); β-thalassemia (India); G6PD deficiency (Iraq, Iran); lactase deficiency; PKU (Yemen)

Source: V. A. McKusick, *Mendelian Inheritance in Man,* 6th ed. (Baltimore: Johns Hopkins Univ Pr, 1983).

*A multifactorial trait. In some instances (e.g., gastric carcinoma among the Japanese) the disease may be environmental (asbestos fibers in polished rice; carcinogenicity of food additives or preparation).

Cultural Differences Usually Outweigh Genetic Differences in Determining Behavior   While a high degree of variation may not be essential for human survival, we might believe that it is precisely this genetic uniqueness which causes cultural differences and individual lifestyles and tastes. From identical twin studies we

know this is not true. The career goals of identical twins depend more on their environment than on their genes. If the twins are encouraged to visit different friends, have different exposures to cultural activities, and attend different colleges, the chances are good that their careers and lifestyles will vary like those of two-egg twins or singleton siblings. If they are always treated by their parents as a single entity and exposed together to major cultural influences, then it is more likely that they will share common careers and lifestyles.

Thus even in the extreme case of identical genotypes, twin studies show an immense cultural variation between the two individuals. Variation is truly worth preserving because the diversity of genotypes provides abundant forms for survival. If all humanity were to be cloned from

Table 29-4 Values and Arguments Used to Debate Eugenics

These are a few of the major pros and cons of eugenics. Excluded here are arguments about racism and spurious elitism, both of which are clearly based on bigotry and naive assumptions about the causes of social failure, such as poverty and crime. Note that some of the pros and cons are mutually exclusive (numbers 1 and 3) and others may both be true (number 4). The arguments may also differ more in underlying conflicts of values than in factual issues (numbers 5 and 6). Most controversial of all is argument number 2, because the degree of genetic involvement (if any) for the traits listed is not known, except in those medical disorders known to be inherited (3300 or more). For this reason, some supporters of eugenics prefer to restrict its application to the medical problems and wait for more evidence on behavioral traits before making eugenic commitments.

Pro-Eugenic View	Anti-Eugenic View
1. The genetic load and its detrimental effects increase in the absence of natural selection.	1. Heterozygosity is beneficial; extremely harmful homozygotes eliminate themselves.
2. Intelligence, talent, longevity, personality, and good health have substantial genetic components.	2. There is no convincing evidence that socially desirable traits have a genetic basis.
3. Modern medicine and civilization permit those with known genetic disorders to survive and reproduce, increasing the genetic load of the population.	3. Since individuals with genetic disorders contribute to the heterozygosity of the next generation, they benefit, rather than harm, the population.
4. An affluent culture and environment increase eugenics through assortative mating.	4. An affluent culture and environment increase well-being by extending social opportunities to more people.
5. Artificial insemination could contribute to eugenics if those using it asked for more information on the donors.	5. Let the physician be limited to matching blood groups and ethnicity; otherwise the variability of the population will be diminished.
6. Conscious, voluntary differential breeding by well-informed individuals can be eugenic.	6. No consideration should be given for genetic load, only actual disorders for which the parents are at risk.

a few individuals, and the world experienced a major disaster such as a new disease, it could prove fatal to a large number of individuals. Fortunately, in humans there are very few clonally derived persons, monozygotic twins constituting less than one percent of all births.

Table 29-4 is a summary of the arguments concerning eugenics. Where then would the loss of variation produced by eugenic practices come from? It could not come from the millions of individuals who might elect to have slightly larger families based on eugenic considerations. Not only do we produce uniquely different sex cells, but we also owe our specific genetic blessings to many different combinations of genes. There is no one genotype for longevity, good health, or special talent. The major threat to variation comes from the possible widespread use of artificial insemination. However, few males have fathered more than 25 children by this means. Physicians use their own judgment on multiple usage, some limiting it to one use, some to five, some not at all. Each of the donor's sperms was unique and each encountered an egg from a different woman. Thus the chance of spreading homozygosity is remote. Even if a donor were spreading unknown undesirable genes the number of couples using artificial insemination is only a minute fraction of one percent of all married couples, and so the effect on the overall gene pool would be minimal.

As long as families choose to reduce genetic load by differential breeding, each new generation can, to some extent, monitor its own medical problems and special talents. Sometimes a family relatively free of problems may encounter a new mutation or evidence of an unlucky high load. It is precisely the continuous use of differential reproduction that, in the long run, will compensate for the spontaneous mutations that will inevitably occur.

Questions to Answer

1. Geneticists have often accepted the political view that "all men are created equal," but they also accept the biological fact that no two zygotes are identical. What are some of the consequences of having both legal equality and biological inequality?
2. What genetic conditions have poor prospects of being perpetuated by their victims? What are the circumstances preventing such persons from having as many offspring as the rest of humanity?
3. What is meant by the term "eugenics"?
4. Can a good environment be eugenic?
5. What assumptions or facts support the view that the genetic load is increasing?
6. What assumptions or facts support the view that no significant change is occurring in genetic load as a result of modern civilization?

7. How would life expectancy be related to genetic load?

8. Does the individual (i.e., you) have any responsibility to the gene pool of the population? Defend your viewpoint.

9. Discuss the ethics of using sperm banks for (a) sterile males, (b) males with a serious genetic defect (e.g., Huntington disease), and (c) males of normal achievement who wish to use the semen from a celebrated artist or scientist. Present the arguments which you believe would justify or prohibit donor sperm in each circumstance.

10. Discuss the ethics of secrecy in artificial insemination. Does a child have a right to know his biological parent's family history if there are hereditary problems in either his adoptive parents' or his biological parent's ancestry?

11. What effect, if any, would differential breeding (eugenics) have on human variation?

12. Is cloning a threat to humanity? Justify your viewpoint.

13. If, on moral grounds, you oppose any differential breeding for any reason, how would you compensate for the spontaneous mutations that arise in the population? Consider your own family and the long-term history of humanity (say, 100 to 1000 generations into the future).

Terms to Master

assortative mating
dysgenic
gene pool

negative eugenics
positive eugenics

Glossary

ABO blood groups. Red blood cell membranes that carry the protein antigens A or B, both, or neither.

Abortion. The removal of an embryo or fetus from the mother's womb (uterus) by artificial means or from natural causes.

Acentric chromosome. A chromosome lacking a centromere (the unit that attaches it to a spindle fiber).

Achondroplasia. An autosomal dominant disorder causing disproportionate dwarfs with normal-sized torsos but short arms, short legs, and altered facial features.

Acrosome. The perforation cap covering the nucleus of a mature sperm and having enzymes that digest the egg surface membrane.

Agglutination. The clumping together of cells.

AID (artificial insemination from a donor). A means of fertilizing a woman whose husband is sterile, using donated semen administered by a physician.

AIH (artificial insemination from the husband). A means of fertilizing a woman by concentrating several samples of the husband's sperm and then having a physician administer it to the man's wife. It is used when the husband has a low sperm count.

Albino. An individual lacking melanin in the eyes or skin. The hair may be white or straw-colored.

Alkaptonuria. An autosomal recessive disorder that produces black urine.

Alkylating agent. A molecule with a reactive side group that removes a hydrogen atom from another molecule and binds the two molecules together.

Allantois. A membrane extending from the embryo that collects metabolic wastes.

Amoeba. A single-celled, irregularly-shaped protozoan that moves by flowing its cytoplasm in a given direction.

Amenorrhoea. The failure to menstruate on a periodic basis.

Amniocentesis. A technique used mainly for diagnosing birth defects in the fetus by means of a needle inserted into the amniotic fluid. The needle can be used to inject substances or withdraw amniotic fluid.

Amnion. A membrane surrounding the fetus which fills with fluid.

Anaphase. A stage of cell division in which the chromatids of a chromosome begin to separate from one another.

Anencephaly. The absence of a major portion of the brain often accompanied by the absence of the cranium.

Aperiodic crystal. A crystal with irregular features such as the sequence of nucleotides in the genes of a chromosome.

Artificial insemination. The fertilization of an egg with sperm by means other than sexual intercourse.

Ashkenazic Jew. A descendant of medieval German Jews or Yiddish-speaking Jews.

Assortative mating. The tendency of like to marry like, for religion, intelligence, and other physical or social traits.

Autolysis. The self-digestion of a cell when it releases its lysosomal enzymes into the cytoplasm.

Autosome. A chromosome that is not a sex chromosome.

α-fetoprotein. A normal protein of the fetus that is not present in excess in the amniotic fluid unless some defect is present such as a neural tube opening.

Bacteriophage. A virus that infects prokaryotes.

Base pair. A complementary association of nitrogenous bases in a DNA duplex, such as adenine with thymine, or guanine with cytosine.

Biochemical pathway. A group of enzymes that collectively synthesize a larger molecule from several smaller components.

Biological determinism. The belief that behavior traits, especially those of humans, have a major genetic component.

Birth control. Any volitional means of preventing the fertilization of an egg or the implantation of an embryo.

Blastocyst. The embryonic mass of cells which has the capacity to implant in the lining of the womb (uterus).

Blastomere. A cell produced by cleavage of another cell after mitosis in the earliest stages following fertilization.

Blue baby. A child with a heart defect present at birth causing reduced oxygen circulation.

Blue-green alga. A bacterium or prokaryotic cell with photosynthetic molecules.

Breakage-fusion-bridge cycle. The process by which dividing cells are killed following the breaking of a chromosome.

Carcinogenicity. The ability of a molecule to induce tumor cell formation.

Cell. The smallest unit of life capable of sustained metabolism.

Cell cycle. A sequence involving growth, DNA synthesis, preparation for cell division, and the mitotic events leading to new cell formation.

Cell membrane. The surrounding membrane which holds the cytoplasm together and regulates movement of molecules into and out of the cell.

Centric chromosome. A chromosome with a centromere (the unit that attaches to a spindle fiber).

Centriole. An organelle of the cytoplasm near the nucleus associated with spindle fiber formation and organization.

Centromere. The unit (a modified gene) on the chromosome to which a spindle fiber is attached.

Cervix. The lower portion of the womb (uterus) forming a junction with the upper end of the vagina.

Chief gene. A gene essential but not fully sufficient for the expression of a trait. Other modifier genes or environmental conditions are required.

Chimera. A composite individual formed from two cell lines of separate origin.

Chromatid. A DNA double helix containing the genes of a chromosome.

Chromatography. A method of separating molecules dissolved in solvent by using paper or columns with cellulose or synthetic resins, through which the molecules pass at different rates.

Chromosome aberrations. Alterations in chromosome length, gene sequence, or composition, as a consequence of chromosome breakage.

Chromosome. One or two chromatids bearing a centromere and a number of genes.

Chromosome banding. The preparation of chromosomes so that a characteristic pattern of light and dark regions emerges distinguishing the chromosome pairs of a karyotype.

Chromosome bridge. The connection, between two centromeres, of a dicentric chromosome as it is pulled to opposite poles of a cell.

Chromosome number. The number of chromosomes present in a fertilized egg.

Cis alignment. In independent alignment, two nonhomologous chromosomes facing the cell's pole with each chromosome bearing two normal alleles or two mutant alleles.

Cleavage. Cell division in the first few divisions of the fertilized egg.

Cleft palate. A condition resulting from the failure of the bones of the palate to close, leaving an opening between the oral cavity and the nasal cavity.

Clinodactyly. A bowing of the small finger.

Clomiphene. A chemical that stimulates ovaries to release eggs.

Club foot. A deformed foot or ankle usually caused by pressure of the uterus on the developing fetus.

Code-script. The term used for the genetic code before the chemical nature of the gene was experimentally worked out.

Coding ratio. In the genetic code, the ratio of bases to amino acids (a coding ratio of 3).

Codon. A triplet or sequence of three bases in a nucleic acid.

Coitus interruptus. Withdrawal of the penis during intercourse just before ejaculation occurs.

Colchicine. An alkaloid produced by plants which can cause mitotic arrest from an absence of spindle fiber synthesis.

Complementation. The normal phenotype obtained when nonalleles are heterozygous or, less often, when a heteroallelic diploid a_1/a_2 leads to a (+) instead of an (a) individual.

Cosanguineous marriage. A marriage between relatives who are genetically related.

Contraceptive. An agent (mechanical or chemical) that prevents fertilization or implantation.

Copulation. The mating or sexual intercourses between two individuals.

Corpus luteum. Follicle after its release of an egg. It then becomes a temporary hormonal gland.

Criss-cross inheritance. A pattern of X-linked inheritance in which sons resemble their mothers, and daughters resemble their fathers.

Crossing-over. A reciprocal exchange of parts of homologous chromatids occurring during the reductional division of meiosis.

Cryptic mosaic. Parents of normal phenotypes who have two or more children with the same trisomy or the same dominant disorder because they have two cell lines in their reproductive tissue.

Cryosurgery. Destruction of tissue by liquid nitrogen.

Cystic fibrosis. An autosomal recessive disorder of the mucous-secreting tissues.

Cytokinesis. The partitioning of a cell as mitosis nears completion.

Deletion. A missing segment of a chromosome.

Demographic transition. The shift from high birth and death rates to low birth and death rates.

Detrimental mutation. An allele that causes diminished viability or fertility.

Diaphragm. A mechanical device that is fitted to the cervix and that prevents passage of sperm from the vagina to the uterus.

Differential breeding. The tendency of beneficial genotypes to leave more surviving progeny than those individuals whose genotypes are detrimental to the survival of their progeny.

Diploid. The number of chromosomes normally present in a fertilized egg.

Dizygotic twins. Twins who arose from separately fertilized eggs.

DNA (deoxyribonucleic acid). The genetic material of chromosomes, containing the nucleotides adenine, thymine, guanine, and cytosine.

Dominant trait. A genetic character expressed by a heterozygote (hybrid).

Double helix. The complementary nucleotide sequences forming the two strands of a DNA molecule.

Down syndrome. Trisomy-21; a semi-lethal disorder characterized by retardation of growth and mental development, and numerous physical abnormalities.

Dysentery. An infection of the intestinal tract causing diarrhea.

Dysgenic trait. A trait which increases genetic load, reduces life expectancy, increases risk of illness, or diminishes intelligence.

Dysmorphia. A condition characterized by abnormal features.

Duplication. The insertion or accidental repetition of a sequence of genes adjacent to the original sequence or, less often, elsewhere in the same chromosome.

Ectoderm. An embryonic tissue from which skin and nerve tissues are derived.

Edward syndrome. Trisomy-18; a lethal condition characterized by multiple birth defects.

Empiric risk. An estimation of the average chance of recurrence of a disorder, arrived at by summing up all the case histories of the disorder.

Emphysema. A disease in which the air sacs of the lungs are destroyed, resulting in diminished oxygen intake.

Endoderm. An embryonic tissue from which the digestive tract, lungs, liver, pancreas, and gall bladder are derived.

Endometrium. The inner lining of the uterus.

Endoplasmic reticulum. An organelle consisting of a spongy network of membranes to which ribosomes are attached.

Environmentalist. A person who advocates that behavioral traits, especially intelligence, are acquired by cultural and familial influences.

Enzyme. A protein that digests or connects molecules.

Enzyme replacement therapy. The attempt to treat genetic disease by injecting cell fragments or synthetic vesicles bearing an enzyme that the individual cannot synthesize.

Epicanthal eye fold. A fold of skin at right angles to the part of the eye closest to the nose.

Equational division. The second meiotic division, which separates the chromatids of a haploid cell.

Erythroblastosis fetalis. The Rh incompatibility disease leading to jaundice, brain damage, or a fatal lysis of the red blood cells shortly after birth.

Erythrocyte. A mature red blood cell.

Estrogen. A hormone released by the ovary which promotes the growth of the uterine lining.

Estrus cycle. In animals (other than humans), the time when a female is receptive to intercourse.

Ethnic disease. A genetic disorder found more frequently in one human population than in other populations.

Eugenics. The promotion of reproductive options favoring desired human genetic traits, especially health, longevity, talent, intelligence, and unselfish behavior.

Eukaryote. A cell with a nucleus and cytoplasm.

Exceptional children. Children whose academic performance is substantially below that of others at their age level; mentally retarded children. Occasionally, it includes the gifted children.

Extrapolation. The use of an argument or finding from one field in another field (e.g., the application of data from mouse experiments to human experiments).

Exon. A segment of a gene's DNA that serves as an informational sequence of nucleotides.

F_1. The first filial generation, or offspring of the parents.

F_2. The second filial generation, or grandchildren of the original parents used in a genetic cross.

Fabry syndrome. An X-linked lysosomal storage disease of cells lining blood vessels.

Factor VIII. A clotting protein absent in the blood of hemophiliacs.

Fallopian tube. The oviduct or passageway carrying the egg from the ovary to the uterus.

Fertilization. The union of an egg with a sperm.

Fetal diffusion syndrome. The partial depletion of the blood supply of the smaller identical twin by the larger identical twin while in the fetal state.

Fetus. The developing embryo; in humans the embryonic stage from the second month to birth.

Frame shift mutation. An increase or decrease of one or two bases (or any number of bases not divisible by 3) in a DNA molecule, resulting in a profoundly altered protein.

FSH, Follicle-Stimulating Hormone. A hormone released by the brain that stimulates follicle formation in the ovaries.

G_1 *phase.* The metabolic growth stage of a cell after mitosis ends and before DNA is synthesized.

G_2 *phase.* The stage of the cell cycle immediately after DNA synthesis in which preparation for cell division occurs.

Gamete. A haploid sperm or egg.

Gene. A sequence of DNA that indirectly produces a protein; a unit of inheritance inferred from breeding analysis.

Genetic counseling. The provision of information to a client concerned with the potential risks of birth defects or genetic disorders.

Genetic death. The prevention of an individual from leaving offspring, by one or more mutations.

Genetic dogma. The belief, before evidence became available, that the sequence of nucleotides in DNA was copied as an RNA sequence and translated into a sequence of amino acids in proteins.

Genetic drift. The tendency of small populations to vary, randomly, in their gene frequencies.

Genetic engineering. The manipulation of chromosomes or genes by biochemical means for the purpose of altering the genotype in a desired direction.

Genetic load. The accumulated detrimental effect of one or more alleles carried in the heterozygous state.

Genetic polymorphism. The perpetration of both homozygous genotypes as well as the heterozygous genotype for two alleles in a population, usually as a consequence of some selective advantage conferred on the heterozygote.

Genotype. The inferred genetic composition of a character trait seen in an individual.

Gestation. Pregnancy.

Giant cell. A dying cell produced by failure of a chromosome bridge to break, with suppression of cell division.

Giemsa. A stain used for chromosome banding studies.

Gloger's rule. The correlation of fur color or skin color with latitude, northern species being lighter than equatorial ones.

Golgi apparatus. An organelle of the cell which buds off lysosomes.

Graafian follicle. The mature egg in a blister on the surface of the ovary.

Haploid. The chromosome number of a sperm or egg.

Hardy-Weinberg law. A mathematical expression for the constancy of gene frequencies in a large population, providing selection and mutation are very low.

Harelip. A cleft lip deformity usually extending to the nostril. It is surgically repairable.

Hemophilia. An X-linked clotting disorder in which the blood protein, factor VIII, is absent.

Hepatitis. Infection of the liver.

Hereditarian. A person who advocates that behavioral traits, especially intelligence, are determined chiefly by genetic constitution.

Heteroalleles. A genotype consisting of two different alleles of the same gene (e.g., a^1a^2).

Heterogeneity. The various genotypes that can result in a given trait or disorder.

Heterosis. The improved size and vigor of an F_1 organism when its parental types come from inbred strains.

Heterozygote. A hybrid, or carrier of a normal gene and its mutant allele.

Hexoseaminidase A. An enzyme which is missing in homozygous Tay-Sachs individuals.

Holist. One who believes that the entirety is greater than the sum of its parts.

Homologous chromosome. One of a pair of chromosomes matching in size and gene content.

Homozygote. An individual with a pair of identical alleles.

Host-mediated assay. The introduction of microbial cells into the peritoneal cavity, or the combination of extracts of eukaryotic cells with prokaryotic cells to test a mutagen whose action may be inhibited or enhanced by the eukaryotic environment.

Huntington disease. An autosomal dominant neurological disorder of late onset, causing paralysis and erratic or psychotic behavior.

Hurler syndrome. An autosomal recessive lysosomal storage disease of mucopolysaccharides, resulting in dwarfed children with multiple abnormalities, and death in late childhood.

Hyaline membrane disease. An immature functioning of the lungs in a premature baby, causing death from improper oxygen intake.

Hybrid. A heterozygote, or carrier of two alleles, one mutant and one normal.

Hydrocephaly. The accumulation of cerebrospinal fluid in the brain because of inadequate drainage from the brain to the spinal cord.

Hydrogen bond. A weak bond in which a hydrogen atom is shared by two other molecules.

Hypospadias. A urethral opening along the shaft of the penis instead of at the tip of the glans.

Hypothalamus. A portion of the brain that controls the pituitary gland.

Implantation. The attachment of the blastocyst to the lining of the uterus.

Impotence. Failure to achieve or maintain an erection of the penis.

Inbreeding. Mating between relatives, including self-fertilization in plants.

Independent assortment. The random segregation of two pairs of heterozygous alleles, each pair carried on a different pair of chromosomes.

Industrial melanism. The appearance of dark-pigmented organisms in sooty environments where, prior to industrialization, light-colored organisms existed.

Inner cell mass. The part of the blastocyst that forms the embryo.

Inovulation. The placing of an egg into a Fallopian tube.

Intelligence quotient (IQ). The ratio of mental age to chronological age multiplied by 100. The measuring tests may be written, mechanical, or perceptual.

Intercalation. The insertion of a molecule into the DNA double helix.

Intron. A segment of a gene's DNA which intervenes, without contributing to the informational content of the gene's product.

Inversion. A 180° rotation and reattachment of a chromosome fragment, reversing the sequence of genes in that fragment.

Ionizing radiation. Atomic particles or energy which alter molecules by imparting an electric charge to them, usually by removal of an electron.

I.U.D. (intrauterine device). A mechanical contraceptive device inserted into the uterus.

Karyotype. A photograph of chromosomes cut up so that the chromosomes are aligned by homologous pairs in order of size.

Keratinocyte. A skin cell which ingests melanosomes and distributes them in the cytoplasm or as a shield around the nucleus.

Kernicterus. A toxic saturation of nerve cells by a hemoglobin breakdown product, bilirubin, resulting in mental retardation.

Klinefelter syndrome. The XXY condition resulting in sterile males with mild abnormalities.

Kwashiorkor. A protein malnutrition resulting in emaciation and a distended abdomen.

LD_{50}. The dose of ionizing radiation needed to kill 50 percent of a population after whole-body exposure.

Leukemia. A cancer of the white blood cells.

LH, Luteinizing Hormone. A pituitary gland hormone which stimulates the ovarian follicles to produce a hormone. In males LH is known as ICSH, interstitial cell stimulating hormone.

Ligase. An enzyme which joins two molecules together.

Lippe's loop. An IUD made of plastic with snake-like twists forming a triangular shape.

Lordosis. Curvature of the spine producing protruding buttocks.

Lymphocyte. A small white blood cell.

Lysis. The bursting of a cell.

Lysosomes. Digestive organelles of a cell.

Lysosomal storage disease. An enzyme defect in a lysosome resulting in the accumulation, without digestion, of a cellular substance, thereby interfering with cell function.

Malthusian law. The viewpoint that organisms reproduce faster than their food supplies or necessities for living.

Mammography. An x-ray of a breast.

Mannosidosis. An autosomal recessive lysosomal storage disease in which mannose-bearing compounds accumulate and cause mental retardation, deafness, and bone deformities.

Marasmus. A progressive wasting away and failure to thrive caused by deficient protein and calorie intake in children.

Marker gene. An allele used to identify a chromosome or a less readily detectable gene on the same chromosome.

Meiosis. The cell divisions converting diploid cells into haploid cells, especially sperm and eggs.

Melanin. A polymer of tyrosine molecules producing a dark-pigmented color.

Melanocyte. A specialized cell in the skin which synthesizes melanin for distribution in skin color.

Malanosome. A lysosome that synthesizes melanin granules.

Menarche. The onset of the first menstrual cycle in puberty.

Meningitis. An infection of the membranes encasing the central nervous system.

Menopause. The cessation of the menstrual cycle, usually in the late middle-age (50s).

Mental age. The average achievement of a large number of children or adolescents of a given age.

Mesoderm. An embryonic tissue from which muscle, bone, connective tissue, and blood are derived.

Messenger RNA (m-RNA). A copy of the DNA sequence in a form suitable for decoding in the cell's cytoplasm.

Mestizo. The term formerly used for a person resulting from the union of an indian with a white.

Metabolism. The enzymatic activities of a cell.

Metaphase. A stage of cell division in which the chromosomes are aligned with their centromeres attached by fibers to the poles of the cell.

Metastasis. The spread of a cancer from its original site to other parts of the body.

Miscegenation. Racial mixture.

Mitochondria. Organelles of the cell that carry out oxidative respiration, producing the bulk of cellular energy.

Mitosis. The process by which one cell produces two cells.

Mitotic nondisjunction. The failure of two chromatids of a chromosome to separate in anaphase of mitosis, resulting in a delayed separation, with one cell increasing its chromosome number by one and the other cell losing one chromosome.

Modifier. A gene that intensifies or diminishes the expression of another gene.

Molecular clone. The multiplication of a nucleic acid molecule in a cell or test tube.

Monozygotic twins. Identical twins arising from a single fertilized egg.

Morula. The embryo as cleavage partitions the fertilized egg into several dozen cells just prior to their differentiation into a blastocyst.

Mosaic. An individual with two strains of cells derived by nondisjunction or mutation after the zygote forms.

Mulatto. A term formerly used to describe the brown-skinned individuals resulting from the union of dark-skinned (black) and caucasian (white) parents.

Multifactorial inheritance. Polygenic traits subject to environmental modification.

Multiple alleles. A normal gene and its mutant forms.

Muscular dystrophy. An X-linked recessive disorder causing muscular degeneration, paralysis, and death before reproductive maturity.

Mutation. An alteration of the gene causing a change in its function.

Mutational distance. The evolutionary time estimated to have occurred in two different organisms based on the number of substituted amino acids in a common protein.

Myelocyte. A white blood cell with characteristic granules in the cytoplasm.

Natural selection. A theory of evolution that claims species change gradually, evolving into new species, according to the survival or elimination of beneficial and detrimental variations in the population.

Neo-Darwinism. The theory of evolution by natural selection based on variations arising from gene mutations and chromosome rearrangements.

Neural crest. The cells forming at the junction of the neural plates when they fold to form the neural tube.

Neurofibromatosis. An autosomal dominant disorder resulting in tumorous distortions of the bones or skin and coffee-colored spots on the back.

Nitrogenous base. A purine or pyrimidine, which are ring-shaped molecules bearing nitrogen in some of their vertices.

Nondisjunction. The failure of one or more chromosomes to separate during mitosis or meiosis.

Nuclear membrane. A porous membrane surrounding the chromosomes in a eukaryotic cell.

Nucleic acid. The sequences of nucleotides composing DNA and RNA.

Nucleolus. A storage vesicle for ribosomal RNA associated with a modified region of one or more chromosomes.

Nucleus. The chromosomes and their surrounding medium enclosed by a porous membrane.

Nystagmus. An involuntary quivering or jerky movement of the eyes.

Oculocutaneous albinism. An autosomal recessive condition in which melanin is absent from the skin, hair, and eyes.

Oocyte. The developing egg prior to completion of meiosis. In humans the oocyte in equational division is called an ovum.

Oogonium. The diploid cell in an ovary prior to its entry into meiosis.

Ommatidium. An eyelet or lens-capped unit of a compound eye.

One gene-one enzyme theory. The view that each gene encodes the information for one enzyme.

Organelle. A specialized structure in a cell that carries out a cellular function.

Organogenesis. The conversion of embryonic tissues into specific organs.

Overdominance. A theory attributing heterosis to the inherent fitness of heterozygotes.

Oviduct. The fallopian tube or passageway transporting an egg from an ovary to the uterus.

Ovulation. The release of an egg from the graafian follicle into the oviduct.

P_1. The first parental generation, or the two individuals starting a genetic cross.

Palindrome. A sequence of letters that is read identically forwards or backwards.

Pancreatitis. An infection of the pancreatic gland.

Patau syndrome. Trisomy-13; a lethal condition characterized by multiple birth defects, especially those associated with failure of the forebrain to develop.

Peritonitis. An infection of the gut cavity caused by perforating ulceration of the gut.

Pessary. A mechanical plug inserted into the cervix, blocking the passage of sperm from the vagina into the uterus.

Petechiae. Rash-like bleeding of the capillaries.

Phagocytosis. The ingestion of external particles by a cell membrane.

Phenotype. The appearance of an individual.

Phenylketonuria (PKU). An autosomal recessive disorder affecting the amino acid phenylalanine and resulting in profound mental retardation if not properly treated.

Phlebitis. An inflammation of a vein.

Phylogenetic tree. An organization of diverse organisms showing their inferred evolutionary origins, each branch or trunk attached to its closest related form.

Piebald. Variegated like a patchwork quilt.

Pill. In popular usage, a hormonal contraceptive taken orally by some women.

Pinocytic vesicle. An inpocketing of the cell membrane bringing liquids into the cell. The pockets pinch off and fuse with lysosomes.

Placebo. An inert medical remedy given to patients with a fantasized illness or to control subjects in an experiment.

Placenta. The membranous organ that is formed when embryonic membranes fuse with the lining of the uterus and is expelled, following birth, as the afterbirth.

Plasma. The fluid or noncellular component of blood.

Plasmid. A small circular DNA fragment associated with a cell as a symbiont (a beneficial parasite).

Platelets. Clotting factors in the blood.

Polar body. A small bit of cytoplasm containing one of the nuclei produced by a meiotic division of an egg.

Polyploid. A chromosome number involving one or more haploid sets greater than the diploid number.

Pompe syndrome. An autosomal recessive lysosomal storage disease in which glycogen accumulates and cannot be digested and broken down to become sugars.

Prenatal diagnosis. The use of invasive or noninvasive techniques to determine the karyotype, biochemistry, or morphology of a fetus.

Primordial germ cell. A diploid cell that arises in the yolk sac, invades the embryonic gonad, and gives rise to reproductive tissue.

Prokaryote. A cell in which the chromosome is in direct contact with the cytoplasm; a bacterium or blue-green alga.

Prophase. A stage of cell division during which chromosomes become compactly coiled.

Propositus. The person being studied in a genetic pedigree.

Protein. A sequence of amino acids functioning as an enzyme, as a subunit of an organelle, or as a regulatory switch in cell metabolism.

Pseudoaging. The induced death of cells causing premature aging.

Punnett square. A checkerboard device to bring sperm genotypes and egg genotypes into their possible zygotic genotypes.

Pure line. A homozygous population.

Quinacrine. An acridine (three-ringed) dye used for chromosome banding studies.

Radioactive labeling. Use of an isotope to identify a molecule, the isotope being detected by a special instrument or by photography.

Radiomimetic agent. A chemical agent whose properties are similar in their genetic effects to ionizing radiation.

Recessive trait. An allele whose function is not noticeable in a heterozygote or hybrid.

Reciprocal cross. The mating of a mutant male with a nonmutant female and of a mutant female with a nonmutant male.

Recombinant DNA. The result of the use of enzymes to cut DNA into fragments and splice these into plasmids so that they may be multiplied in appropriate cells.

Reductional division. The first meiotic division in which a diploid cell produces two haploid cells.

Reductionist. A scientist who believes the analysis of isolated parts of an organism or cell is a proper procedure to reveal the mechanisms, processes, and components of the whole.

Regulation. The turning on or off of genes participating in a metabolic or developmental activity.

Repair enzyme. A protein that recognizes and repairs chromosome breaks or abnormal molecular configurations.

Replication. The copying of a nucleic acid to make two identical copies.

Restituted chromosome. A chromosome repaired so that its original sequence of genes is preserved.

Restriction enzymes. Enzymes that recognize a specific sequence in DNA and digest the molecule by cleaving in that specific region.

Retinitis pigmentosa. An autosomal recessive disease of the retina causing progressive blindness.

Retinoblastoma. A cancer of the eye in children, transmitted as an autosomal dominant mutation.

Rh blood groups. Protein antigens on the red blood cell membrane which, if present, provoke serum antibodies in a person whose own red blood cells lack these antigens.

Rhogam. The antibody against the Rh positive antigen; it is administered to Rh negative mothers who deliver an Rh positive baby.

Rhythm method. A birth control technique using abstinence from intercourse during the female's fertile period.

Ribosomal RNA (r-RNA). Nucleic acid copied from genes that serves as a mechanical device in ribosomes to bring about protein synthesis.

Ribosome. A structure composed of proteins and r-RNA that permits protein synthesis to take place.

Rickets. A disorder of calcium metabolism producing deformed bones and stunted growth, often from low production and intake of vitamin D.

RNA. A nucleic acid whose nucleotides contain the nitrogenous bases, guanine, adenine, cytosine, and uracil. The sugar connecting these bases to phosphate is ribose.

Roentgen. A unit of radiation measured by the number of ions induced in a cubic centimeter of air that impart a one-volt charge.

S phase. The part of the cell cycle in which DNA synthesis occurs.

Sacral spot. The bluish spot at the base of the spinal region seen, at birth, in infants of oriental or black ancestry.

Segregation. The separation of a pair of homologous chromosomes carrying a pair of heterozygous alleles.

Semi-conservative replication. Separation of complementary strands, each of which synthesizes a new complementary strand.

Seminal fluid. The nutrients and buffers surrounding the sperm in an ejaculate.

Serum. The fluid portion of blood remaining after a clot is formed.

Sex chromosome. An X or Y chromosome.

Sex ratio. The excess or deficit of males relative to females.

Shotgun experiment. The fragmentation of an entire genome so that one or more of the fragments may bear a desirable gene which can be cloned and later isolated by selective techniques.

Sickle cell anemia. An autosomal recessive condition producing severe anemia and red blood cells whose deformed membranes collapse into a sickled shape.

Simian crease. The single palmar crease often seen in trisomy-21 patients but which is also present in about one percent of normal persons.

Somatic cell hybridization. The fusion of two unrelated cell lines to produce cells capable of growing in tissue culture which bear one or more chromosomes of each species.

Species. A population of organisms that normally breed freely with each other.

Sperm. Mature spermatozoa or male reproductive cells.

Spermatid. The male gamete upon completion of meiosis but prior to its streamlining into a functional sperm.

Spermatocyte. A male reproductive cell from the time it enters meiosis until it completes the second meiotic division.

Spermatogenesis. The entire process, from the time a diploid spermatogonium enters meiosis until mature sperm are formed in the testes.

Spermatogonium. A diploid male reproductive cell that retains the capacity to divide by mitosis.

Spermatozoan. The mature sperm.

Spermiogenesis. The post-meiotic process by which a spermatid is converted into a spermatozoan.

Spina bifida. A neural tube defect involving incomplete closure and resulting in an open or covered defect of the spinal cord.

Spindle apparatus. The organelles and the microtubular fibers they form which attach to centromeres of chromosomes during cell division.

Strabismus. A cross-eyed condition in which one or both eyes may wander and come to rest at the temple or nasal side of the eye.

Substitution mutation. A substitution of a single base by any of the other three bases resulting in a single amino acid replacement in a protein.

Symbiont. An organism that lives with another organism (or within a cell) so that both organisms benefit from the association.

Syncytium. Many nuclei embedded in a common cytoplasm with only one surrounding cell membrane.

Tampon. A device inserted into the vagina (or other body cavity) to absorb blood or secretions.

Tay-Sachs disease. An autosomal recessive lysosomal storage disease of the neurons resulting in blindness, paralysis, seizures, and death in infancy.

Telophase. The mitotic stage in which chromosomes move to the poles of a cell and the cytoplasm is partitioned into two cells.

Thanatophoric dwarf. A lethal condition in which the baby has very short limbs and a narrow thorax; it is an autosomal recessive trait.

Thymine dimer. The fusion of adjacent thymines along a DNA strand, usually induced by ultraviolet radiation.

Tissue. A group of cells performing a common function.

Toxicity. Capacity of a substance to destroy cells or interfere with their metabolism.

Trans alignment. In independent assortment two nonhomologous chromosomes facing a cell's pole with one chromosome bearing a mutant allele and the other nonhomolog bearing a normal allele.

Transfer RNA (t-RNA). A nucleic acid that carries an amino acid and has a specific triplet sequence to pair with a complementary sequence of m-RNA during protein synthesis.

Transformation. The alteration of a cell by replacement of a gene from a fragment of DNA introduced into the cell.

Translocation. The exchange of nonhomologous chromosome parts during the repair of two broken chromosomes.

Trimester. One third of the nine months of pregnancy, measured as first, second, and third trimesters.

Triplet. A codon or sequence of three bases in a nucleic acid.

Triploid. A cell or individual with three haploid sets of chromosomes for its chromosome number.

Trisomic. A diploid cell bearing one extra chromosome homologous to a pair normally present.

Trisomy. A condition in which all or some of the cells in an individual contain an extra chromosome, resulting in a syndrome characteristic of that particular chromosome excess.

Trisomy-13. Patau syndrome; a lethal condition with multiple birth defects, especially those involving failure of forebrain development.

Trisomy-18. Edward syndrome; a lethal condition with multiple birth defects and characteristic skeletal defects.

Trisomy-21. Down syndrome; a condition involving short stature, mental retardation, and characteristic facial abnormalities.

Trophoblast. The tissue of a blastocyst which implants into the uterine lining.

Tubal ligation. The cutting or blocking of the oviducts to prevent the passage of an egg to the uterus.

Turner syndrome. The XO condition producing females with dwarf stature, webbed neck, and prepubertal body features associated with the absence of ovaries.

Ultrasound. A high-frequency wavelength of sound whose echoes are converted by computer analysis into a visual image on a TV screen revealing internal organs, a tumor, or a fetus in a patient.

Uterus. The womb or organ that houses the embryo and fetus during pregnancy.

Vas deferens. The male sperm duct carrying sperm from the testes (or epididymis) to the penis.

Vasectomy. A cutting or blocking of the sperm ducts in the scrotum to prevent sperm (but not seminal fluid) from being released in an ejaculate.

Virulent. An organism which causes illness in its infected host.

Virus. A noncellular life form consisting of a nucleic acid and protective protein coat.

Whole-body dose. The amount of radiation administered to the entire individual.

X-inactivation. A mechanism which keeps one X coiled and one X unwound in a mammalian cell, resulting in the functioning of only one X chromosome in the diploid cell.

X-linked inheritance. Traits whose genes are on the X chromosome and thus expressed in a male. Expression in the female usually requires homozygosity if the trait is recessive.

X-linked lethal. An allele on the X chromosome affecting a vital process and leading to death of the male embryo, fetus, or child.

x ray. Energy released by atoms which can cause other atoms to ionize.

Yolk sac. A part of the embryo in egg laying organisms composed of endoderm and containing massive amounts of nutrients.

Zambo. The term formerly used for an individual resulting from the union of a black with an indian.

ZPG. Zero population growth; the state of population in which birth rates and death rates are identical.

References

Abraham, Morris B., ed. 1982. *Splicing Life: The Social and Ethical Issues of Genetic Engineering with Human Beings,* President's Commission for the Study of Ethical Problems in Medicine and Biomedical and Behavioral Research, Superintendent of Documents, Washington, D. C.

Altman, P. L. and Dittmer, D. S. 1972. *Biology Data Book.* Bethesda: Federation of American Societies for Experimental Biology.

Altenburg, E. and Muller, H. J. 1920. The Genetic Basis of Truncate Wings—An Inconstant and Modifiable Character in Drosophila. *Genetics* 5:1–59.

Auerbach, C. 1976. *Mutation Research.* London: Chapman and Hall.

Austin, C. R. and Short, R. V. 1976. *Mammalian Reproduction and Development.* New York: Cambridge Univ. Pr.

Ayala, F. J. and Valentine, J. W. 1979. *Evolving: The Theory and Processes of Organic Evolution.* Menlo Park, California: Benjamin/Cummings.

Bajema, C., ed. 1976. *Eugenics: Then and Now.* Illinois: Halsted Pr.

Balinky, B. I. 1981. *An Introduction to Embryology, 5th ed.* Philadelphia: Saunders.

Barron, M. L., ed. 1972. *The Blending American: Patterns of Intermarriage.* Chicago: Quadrangle.

Bergsma, D., ed. 1979. *Birth Defects Compendium, 2nd ed.* New York: Alan R. Liss.

Bishop, J. M. 1982. Oncogenes. *Scientific American* 246:80–92.

Blandau, R. J., ed. 1971. *The Biology of the Blastocyst.* Chicago: Univ. of Chicago Pr.

Blechschmidt, E. 1978. *Biokinetics and Biodynamics of Human Differentiation.* Springfield, Ill.: Thomas.

Borgatta, E. F. and Evans, R. R., eds. 1968. *Smoking, Health, and Behavior.* Chicago: Aldine

Brace, C. L. 1979. *Atlas of Human Evolution.* New York: Holt, Rinehart, and Winston.

Brent, P. L. 1981. *Charles Darwin: A Man of Enlarged Curiosity.* New York: Harper & Row.

Bulmer, M. G. 1970. *The Biology of Twinning in Man.* New York: Oxford Univ. Pr.

Bytwerk, R. L. 1983. *Julius Streicher.* New York: Stein and Day.

Cairns, J. 1978. *Cancer: Science and Society.* San Francisco: W. H. Freeman.

Carlson, E. A. 1973; Eugenics Revisited: The Case for Germinal Choice. *Stadler Genetics Symposia* 5:13–34.

Cavalli-Sforza, L. and Bodmer, W. F. 1971. *The Genetics of Human Populations.* San Francisco: W. H. Freeman.

Chambon, P. 1981. Split Genes. *Scientific American* 244:60–71.

Chase, A. 1976. *The Legacy of Malthus.* New York: Knopf.

Chiarelli, A. B. 1973. *Evolution of the Primates.* New York: Academic Pr.

The Committee for the Compilation of Materials on Damage Caused by the Atomic Bombs in Hiroshima and Nagasaki. 1981. *Hiroshima and Nagasaki: The Physical, Medical, and Social Effects of the Atomic Bombings.* New York: Basic Books.

Cox, Catherine. 1929. *Genetic Studies of Genius: The Early Mental Traits of Three Hunded Geniuses.* Stanford, Calif.: Stanford Univ. Pr.

Crosignani, P. G. and Mishell, D. R. 1976. *Ovulation in the Human.* New York: Academic Pr.

Darwin, C. 1839. Reprint. *The Voyage of the Beagle.* Garden City, N.Y.: Doubleday, 1962.

———. 1859. *On the Origin of Species by Means of Natural Selection or the Preservation of Favored Races in the Struggle for Life.* London: J. Murray.

Davenport, C. B. 1913. *Heredity of Skin Color in Negro-White Crosses.* Carnegie Institution of Washington, D. C.

deGrouchy, J. and Turleau, C. 1977. *Clinical Atlas of Human Chromosomes.* New York: Wiley.

DeRobertis, E.D.P., Saez, F. A., and DeRobertis, E.M.F. 1975. *Cell Biology, 6th Ed.* Philadelphia: Saunders.

DeSerres, F. D. and Hollaender, A., eds. 1971–1982. *Chemical Mutagens: Principles and Methods for Their Detection* (7 volumes). New York: Plenum Pr.

Drake, J. W. 1970. *The Molecular Bases of Mutation.* San Francisco: Holden-Day.

Dublin, L. and Lotka, A. 1935. *Length of Life.* New York: Ronald Pr.

DuPraw, E. J. 1970. *DNA and Chromosomes.* New York: Holt, Rinehart, and Winston.

Dyer, K. F. 1974. *The Biology of Racial Integration.* Bristol: Scientechnica.

East, E. M. and Jones, D. F. 1919. *Inbreeding and Oubreeding.* Philadelphia: Lippincott.

Edwards, R. and Steptoe, P. 1980. *A Matter of Life.* New York: Morrow.

Epstein, S. S. 1978. *The Politics of Cancer.* San Francisco: Sierra.

Erhardt, C. L. and Berlin, J. E., eds. 1974. *Mortality and Morbidity in the United States.* Cambridge, Mass.: Harvard Univ. Pr.

Fawcett, D. W. and Bedford, J. M., eds. 1979. *The Spermatozoon.* Baltimore: Urban and Schwarzenberg.

Federman, D. 1967. *Abnormal Sexual Development*. Philadelphia: Saunders.

Fitzpatrick, T. B., Quevedo, W. C., Jr., Szabo, G., and Seiji, M. 1971. Biology of the Melanin Pigmentory System. In Fitzpatrick, T. B., *et al. Dermatology in General Medicine*, pp. 117–146. New York: McGraw-Hill.

Fuchs, F. 1980. Genetic Anmiocentesis. *Scientific American* 242:47–53.

Fuhrman, W. and Vogel, F. 1983. *Genetic Counseling, 3rd ed.* New York: Springer-Verlag.

Futyuma, D. J. 1979. *Evolutionary Biology*. Sonderland, Mass.: Sinauer Assoc.

Galton, F. 1869. *Hereditary Genius: An Inquiry into Its Laws and Consequences*. London: MacMillan.

Goertzel, V. and Goertzel, M. G. 1962. *Cradles of Eminence*. Boston: Little, Brown.

Gorlin, R. J., Pindborg, J. J., and Cohen, M. M. 1976. *Syndromes of the Head and Neck, 2nd ed.* New York: McGrawHill.

Gould, S. J. 1981. *The Mismeasure of Man*. New York: Norton.

Guttmacher, A. F. 1973. *Pregnancy, Birth, and Family Planning*. New York: Viking Press.

Hafez, E.S.E., ed. 1976. *Human Semen and Fertility Regulation in Men*. Saint Louis, Mo.: Mosby.

Haller, M. H. 1968. *Eugenics: Hereditarian Attitudes in American Thought*. New Brunswick, N.J.: Rutgers Univ. Pr.

Harrison, R. G. 1959. *A Textbook of Human Embryology*. New York: Oxford Univ. Pr.

Hook, E. B. and Hamerton, I. H. 1977. The Frequency of Chromosome Abnormalities Detected in Consecutive Newborn Studies—Differences Between Studies—Results by Sex and Severity of Phenotypic Involvement. In *Population Cytogenetics, Studies in Humans*, E. B. Hook and I. H. Porter, eds, pp. 63–79. New York: Academic Pr.

Hook, E. B. 1973. Behavioral Implications of the Human XYY Genotype. *Science* 179:139–151.

The Insight Team of the Sunday Times of London. 1979. *Suffer the Children: The Story of Thalidomide*. New York: Viking Pr.

Jensen, A. R. 1969. How Much Can We Boost IQ and Scholastic Achievement? *Harvard Educational Review* 39:1–123.

Jeon, K. W. and Danielli, J. F. 1971. Micrugical Studies with Large Free-Living Amoebas. *International Revue of Cytology* 30:49–90.

Johannsen, W. (1903) 1959. Genotype and Phenotype. *Classic Papers in Genetics*. Englewood Cliffs, N.J.: Prentice-Hall.

John, B. and Lewis, K. R. 1968. *The Chromosome Complement*. New York: Springer-Verlag.

Jones, H. W. and Scott, W. W. 1971. *Hermaphroditism, Genital Anomalies and Related Endocrine Disorders, 2nd ed.* Baltimore: Williams and Wilkins.

Judson, H. F. 1979. *The Eighth Day of Creation: Makers of the Revolution in Biology*. New York: Simon & Schuster.

Kamin, L. J. 1974. *The Science and Politics of IQ*. New York: Erlbaum Assoc.

Lack, D. L. 1947. *Darwin's Finches.* London: Cambridge Univ. Pr.

Langman, J. 1981. *Medical Embryology, 4th ed.* Baltimore: Williams and Wilkins.

Leaky, R. E. 1978. *People of the Lake: Mankind and Its Beginnings.* Garden City, New York: Anchor Pr.

Lear, John. 1978. *Recombinant DNA: The Untold Story.* New York: Crown.

Lifton, R. J. 1968. *Death in Life: Survivors of Hiroshima.* New York: Random House.

Ludmerer, K. M. 1972. *Genetics and American Society.* Baltimore: Johns Hopkins Univ. Pr.

MacGillivray, I., Nylander, P.O.S., and Corney, G. 1975. *Human Multiple Reproduction.* Philadelphia: Saunders.

Magenis, R. E., Overton, K. M., Chamberlin, J., Brady, T., and Lorrein, E. 1977. Parental Origin of the Extra Chromosome in Down's Syndrome. *Human Genetics* 37:7–16.

Martin, A. D. and Harbison, S. A. 1972. *An Introduction to Radiation Protection.* London: Chapman and Hall.

McGrath, J. and Solter, D. 1983. Nuclear transplantation in the mouse embryo by microsurgery and cell fusion. *Science* 220:1300–1302.

McKusick, V. 1980. *Mendelian Inheritance in Man, 5th ed.* Baltimore: Johns Hopkins Univ Pr.

McKusick, V. A. 1975. The Growth and Development of Human Genetics as a Clinical Discipline. *Am. J. Human Genetics* 27:261–273.

McLaren, A. 1981. *Germ Cells and Soma: A New Look at an Old Problem.* New Haven: Yale Univ. Pr.

Mendel, G. 1866. Experiments on Plant Hybrids. Translated in: Stern, C. and Sherwood, E., eds. 1966. *The Origin of Genetics.* San Francisco: Freeman.

Mitchison, J. M. 1971. *The Biology of the Cell Cycle.* New York: Cambridge Univ. Pr.

Money, J. W. and Ehrhardt, A. A. 1972. *Man and Woman, Boy and Girl: Differentiation and Dimorphism of Gender Identity from Conception to Maturity.* Baltimore: Johns Hopkins Univ. Pr.

Montagu, A. 1969. *Man: His First Two Million Years.* New York: Columbia Univ. Pr.

Moore, K. L. 1974. *Before We are Born: Basic Embryology and Birth Defects.* Philadelphia: Saunders.

Moriyama, I. W. 1977. Capsule Summary of Results of Radiation Studies on Hiroshima and Nagasaki Atomic Bomb Survivors. Hiroshima: Radiation Effects Research Foundation.

Mörner, M. 1967. *Race Mixture in the History of Latin America.* Boston: Little Brown.

Moser, H. W. and Wolf, P. A. 1971. The nosology of mental retardation: including the report of a survey of 1378 mentally retarded individuals at the Walter E. Fernald State School. *Birth Defects Original Articles Series* vii:117–134.

Mourant, A. E. 1976. *The Distribution of the Human Blood Groups and Other Polymorphisms.* London: Oxford Univ. Pr.

Muller, H. J. 1961. Human Evolution by Voluntary Choice of Germ Plasm. *Science* 134:643–649.

———. 1950. Our Load of Mutations. *American Journal of Human Genetics* 2:111–176.

———. 1932. The Dominance of Economics over Eugenics. In *A Decade of Progress in Eugenics.* The Third International Congress of Eugenics, pp. 138–144. Baltimore: Williams and Wilkins.

Neel, J. V. and Schull, W. J. 1956. *The Effect of Exposure to the Atomic Bombs on Pregnancy Termination in Hiroshima and Nagasaki.* Washington: Atomic Bomb Casualty Commission, Hiroshima.

Neumann, M., ed. 1978. *The Tricentennial People: Human Applications of the New Genetics.* Ames, Iowa: Iowa State Univ. Press.

Nilsson-Ehle, H. 1959. Breeding Analysis in Oats and Wheat (1909). In Peters, J. A., ed. *Classic Papers in Genetics.* Englewood Cliffs, New Jersey: Prentice-Hall.

Novitski, E. 1982. *Human Genetics, 2nd ed.* New York: MacMillan.

Ohno, S. 1979. *Major Sex Determining Genes.* New York: Springer-Verlag.

Olby, R. C. 1974. *The Path to the Double Helix.* London: MacMillan.

———. 1965. *The Origins of Mendelism.* London: Constable.

Pochin, Edward E. 1976. *Estimated Population Exposure from Nuclear Power Production and Other Radiation Sources.* Paris: Nuclear Energy Agency, Organization for Economic Co-operation and Development.

Prescott, David M. 1976. *Reproduction of Eukaryotic Cells.* New York: Acad Pr.

Race, R. R. and Sanger, R. 1975. *Blood Groups in Man, 6th ed.* New York: Lippincott.

Rost, T. L. and Gifford, E. M. eds. 1977. *Mechanisms and Control of Cell Division.* New York: Halsted Pr.

Salmon, M. A. 1978. *Developmental Defects and Syndromes.* Aylesbury, England: H. M. & M.

Sayre, A. 1975. *Rosalind Franklin and DNA.* New York: Norton.

Schmeltz, Irwin, ed. 1972. *The Chemistry of Tobacco and Tobacco Smoke.* New York: Plenum Pr.

Schottenfeld, D. and Fraumeni, J. F. 1982. *Cancer Epidemiology and Prevention.* Philadelphia: Saunders.

Schrödinger, Erwin. 1945. *What is Life?* New York: Cambridge Univ. Pr.

Singer, B. and Kusmierek, J. T. 1982. Chemical Mutagenesis. *Annual Review of Biochemistry* 51:655–694.

Smith, D. W. 1982. *Recognizable Patterns of Human Malformation, 3rd Ed.* Philadelphia: Saunders.

Smith, J. M., ed. 1982. *Evolution Now: A Century After Darwin.* San Francisco: Freeman.

Snyder, L. H. 1973. *Blood Groups.* Minneapolis: Burgess.

Solecki, R. 1971. *Shanidar: The First Flower People.* New York: Knopf.

Spector, W. S. 1956. *Handbook of Biological Data.* Philadelphia: Saunders.

Stanbury, J. B., Wyngaarden, J. B., and Frederickson, D. S., eds. 1978. *The Metabolic Basis of Inherited Disease, 4th ed.* New York: McGraw-Hill.

Steer, Martin W. 1981. *Understanding Cell Structure.* New York: Cambridge Univ. Pr.

Stern, C. 1973. *Principles of Human Genetics, 3rd ed.* San Francisco: Freeman.

Streissguth, P., et al. 1980. Teratogenic Effects of Alcohol in Humans and Laboratory Animals. *Science* 209:353–361.

Strong, S. J. and Corney, G. 1967. *The Placenta in Twin Pregnancy*. New York: Pergamon.

Taylor, I. and Knowelden, J. 1964. *Principles of Epidemiology*. Boston: Little, Brown.

Terman, L. M. 1925. *Genetic Studies of Genius: Mental and Physical Traits of a Thousand Gifted Children*. Stanford: Stanford Univ Pr.

Therman, E. 1980. *Human Chromosomes: Structure, Behavior, Effects*. New York: Springer-Verlag.

Titmuss, R. M. 1971. *The Gift Relationship: From Human Blood to Social Policy*. New York: Pantheon.

U.S.D.H. Government Printing Office. 1979. *Smoking and Health: Report of the Surgeon General*. Washington, D. C.

Wallace, Bruce. 1970. *Genetic Load, Its Biological and Conceptual Aspects*. Englewood Cliffs, New Jersey: Prentice-Hall.

Wasserman, H. P. 1974. *Ethnic Pigmentation: Historical, Physiological, and Clinical Aspects*. New York: American Elsevier.

Watson, J. D. 1984. *Molecular Biology of the Gene, 4th ed*. Menlo Park, California: Benjamin/Cummings.

———.1969. *The Double Helix*. New York: Antheneum.

Watson, J. D. and Tooze, J. 1981. *The DNA Story: A Documentary History of Gene Cloning*. San Francisco: Freeman.

Williamson, R., ed. 1981. *Genetic Engineering*. London: Acad Pr.

Woese, C. R. 1967. *The Genetic Code*. New York: Harper and Row.

INDEX